Process-Structure-Property Relationships in Metals

Special Issue Editor
Soran Birosca

MDPI • Basel • Beijing • Wuhan • Barcelona • Belgrade

MDPI

Special Issue Editor
Swansea University
Swansea
UK

Editorial Office
MDPI AG
St. Alban-Anlage 66
Basel, Switzerland

This edition is a reprint of the Special Issue published online in the open access journal *Metals* (ISSN 2075-4701) from 2016–2017 (available at: http://www.mdpi.com/journal/metals/special_issues/process_structure_property).

For citation purposes, cite each article independently as indicated on the article page online and as indicated below:

Author 1; Author 2. Article title. *Journal Name* **Year**, *Article number*, page range.

First Edition 2017

ISBN 978-3-03842-456-7 (Pbk)
ISBN 978-3-03842-457-4 (PDF)

Table of Contents

About the Special Issue Editor

Soran Birosca is an associate professor in material engineering at Swansea University—College of Engineering. Prior to his appointment at Swansea he was conducting his research at the University of Cambridge—Department of Materials Science and Metallurgy and the University of Manchester—School of Materials. He also spent a year as a research associate in Pohang University of Science and Technology—Materials Design Laboratory—Pohang (South Korea). Soran has a PhD degree in materials science and engineering from Loughborough University, an MSc degree from the University of Manchester in corrosion science and engineering and BSc degree from Salahaddin University in mechanical engineering. Soran research portfolio over years focused on material characterisation, physical metallurgy, materials processing, materials performance and integrity, phase transformation and heat treatment, low and high temperature corrosion, structure/property relationship, mechanical property of materials, fracture mechanics, fatigue, microtexture and texture analysis for materials used in power generation, nuclear, automotive and aero- engine applications.

![metals logo] *metals*

MDPI

Article

Atmospheric-Induced Stress Corrosion Cracking of Grade 2205 Duplex Stainless Steel—Effects of 475 °C Embrittlement and Process Orientation

Cem Örnek [1,2,*], Safwan A.M. Idris [3], Pierfranco Reccagni [2] and Dirk L. Engelberg [1,2]

[1] Materials Performance Centre, School of Materials, The University of Manchester, Sackville Street, Manchester M13 9PL, UK; dirk.engelberg@manchester.ac.uk

[2] Corrosion and Protection Centre, School of Materials, The University of Manchester, Sackville Street, Manchester M13 9PL, UK; pierfranco.reccagni@postgrad.manchester.ac.uk

[3] Brunei Shell Petroleum Sendirian Berhad, Jalan Utara, Panaga, Seria KB2933, Negara Brunei Darussalam; a.pgidris@shell.com

* Correspondence: cem.oernek@manchester.ac.uk or cem_oernek@hotmail.de; Tel.: +44-161-306-4838

Academic Editor: Soran Birosca

Received: 30 April 2016; Accepted: 12 July 2016; Published: 19 July 2016

Abstract: The effect of 475 °C embrittlement and microstructure process orientation on atmospheric-induced stress corrosion cracking (AISCC) of grade 2205 duplex stainless steel has been investigated. AISCC tests were carried out under salt-laden, chloride-containing deposits, on U-bend samples manufactured in rolling (RD) and transverse directions (TD). The occurrence of selective corrosion and stress corrosion cracking was observed, with samples in TD displaying higher propensity towards AISCC. Strains and tensile stresses were observed in both ferrite and austenite, with similar magnitudes in TD, whereas, larger strains and stresses in austenite in RD. The occurrence of 475 °C embrittlement was related to microstructural changes in the ferrite. Exposure to 475 °C heat treatment for 5 to 10 h resulted in better AISCC resistance, with spinodal decomposition believed to enhance the corrosion properties of the ferrite. The austenite was more susceptible to ageing treatments up to 50 h, with the ferrite becoming more susceptible with ageing in excess of 50 h. Increased susceptibility of the ferrite may be related to the formation of additional precipitates, such as R-phase. The implications of heat treatment at 475 °C and the effect of process orientation are discussed in light of microstructure development and propensity to AISCC.

Keywords: duplex stainless steel; 475 °C embrittlement; atmospheric-induced stress corrosion cracking; electron backscatter diffraction; X-ray diffraction; residual stress

---

1. Introduction

Duplex stainless steels (DSSs) are used as corrosion-resistant materials for marine, off-shore, and nuclear applications [1,2]. DSSs provide superior mechanical and electrochemical properties to most of their austenitic and ferritic counterparts, which is related to the complementary nature of both crystallographic phases [3,4]. However, service temperatures in excess of 250 °C or welding can result in embrittlement and loss of corrosion resistance [5–7].

Exposure to the 250–550 °C low temperature embrittlement window, also known as "475 °C embrittlement", can cause loss of toughness and ductility, primarily related to phase reactions in the ferrite [5,7,8]. Spinodal decomposition in ferrite (δ) to form Fe-enriched (Cr-depleted) α' and Cr-enriched (Fe-depleted) α'' structures, or the nucleation and growth of α''-particles embedded in an α'-matrix can occur, typically followed by the formation of further precipitates, such as χ, Frank-Kasper R-phase, τ, or Mo-Si enriched G-phase [5,7,9–15]. The phase separation products ($\alpha' + \alpha''$), in general, do not exceed dimensions larger than 100 nm, whereas precipitates such as G- or R-phase can grow to sizes larger than 500 nm [6,10–14]. It should be noted that similar phase reactions have also been reported in austenite, which, however, requires further investigations to make clear statements about their origin [8,9,16,17].

There has been extensive research on the characterisation of 475 °C embrittlement, which predominantly focused on investigating microstructure development and mechanical property changes [6,7,14,18–20]. It is common that embrittlement gradually increases with ageing, reaching saturation after prolonged exposure times. Corrosion properties and associated atmospheric chloride-induced stress corrosion cracking (SCC) performance of 475 °C embrittled microstructures, however, have not been studied in detail, and the understanding of their SCC propensity is lacking. So far, most research to better understand SCC was carried out in extremely aggressive environments [21–31], often at temperatures in excess of 100 °C [23,29,30,32,33]. The low-temperature SCC performance (<100 °C), in particular under atmospheric exposure conditions, has not been elucidated, but is certainly important for engineering applications.

In this work grade 2205 DSS was heat treated to investigate the effect of 475 °C embrittlement on the atmospheric chloride-induced SCC behaviour. U-bend samples were manufactured along RD and TD to assess the effect of process orientation on microstructure AISCC susceptibility. The key focus of this study was the effect of 475 °C heat treatment on microstructure propensity to AISCC, and the work reported here is part of a larger research project on understanding the effect of 475 °C embrittlement on DSS performance [34–38].

2. Experimental Section

A solution-annealed (as-received) sheet of grade 2205 DSS (EN 1.4462, UNS S32205) manufactured by OutoKumpu (Torshälla, Sweden) of 2 mm thickness was used in this study. The sheet had a chemical composition (in wt. %) of 22.44Cr, 5.75Ni, 3.32Mo, 1.41Mn, 0.42Si, 0.015C, 0.155N, 0.006Nb, 0.21Cu, 0.12Co, and Fe (bal.). The sheet was then cut along RD and TD to produce rectangular strips with dimensions of 70 mm × 20 mm × 2 mm (L × W × T). Two holes of 5.2 mm diameter each were drilled on both sides of the strips, with samples exposed to heat treatments at 475 °C ± 5 °C for 5, 10, 20, 50, and 255 h, followed by water quenching. The surface of all specimens was ground to 4000-grit using SiC paper. A schematic illustration of the strips, including their orientation with respect to RD/TD, and how these were further shaped into U-bend samples is shown in Figure 1. Bending was carried out using an Instron 5569 testing machine (Instron, Norwood, MA, USA) under compression with a load cell of 50 kN and a crosshead displacement rate of 10 mm/min. The U-bend samples were tightened using stainless steel bolts and nuts.

2.1. Microstructure Characterisation

Two as-received strip specimens were cut along RD and TD. The samples were then ground and polished to $1/4$ μm finish using diamond paste, followed by an OP-S end-polishing treatment for one hour. Microstructure characterisation was carried out on both specimens before and after bending using electron backscatter diffraction (EBSD).

EBSD analysis was performed over a scan area of 711 μm × 622 μm of the samples before bending, and 728 μm × 728 μm of the bent specimens using an FEI Quanta 650 scanning electron microscope (SEM, Hillsboro, OR, USA) interfaced with a Nordlys EBSD detector (Oxford Instruments, Abingdon, Oxfordshire, UK) and AZtec V2.2 software for data acquisition from Oxford Instruments (Abingdon, Oxfordshire, UK). A 20 kV accelerating voltage with step sizes of 527 nm and 773 nm was used for microstructure analyses. Both EBSD maps were acquired over approximately the same area in both samples, located towards the centre of the strips, schematically shown in Figure 1b.

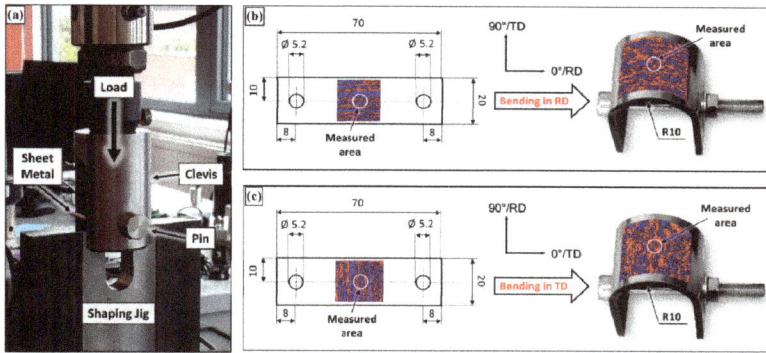

Figure 1. (a) U-bend shaping jig with (b) sheet dimensions and bending orientation in RD and (c) in TD. The circle in the centre of each specimen indicates the position of XRD stress measurements before and after bending. The angles indicate the measurement direction in 0° and 90°.

EBSD data processing was undertaken with HKL Channel 5 software (version 5.12.62.0 64-bit, Oxford Instruments, Abingdon, Oxfordshire, UK). High-angle grain boundaries (HAGBs) were defined with misorientation $\geqslant 15°$ and low-angle grain boundaries (LAGBs) between >1° and <15°. The grain size was determined by the mean linear intercept method, as the mean of vertical and horizontal directions, with 50 intercept lines used for grain boundary detection. The aspect ratio of the average grain size of ferrite and austenite was determined for the as-received and U-bend microstructures. The aspect ratio was defined as the quotient of the average grain sizes obtained from vertical linear intercepts (V) and those obtained from horizontal linear interception (H). The phase fractions of ferrite and austenite were also extracted.

Local misorientation (LMO) maps (Kernel average misorientation) were generated by using 3×3 pixel binning with a 5° sub-grain angle threshold. This analysis gives the average LMO below the pre-determined sub-grain angle threshold, and this method was used to locate regions with higher concentrations of misorientation in the microstructure. The latter is typically associated with local micro-deformation in the form of elastic/plastic strain [39].

2.2. Hardness Measurements

Macro-hardness measurements were carried out on a Georg Reicherter Briviskop 1875 Vickers hardness machine (Esslingen a.N., Baden-Württemberg, Germany) on all samples. Measurements were undertaken on flat strips and U-bend samples. For the samples before testing, 10 hardness indentations were produced with five indentations on each end of the samples, as seen in Figure 2a. Three indentations were placed on the apex of the U-bend samples after bending, as shown in Figure 2b. The corresponding hardness was then calculated from the mean of the hardness measurements.

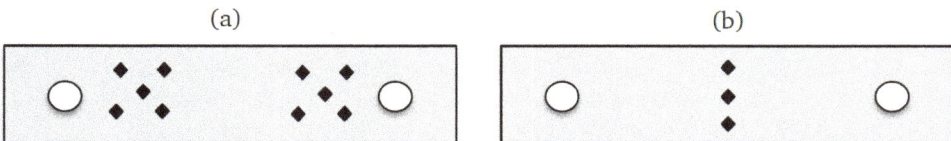

Figure 2. Schematic representation of the location of hardness indentations in (**a**) flat strips and (**b**) U-bend samples.

2.3. XRD Stress Measurements

XRD stress measurements were carried out using the Proto iXRD Combo testing machine (Proto Manufacturing Inc., Oldcastle, ON, Canada), which is equipped with a two-detector system. Cr and Mn X-ray sources were used to measure the strain in ferrite and austenite, respectively. All setup parameters are listed in Table 1. Prior to each test, the X-ray diffractometer was calibrated to determine the zero stress positions of the peaks of the diffracted X-rays. Stress-free and stressed standard samples, provided by the manufacturer, were used for calibration. The multiple exposure technique with 11 angles was applied for inter-planar d-spacing measurements. Two X-ray measurement orientations, i.e., $0°$ and $90°$, aligned to the RD and TD process orientation, were chosen. Each measured orientation corresponds to the stress direction specified in Figure 1b,c. The measurement position was in the centre of each specimen, measured at the same position before and after bending.

Table 1. X-ray measurement conditions.

Parameter	Measure	
X-ray type	Cr-Kα	Mn-Kα
Source voltage, current	20 kV, 4 mA	20 kV, 4 mA
Aperture size	2 mm	2 mm
Bragg angle, 2θ	156.4	152.8
Diffraction plane	(2 1 1)	(3 1 1)
Wavelength	2.291 Å	2.1034 Å
Max. measurement angle, β	$27°$	$27°$
Number of β angles	11	11
β angles	27	27
β oscillation angle	$3°$	$3°$
Phi angles	$0°$ and $90°$	$0°$ and $90°$
Exposure time	2 s	2 s
Number of exposure profiles	10	10
Number of exposures gain	30	30
X-ray elastic constant $S_1^{(hkl)}$	1.28×10^{-6} MPa	1.2×10^{-6} MPa
X-ray elastic constant $^1/_2 S_2^{(hkl)}$	5.92×10^{-6} MPa	7.18×10^{-6} MPa
Peak fit	Gaussian	Gaussian

2.4. Atmospheric-Induced Stress Corrosion Cracking Tests

AISCC tests were performed by exposing samples at a temperature of 75 °C $\pm$ 3 °C and a relative humidity (RH) of 35% $\pm$ 5% for 35 days. Three water droplets containing $MgCl_2$ were applied onto the surface of all U-bend samples, yielding an initial $MgCl_2$ deposition density of 20 µg/cm^2 (Drop 1), 332 µg/cm^2 (Drop 2), and 3835 µg/cm^2 (Drop 3). Drops 1, 2, and 3 correspond to a chloride deposition density of 15 µg/cm^2, 247 µg/cm^2, and 2856 µg/cm^2, respectively. An Eppendorf micropipette was used to dispense the droplets with volumes of 0.5, 1.5, and 2.5 µL, producing an overall droplet radius of 1.8, 2.3, and 2.8 mm, respectively. The effect of secondary spreading of the droplet during the test was not considered, therefore the exposed chloride concentrations should be regarded as "initial" chloride densities. The samples were placed in a Perspex box, and the humidity controlled by a saturated solution of $MgCl_2$, which was placed into the box together with the specimens. The box was sealed with a special temperature-resistant sealant and placed in a heating cabinet at 75 °C. The temperature and RH over the entire testing time was recorded with an EL-USB-2+ data logger from Lascar (Lascar Electronics Ltd., Salisbury, UK).

At the end of all tests, the exposed samples were imaged with a Zeiss stereo microscope (Jena, Thüringen, Germany). The samples were then cleaned in $\approx$70 °C citric acid solution for 2 h to remove all corrosion products without attacking the metal substrate. A photograph of a U-bend specimen deposited with salt-laden droplets and a stereo-microscopic image of a deposit after 35 days exposure is shown in Figure 3.

The corrosion morphology was analysed with an FEI Quanta 650 SEM and Zeiss Ultra V55 SEM (Jena, Thüringen, Germany). Ferrite could easily be distinguished from austenite, by its darker channelling contrast and morphological appearance. When this was not obvious, energy-dispersive X-ray spectroscopy (EDX, Oxford Instruments, Abingdon, Oxfordshire, UK) was used to inform about the chemical composition of each phase. Ferrite is typically richer in Cr and Mo, whereas the austenite is richer in Ni and Mn. The corrosion area and volume of each deposit was measured with a Keyence VK-X200 3D laser scanning confocal microscope (LSCM, Keyence Corporation, Osaka, Japan) using 20× and 50× lenses in stitching mode. Corrosion area calculations were further performed on SEM micrographs by converting the image into a binary matrix and determining the area of corrosion by threshold segmentation using the open source ImageJ image processing programme (version 1.6.0_24 64-bit, Wayne Rasband from National Institutes of Health, Bethesda, MD, USA).

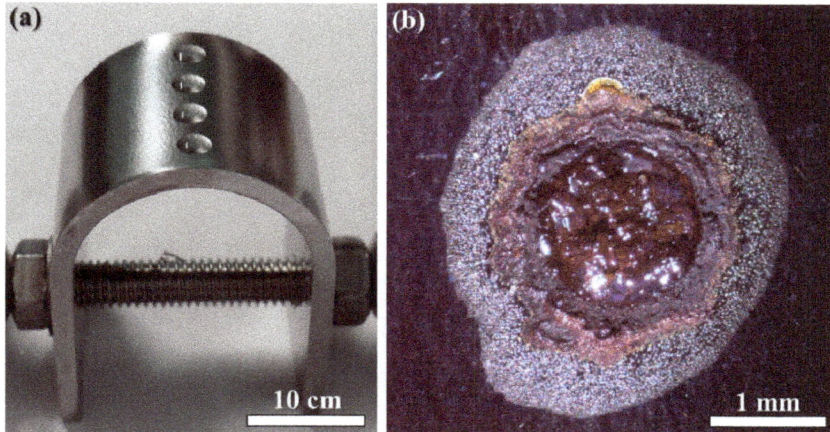

Figure 3. (a) Photograph of a U-bend specimen with salt-laden droplets containing $MgCl_2$; (b) stereo-microscopic image of one of the deposits after AISCC testing.

The microstructure propensity towards AISCC was assessed without disassembling the U-bend samples. The extent of SCC damage was determined as crack length per measured nominal corrosion area, and plotted against 475 °C exposure time. For the latter analysis, only the longest crack was considered, which was measured using SEM images and the ImageJ analysis program.

3. Results

3.1. Microstructure Characterisation

The as-received microstructure consisted of 54% ± 1% austenite (γ) and 46% ± 1% ferrite (δ). The average grain size of ferrite and austenite was 4.2 μm ± 2.3 μm and 2.2 μm ± 1.4 μm, respectively. The austenite grains are arrayed in a band-like structure and embedded in the ferritic matrix, as shown by the EBSD phase map in Figure 4a. The microstructure has elongated grains along RD, with no intermetallic phases observed. The aspect ratio of ferrite and austenite was 0.84 and 0.93, respectively. Numerous LAGBs were present in the microstructure, and a summary of all EBSD results is listed in Table 2.

After bending the sample in TD, the phase fraction of austenite to ferrite changed to 48% ± 1% versus 52% ± 1%, highlighting a clear reduction of the austenite content. The apex microstructure was heavily deformed with elongated ferrite and austenite grains towards the bending direction, as shown in Figure 4b. The band-like austenitic structure seemed disrupted after the bending process.

The average grain size of ferrite and austenite increased to 5.1 μm ± 2.3 μm and 3.7 μm ± 1.7 μm, respectively, with the aspect ratio of ferrite and austenite increasing to 1.61 and 1.71, respectively. This shows that austenite and ferrite grains were highly deformed in the bending direction (TD). The microstructure bent in TD shows far more disrupted grain morphologies, compared to the microstructure bent in RD.

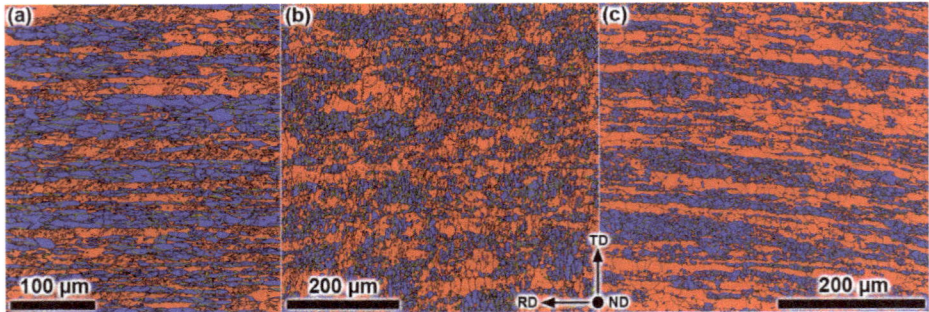

Figure 4. EBSD phase maps of (**a**) as-received 2205 DSS, and (**b**) after bending in TD and (**c**) in RD. Black lines are phase boundaries and HAGBs. The arrows indicate normal direction (ND), rolling direction (RD), and transverse direction (TD).

Table 2. EBSD grain geometry and morphology data.

Sample	Phase	Fraction (%)	G_S (μm)	σ (μm)	Aspect Ratio (X_V/X_H)
	δ	46	4.2	2.3	0.84
As-received	γ	54	2.2	1.4	0.93
	Total	100	2.8	1.8	0.91
	δ	53	5.1	2.3	0.80
Bent in RD	γ	47	2.5	1.4	0.92
	Total	100	3.4	1.9	0.89
	δ	52	5.1	2.2	1.61
Bent in TD	γ	48	3.7	1.7	1.71
	Total	100	4.3	1.7	1.67

G_S: Average grain size; σ: standard deviation.

After bending in RD, the phase fraction showed a similar change to 47% ± 1% austenite versus 53% ± 1% ferrite, supporting observations of a clear reduction in the austenite content. Large deformations can be noticed along the bending direction, as shown in Figure 4c. The band-like austenitic structure had an even more elongated appearance. The average grain size of ferrite and austenite slightly increased to 5.1 μm ± 2.3 μm and 2.5 μm ± 1.4 μm. The aspect ratio of ferrite and austenite decreased slightly to 0.80 and 0.92, respectively, indicating no significant change in grain morphology of both ferritic and austenitic grains.

The LMO analysis of both U-bend samples showed the presence of elastic-plastic strain and deformation in austenite and ferrite, summarised in Figure 5a–c. The degree of local misorientation changed to values up to 5°, indicating the presence of large strains across the entire microstructure due to the bending deformation. The microstructure bent in TD had nearly twice the degree of LMO compared to RD, indicating a far stronger effect on microstructure changes when samples are bent in TD. The austenite seemed to have a somewhat larger LMO distribution in comparison to ferrite, which was observed in both bending directions. In the microstructure bent in RD, a local region with accumulation of high LMOs, indicative of elastic/plastic strain, was observed in austenite.

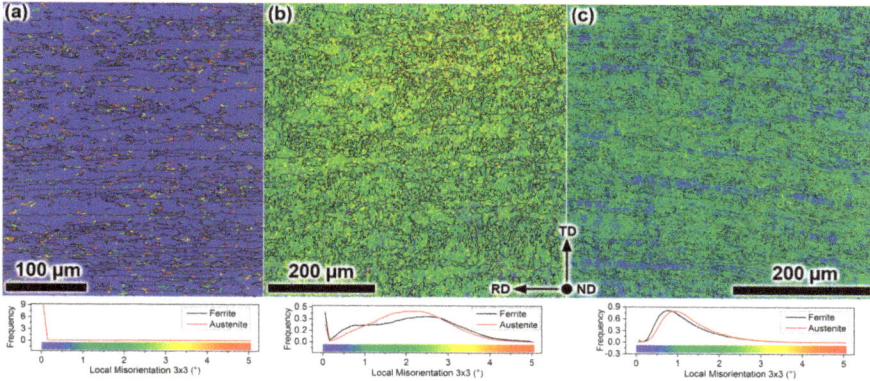

Figure 5. EBSD local misorientation maps of (**a**) the as-received microstructure, (**b**) after bending in TD, and (**c**) after bending in RD. Black lines are phase boundaries only. Note that the misorientation profiles of ferrite and austenite in (**a**) overlap with each other. The arrows indicate normal direction (ND), rolling direction (RD), and transverse direction (TD).

3.2. Hardness Behaviour

The hardness results of all samples as a function of ageing time at 475 °C are shown in Figure 6. The hardness increased from 285 HV30 ± 5 HV30 to 369 HV30 ± 5 HV30 after 255 h ageing. The same trend, but with significantly higher hardness values, was also observed for all U-bend specimens. The hardness after bending in RD and TD was 510 HV30 ± 15 HV30 and 529 HV30 ± 5 HV30, respectively. The hardness measured on U-bend specimens with 475 °C ageing treatment up to 10 h followed similar trends, as the hardness measured on as-received samples with the same heat treatment. This indicates only a minor effect of bending deformation on the hardness for short-term ageing treatments at 475 °C. Heat treatments in excess of 10 h showed a clear difference in hardness after bending.

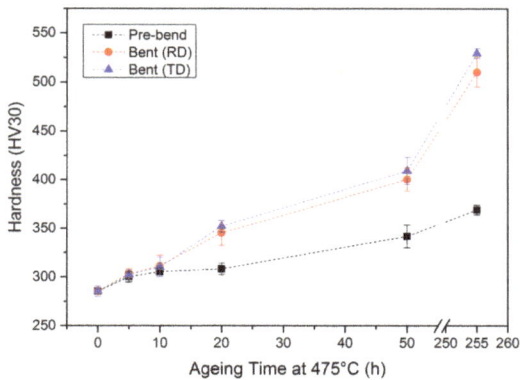

Figure 6. Hardness of flat strips (pre-bend) and U-bend samples in (RD) and (TD) as function of ageing exposure at 475 °C.

3.3. XRD Surface Stress Measurements

The surface stresses of ferrite and austenite before and after bending in RD and TD as a function of ageing time at 475 °C are shown in Figure 7a–d. Bending deformation of the as-received specimens led to the formation of tensile stresses in austenite with respect to both measurement directions (0° and 90°).

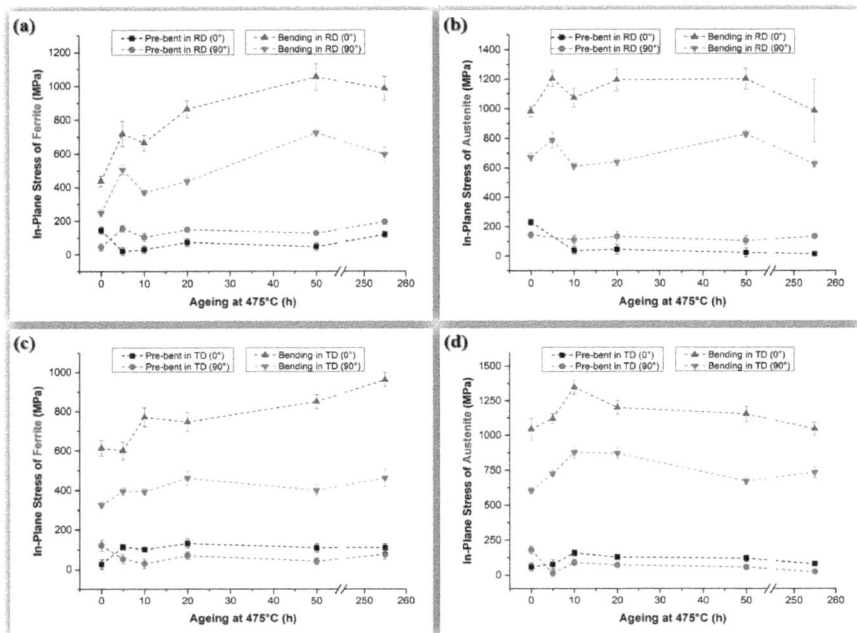

Figure 7. In-plane surface stress development in ferrite and austenite as a function of ageing at 475 °C: (**a–b**) before (pre-) and after bending in RD; (**c–d**) before (pre-) and after bending in TD.

The surface stresses of austenite of the flat specimen (Figure 7b) decreased with ageing from 232 MPa ± 20 MPa in 0° to 15 MPa ± 10 MPa, whereas in 90° no significant stress changes were noticed. The in-plane stress of ferrite increased from 44 MPa ± 23 MPa to 194 MPa ± 15 MPa in 90°, with no significant changes in 0°. The measurement orientation 0° in RD corresponds to 90° in TD and vice versa, and obtained data in the different process orientations always showed similar trends with only small variations. In summary, the austenite seemed to be the phase with the highest stress in the as-received microstructure, but the stress in ferrite grew with ageing to values similar to those measured in the austenite after long-term ageing.

After U-bending, the surface stresses of all microstructures increased significantly. The stresses in austenite after bending in RD were 980 MPa ± 33 MPa and 671 MPa ± 29 MPa in 0° and 90°, respectively. The stress in austenite in the microstructure aged for 5 h was higher than the stress measured in the as-received condition, reaching tensile stresses of 1203 MPa ± 55 MPa (0°) and 787 MPa ± 51 MPa (90°). No significant change of the stress in austenite was seen with longer ageing times. The stress of ferrite, however, showed a steady increase as a function of ageing time, increasing from 437 MPa ± 31 MPa (0°) and 247 MPa ± 24 MPa (90°) to 984 MPa ± 73 MPa (0°) and 595 MPa ± 44 MPa (90°) after 255 h ageing, in line with expected microstructure changes in ferrite with 475 °C embrittlement. The highest stress was clearly concentrated along the bending direction for both, ferrite and austenite.

The stress of austenite and ferrite after bending in TD showed a similar increase with the austenite reaching slightly higher stress values than ferrite. The stress measured in austenite showed a steady increase from 1042 MPa ± 80 MPa (0°) and 605 MPa ± 20 MPa (90°) to 1348 MPa ± 51 MPa (0°) and

878 MPa $\pm$ 44 MPa (90°), as a function of ageing up to 10 h. This was followed by a decrease to 1046 MPa $\pm$ 47 MPa and 735 MPa $\pm$ 39 MPa with longer ageing (255 h). The in-plane stress of ferrite rose steadily from 612 MPa $\pm$ 40 MPa (°0) and 325 MPa $\pm$ 15 MPa (90°) to 962 MPa $\pm$ 37 MPa (0°) and 463 MPa $\pm$ 44 MPa (90°) with ageing exposure.

3.4. Atmospheric-Induced Stress Corrosion Cracking Behaviour

3.4.1. As-Received (RD vs. TD)

The corrosion morphology in (RD) after exposure to 20 $\mu g/cm^2$ $MgCl_2$ (Drop 1) is shown in Figure 8a. Numerous nanometre-sized corrosion sites were present on deformed austenitic sites, whereas, the ferrite seemed unaffected. The deposit with 332 $\mu g/cm^2$ of $MgCl_2$ (Drop 2) also resulted in nanometre-sized corrosion sites, which selectively nucleated in the austenite, possibly associated with deformation sub-structures (slip bands), as can be seen in Figure 8b,c. Furthermore, cracks were observed in the ferrite of 5–20 μm length, either oriented parallel or ~45° inclined to the stress axis, as shown in Figure 8b,c. Under the droplet containing 3835 $\mu g/cm^2$ $MgCl_2$ (Drop 3) far more corrosion attack was observed as well as small micro-cracks (Figure 8d–f). Selective dissolution of ferrite occurred at the droplet periphery, with dissolution of both ferrite and austenite observed in the droplet centre. Most cracks propagated along grain boundaries and stopped after encountering other grains, shown in Figure 8e. However, connected cracks of up to 100 μm in length were found.

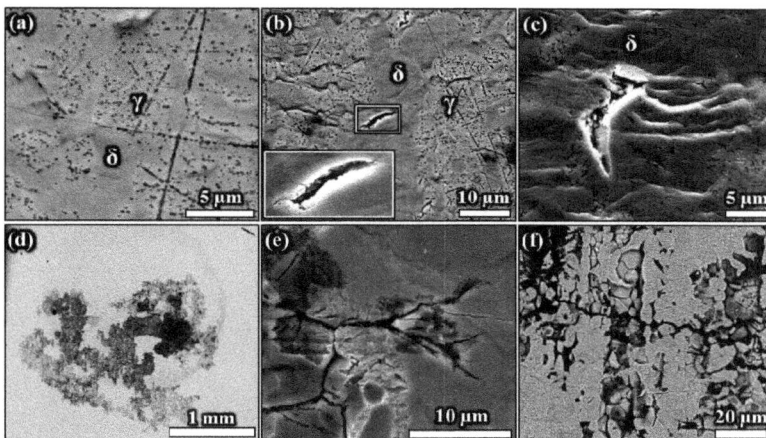

Figure 8. Corrosion morphology of as-received microstructure (in RD) with (**a**) "nano-pits" preferentially nucleated in austenite under Drop 1 (BSE-SEM image); (**b**) a micro-crack in ferrite with "nano-pits" in austenite under Drop 2 (SE-SEM image); (**c**) another micro-crack in ferrite under Drop 2 (SE-SEM image); (**d**) the entire attacked area under Drop 3; (**e,f**) small SCC cracks and crack-like regions under Drop 3. Bending direction is vertical to the images.

On the U-bend specimen (TD), two corrosion sites with 50–60 μm diameters were observed beneath Drop 1 (Figure 9a). Selective dissolution was identified as the main corrosion mechanism at those sites, shown in Figure 9a, especially at the circumference of the corroded region. Some intergranular corrosion and micro-cracks were also observed. These micro-cracks were branched with very narrow crack openings, and a maximum length of 10 μm. Under Drop 2, minor selective attack on ferrite with disintegration and fracture of austenite grains was observed, as shown in Figure 9b. Corrosion and SCC was observed under Drop 3, with dissolution of both phases and numerous cracks in austenite. The micro-cracks in austenite were indicative of chloride-induced SCC, showing transgranular cracks with branching (Figure 9d). The cracks were perpendicular to the stress axis.

In ferrite, however, multiple closely-spaced, parallel-arrayed crack-like morphologies were observed, indicating a hydrogen-embrittlement type of attack, as shown in Figure 9c.

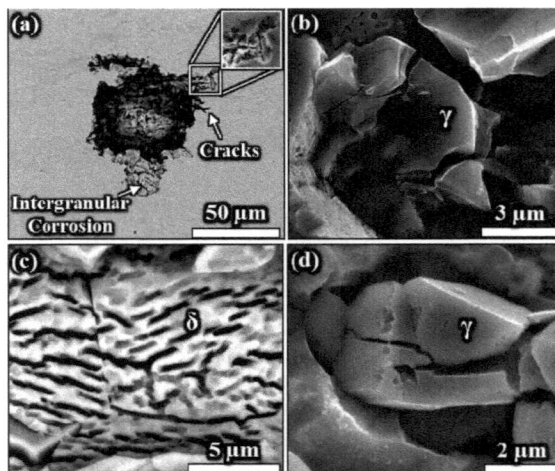

Figure 9. Corrosion morphology of the as-received microstructure (in TD) showing (**a**) corrosion sites with intergranular corrosion and some cracks emanating from the circumference of the pit wall under Drop 1 (BSE-SEM image); (**b**) cracked, disintegrated, and possibly fractured austenite grains under Drop 2 (SE-SEM image); (**c**) closely-spaced crack-like corrosion morphology in ferrite (BSE-SEM image), and (**d**) a branched transgranular micro-crack in austenite (SE-SEM image), both under Drop 3. Bending direction is vertical to all images.

3.4.2. Ageing at 475°C for 5 h (RD)

Multiple closely-spaced corrosion and crack sites were observed, primarily in the austenite under Drop 1, as can be seen in Figure 10a. No selective dissolution was observed, indicating no difference between the anodic and cathodic corrosion behaviour of both phases. A filiform-like corrosion pattern was found under Drop 2, as shown in Figure 10b. This corrosion morphology was neither associated with microstructure nor process orientation. Dissolution of ferrite and austenite with SCC cracks in austenite were observed under Drop 3 (Figure 10c). The maximum crack length found was ~120 µm.

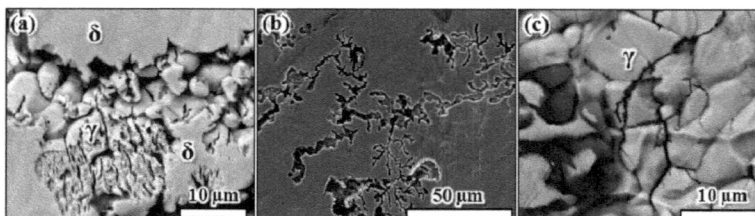

Figure 10. Corrosion morphology of microstructure aged at 475 °C for 5 h (in RD) with (**a**) closely-spaced micro-cracks and corrosion sites in austenite and ferrite under Drop 1 (BSE-SEM image), (**b**) filiform-type corrosion appearance under Drop 2 (SE-SEM image), and (**c**) corrosion with cracks in austenite under Drop 3 (BSE-SEM image). Bending direction is horizontal to all images.

3.4.3. Ageing at 475 °C for 10 h (RD)

Neither corrosion nor cracking was observed under Drop 1 and 2. Selective corrosion of ferrite had occurred with preferential dissolution along RD under Drop 3. Transgranular SCC was observed in austenite, with examples given in Figure 11a. Some SCC cracks had penetrated into ferrite grains, as shown in Figure 11b. Localised corrosion on slip bands and fractured austenite grains were observed (Figure 11c), suggesting slip-band dissolution-assisted SCC of the austenite.

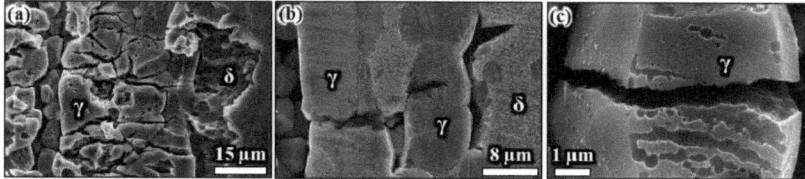

Figure 11. Corrosion morphology of microstructure aged at 475 °C for 10 h (in RD) under the deposit containing 3835 μg/cm^2 magnesium chloride showing (**a**) multiple micro-cracks in austenite with selective dissolution of ferrite; (**b**) selective dissolution of ferrite and attack on γ/δ interface; (**c**) a fractured austenite grain (all SE-SEM images). Bending direction is vertical to these images.

3.4.4. Ageing at 475 °C for 20 h (RD)

A few corroded nano-sites were found on ferrite and austenite grains with exposure to Drop 1. Under Drop 2, both austenite and ferrite seemed to have dissolved, as can be seen in Figure 12a. Multiple closely spaced and densely arranged micro-cracks were observed in the austenite. The dissolution of ferrite occurred beneath Drop 3 (Figure 12b). Multiple micro-cracks with transgranular pathways were observed in austenite, with a maximum length of ~100 μm.

Figure 12. Corrosion morphology of microstructure aged at 475 °C for 20 h (in RD) showing (**a**) multiple closely-spaced fine-scale cracks formed in austenite under Drop 2 (SE-SEM image); (**b**) micro-cracks in austenite and selective dissolution of ferrite under Drop 3 (BSE-SEM image). Bending direction is horizontal to the images.

3.4.5. Ageing at 475 °C for 50 h (RD)

Corrosion and cracking in ferrite and austenite was observed under Drop 1 (Figure 13a). Despite the large number of micro-cracks, a maximum crack length of only 5 μm was measured. In contrast, longer cracks were seen in the austenite under Drop 2, as shown in Figure 13b. The cracks seemed to be primarily of an intergranular nature, with a maximum crack length of 25 μm. Severe SCC with multiple transgranular cracks were found all over the corroded region under Drop 3 (Figure 13c), with a maximum length of ~100 μm.

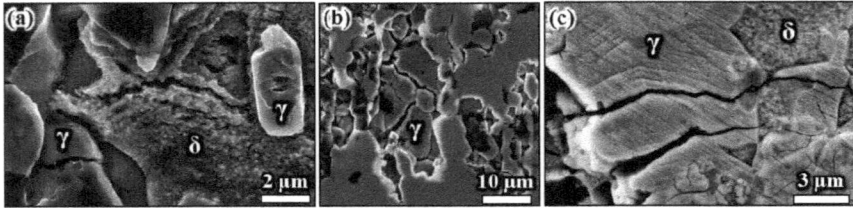

Figure 13. Corrosion morphology of microstructure aged at 475 °C for 50 h (in RD) showing (**a**) small cracks in ferrite and austenite under Drop 1 (SE-SEM image); (**b**) intergranular attack in austenite under Drop 2 (SE-SEM image); (**c**) transgranular micro-cracks with γ-δ-γ crack path (SE-SEM image). Bending direction is vertical to the images.

3.4.6. Ageing at 475 °C for 255 h (RD)

The specimen aged for 255 h showed the strongest corrosion attack response. The corrosion morphology observed under Drop 1 is shown in Figure 14a–d, with localised corrosion on austenite grains, in the vicinity of partially dissolved ferrite regions (Figure 14a), indicating preferential dissolution of ferrite. However, an increased susceptibility to localised corrosion was noticed in the austenite, associated with long-term ageing at 475 °C.

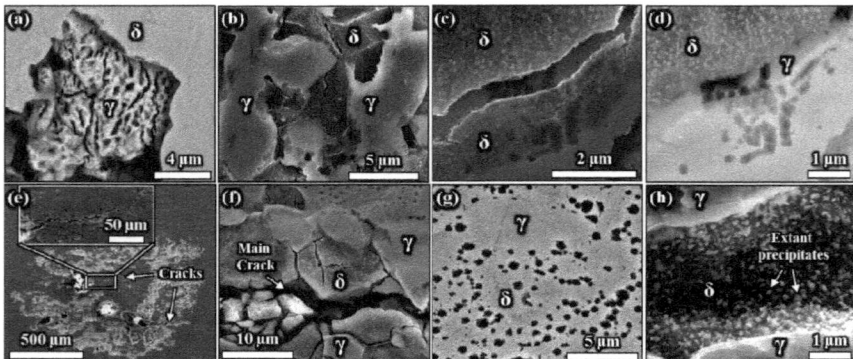

Figure 14. Corrosion morphology of microstructure aged at 475 °C for 255 h (in RD) showing (**a**) localised corrosion in austenite adjacent to a partially dissolved ferrite grain under Drop 1 (BSE-SEM image); (**b**) partially attacked austenite and embrittlement of ferrite under Drop 1 (SE-SEM image); (**c**) fractured ferrite grains (SE-SEM image); (**d**) localised corrosion on austenite underneath a ferrite grain under Drop 1 (BSE-SEM image); (**e**) cracks under Drop 2 (SE-SEM image); (**f**) multiple cracks in ferrite grains (SE-SEM image); (**g**) numerous corrosion pits in ferrite (BSE-SEM image); (**h**) precipitates at the crack wall under Drop 2 (SE-SEM image). Bending direction is vertical to all images.

In some areas, partially dissolved austenite grains and cracked ferrite regions were observed (Figure 14b). These cracks are clear evidence of the embrittlement of ferrite, with a preferred crack orientation towards the processing orientation. Moreover, in some austenite grains, local corrosion sites were observed beneath fractured ferrite grains (Figure 14c). This observation is supported by the SEM image in Figure 14d. The maximum crack length was ~250 μm. The extent of corrosion and SCC was more pronounced under Drop 2. The maximum crack length was ~700 μm, orientated perpendicular to the bending axis. The cracks could be noticed at low magnification, which can be seen in Figure 14f. Cracks were preferentially formed in the ferrite. Fractured ferrite grains were seen with multiple cracks, indicating microstructure embrittlement. The number and severity of cracks in ferrite were higher than those formed in austenite. Numerous corrosion pits also nucleated in the ferrite and at interphase boundaries, indicating enhanced local corrosion propensity (Figure 14g). No SCC could be seen in the austenite. At the bottom of the main crack, protruding precipitates were seen at the surface of ferrite regions in the form of bright speckles, shown in Figure 14h. These precipitates seemed to have net cathodic character with respect to the decomposed ferrite matrix, and are believed to be associated with R-phase. Severe corrosion and cracking was observed under Drop 3. General dissolution of both ferrite and austenite beneath the entire exposed area was seen. Wide and long cracks with a maximum crack length of ~1500 µm were found.

3.4.7. Ageing at 475 °C for 5 h (TD)

No corrosion attack was observed under Drop 1, clearly showing better corrosion resistance of the specimen aged for 5 h only. Beneath Drop 2, only nanometre-sized corrosion sites were observed in both ferrite and austenite. However, SCC was found under Drop 3 (Figure 15a,b), with a preferential pathway of cracks aligned along the microstructure process orientation. Cracks seemed to have nucleated in ferrite, and either propagated transgranularly through ferrite grains or along interphase boundaries. No selective dissolution of either phase was observed, indicating a more balanced net anodic and net cathodic character of both phases. Long SCC cracks were observed with a maximum crack length of ~1700 µm, oriented perpendicular to the stress axis.

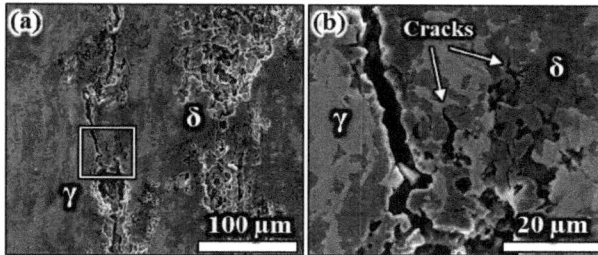

Figure 15. Corrosion morphology of 2205 DSS aged at 475 °C for 5 h (in TD) showing (**a**,**b**) long cracks in ferrite under Drop 3 (both SE-SEM images). Note that (**b**) is a magnified view of the highlighted region in (**a**). Bending direction is horizontal to the images.

3.4.8. Ageing at 475 °C for 10 h (bent in TD)

Corrosion and SCC was observed beneath Drop 1, but with only shallow corrosion sites. A number of transgranular micro-cracks were found in the austenite with a maximum crack length of ~30 µm. Filiform-type corrosion morphologies were observed under Drop 2 (Figure 16a), with transgranular SCC in austenite under Drop 3 (Figure 16b,c). The maximum crack length was ~50 µm, oriented perpendicular to the stress axis.

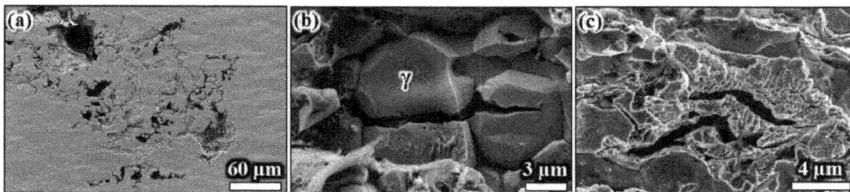

Figure 16. Corrosion morphology of microstructure aged at 475 °C for 10 h (in TD) showing (**a**) filiform-like corrosion of ferrite and austenite under Drop 2; (**b**,**c**) micro-cracks in austenite under Drop 3 (all SE-SEM images). Bending direction is vertical to the images.

3.4.9. Ageing at 475 °C for 20 h (TD)

On the specimen aged for 20 h two corrosion sites with maximum lateral sizes of 10 μm were seen underneath Drop 1 (Figure 17a). Significant surface attack was observed under Drop 2, indicating corrosion of both ferrite and austenite (Figure 17b). A few fine-scale micro-cracks were seen in austenite, but no cracks in ferrite. Under Drop 3, numerous fine-scale cracks were also found in austenite, whereas the ferrite seemed to have corroded only superficially (Figure 17c). In austenite, multiple SCC-containing sites were found (Figure 17d).

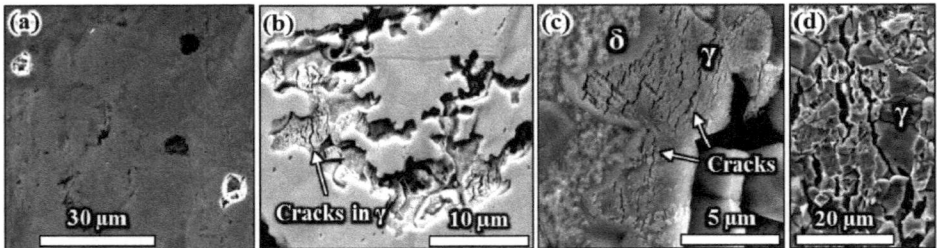

Figure 17. Corrosion morphology aged at 475 °C for 20 h (in TD) showing (**a**) pits formed under Drop 1 (SE-SEM image); (**b**) filiform-like corrosion with micro-cracks in austenite under Drop 2 (BSE-SEM image); (**c**) numerous closely-spaced fine-scale cracks in austenite under Drop 3 (SE-SEM image); (**d**) cracks in austenite under Drop 3 (SE-SEM image). Bending direction is horizontal to all images.

3.4.10. Ageing at 475 °C for 50 h (TD)

Selective corrosion occurred on austenite grains under Drop 1 with slip planes acting as preferential nucleation sites, as shown in Figure 18b. General corrosion of ferrite and austenite was observed under Drop 2, with numerous crack-like features in the austenite (Figure 18b). Severe corrosion was observed under Drop 3, with cracks in ferrite and austenite (also summarised in Figure 18c). Bright features at the surface of ferrite were observed, indicating R-phase precipitates.

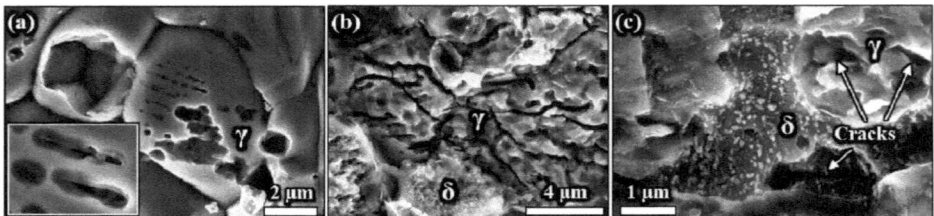

Figure 18. Corrosion morphology of microstructure aged at 475 °C for 50 h (in TD) showing (**a**) localised corrosion along slip planes on austenite under Drop 1 (SE-SEM image); (**b**) crack-like features in austenite under Drop 2 (SE-SEM image); (**c**) precipitates in ferrite (bright speckles) with cracks in ferrite and austenite under Drop 3 (SE-SEM image). Bending direction is vertical to the images.

3.4.11. Ageing at 475 °C for 255 h (TD)

The specimen aged for 255 h showed localised corrosion in austenite under Drop 1 (Figure 19a), and selective dissolution of ferrite along the process orientation was also observed. This observation was different to the corrosion morphology observed on the specimen aged for 255 h and bent in RD. Multiple closely-spaced crack-like features were seen in austenite, as shown in Figure 19a, with numerous transgranular micro-cracks under Drop 2 (Figure 19b). Corrosion was observed under Drop 3, with dissolution of both ferrite and austenite, as well as cracks seen in both ferrite and austenite (Figure 19d).

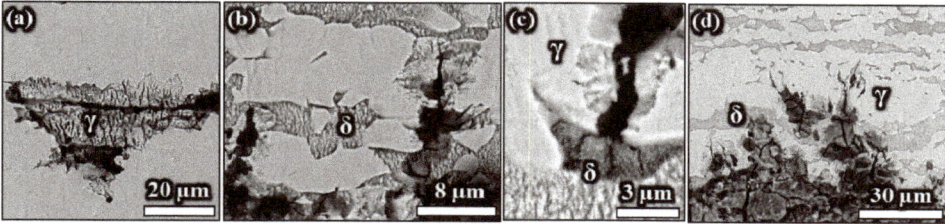

Figure 19. Corrosion morphology aged at 475 °C for 255 h (in TD) showing (**a**) selective dissolution of ferrite and multiple closely-spaced crack-like features in austenite under Drop 1 (BSE-SEM image); (**b**) transgranular cracks through ferrite and austenite under Drop 2 (BSE-SEM image); (**c,d**) cracks in ferrite and austenite under Drop 3 (both BSE-SEM image). Bending direction is horizontal to the images.

4. Comparison and Quantification

The measured corrosion area and corrosion volume as a function of ageing time and process direction are summarised in Figures 20 and 21, respectively. In general, the area and volume of corrosion increased with increasing chloride deposition density, and with 475 °C embrittlement heat treatment exposure time. Most corrosion was observed on specimens aged for 255 h. Furthermore, corrosion attack on specimens bent in TD was in general larger than those bent in RD. No corrosion was observed under Drop 1 for all heat treatment conditions and both process orientations. Only very little attack was present under Drop 2, with corroded areas in the order of ~10^5 μm^2 and corrosion volume of ~10^6 μm^2. The area of corrosion attack beneath Drop 3 was typically ~10^6–10^7 μm^2, with a volume of ~10^7 μm^3.

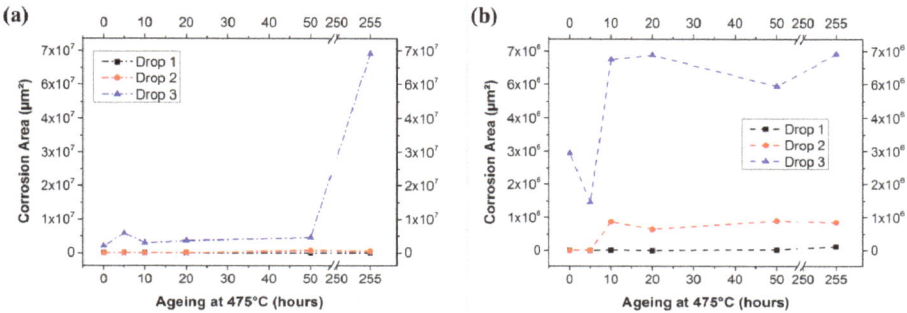

Figure 20. Measured surface area of corrosion as a function of ageing time at 475 °C of microstructures bent in (**a**) RD and (**b**) TD.

The extent of corrosion showed the lowest volume of corrosion for the 10-h-aged specimens bent in RD and the five-hour-aged specimen bent in TD. Corrosion under Drop 3 did not seem to have stopped or saturated, even for a microstructure aged for 255 h. The corrosion volume on the 255 h aged specimens was ~5-times larger compared to the as-received samples.

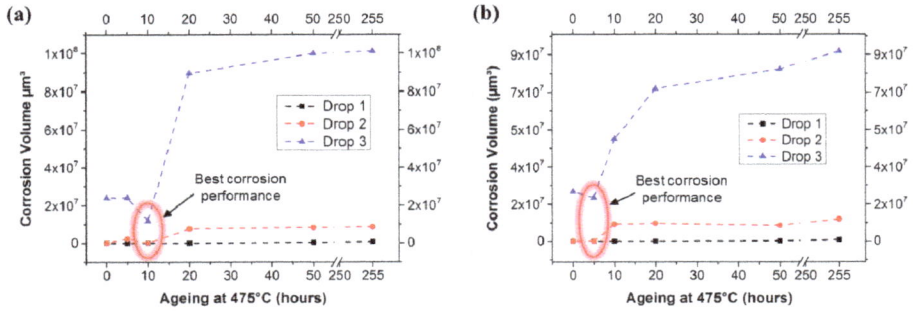

Figure 21. Measured volume of corrosion as a function of ageing time at 475 °C bent in (**a**) RD and (**b**) TD.

SCC cracks were also measured as a function of ageing time (Figure 22). The shortest cracks were found on the specimen aged for 10 h (in RD), indicating the best SCC performance. Severe SCC occurred on the 255-h-aged specimen, with maximum crack length measured under Drops 1, 2, and 3 of ~250, ~700, and ~1500 μm, respectively. However, the U-bend specimens in TD showed a somewhat different crack length distribution. The longest crack with 1700 μm length was measured under Drop 3 for the specimen aged for 5 h.

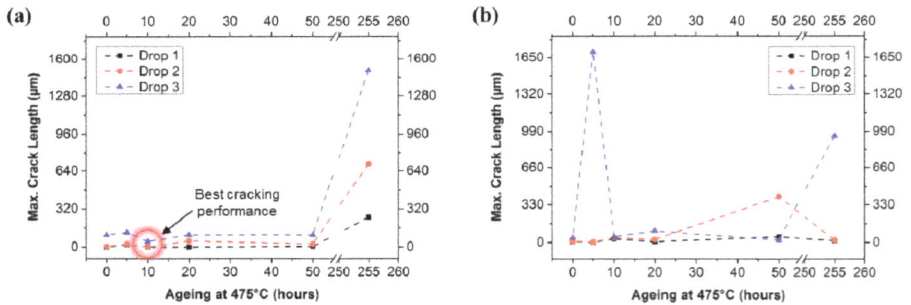

Figure 22. Measured maximum SCC crack length of 2205 DSS microstructure bent (**a**) in RD and (**b**) TD.

5. Discussion

5.1. Microstructure Characterisation

The fraction of austenite decreased from 54% to 47% in the microstructure with bending in RD and to 48% in TD. This can possibly be attributed to the formation of deformation-induced martensite, which is typically indexed as ferrite due to its similar crystallographic structure. Deformation-induced martensite in 2205 DSS has been reported after plastic deformation by cold rolling ($\varepsilon = 3.38$), which supports this assumption [40–42].

The bending deformation resulted in an increase of grain sizes of both ferrite and austenite, with the latter being more affected in the TD direction. All grain structures were apparently more affected after bending in TD than after bending in RD. The microstructure resembled a sandwich-like appearance consisting of bands of ferrite and austenite. As a consequence of the deformation of the ferrite, the surrounding austenite grains were also constraint.

LMO development clearly showed that the austenite contained larger proportions of elastic-plastic strain than the ferrite after bending in RD, whereas the opposite was apparent after bending in

TD. The structural change and strain response clearly demonstrated the effect of microstructure on bending. Similar results have been recently reported in cold-rolled 2205 DSS in which the fraction of HAGBs and LAGBs increased with cold reduction, causing micro-deformation and, hence, strain in the microstructure [36,43]. Severe strain localisation in austenite grains was observed, resulting in localised corrosion [36,43].

5.2. Hardness Behaviour

Bending increased the hardness of the stainless steel strips resulting in a higher dislocation density in the microstructure. The hardness also increased with ageing, indicating an effect of the 475 °C embrittlement treatment. The latter is known to impede dislocation motion in ferrite [44]. The hardness increased gradually up to 20 h of ageing exposure, with a steep rise after 50 h, reaching a maximum hardness after 255 h. Larger precipitates were seen after 255 h of ageing. These are also believed to be responsible for the additional hardness increase, and they may even play a larger role than spinodal decomposition for the observed hardness changes [45,46].

5.3. XRD Stress Development

The surface stress of austenite was not substantially affected by ageing alone, whereas the stress in the ferrite became more tensile after 255 h of ageing. The latter was even more pronounced after bending in RD. Precipitates formed at the interface between austenite and ferrite can hinder dislocation motion, forcing ferrite grains to distort along RD. The effect of 475 °C embrittlement on the stress state of both phases can clearly be noticed after bending in RD and TD. Both phases became tensile in all directions with ageing exposure, which is the response of the microstructure due to the applied strain. The observed stress in austenite was higher than the stress in ferrite in the as-received microstructure, as well as in all conditions aged for up to 50 h. This is related to larger plastic strain, which can be accommodated in austenite, most likely accompanied by strain-induced martensite formation. Martensite formation is typically accompanied by a volume expansion resulting in strain hardening, which in turn is associated with retardation of dislocation motion. Both phases had similar tensile strength after 255 h of ageing, indicating a balanced stress state.

5.4. Corrosion and AISCC Behaviour

Temperature and RH play a key role in atmospheric corrosion, and the highest susceptibility to AISCC has been reported to occur on stainless steels at RHs at or close to salt deliquescence [47]. Up to 12 M chloride can be formed at the deliquescence point of $MgCl_2$ at room temperature [48,49], and this can be 15 M or even higher for $\geqslant$50 °C [24,50]. Large droplet sizes (diameter, thickness) have also been reported to increase the extent of atmospheric corrosion [51,52]. There is a relationship between the extent of corrosion with the droplet size (ranking order Drop 3 > Drop 2 > Drop 1). Increasing the chloride deposition density from 15 μg/cm^2 (Drop 1) to 247 μg/cm^2 (Drop 2) increased the corrosion by more than six orders of magnitude, with a further increase of the chloride deposition density to 2856 μg/cm^2 (Drop 3) the extent of corrosion increased approximately by one order of magnitude (see Figures 20 and 21).

Ferrite and austenite have different electrochemical corrosion potentials, therefore a different form and extent of corrosion attack can be expected [53–56]. In most 2205 DSSs, ferrite usually dissolves in chloride-bearing media [34,35,54–59], which is due to a higher corrosion potential of austenite deriving from its higher nickel and nitrogen contents [54–56]. Furthermore, ferrite and austenite show inherently different mechanical behaviour during plastic deformation due to their different crystal structure [3,36,38,60]. The corrosion mechanism was clearly affected by the ageing time, bending deformation, and also process orientation (RD, TD).

Short-term ageing in the 475 °C embrittlement temperature window of duplex stainless steels is typically associated with an increase of the corrosion resistance, but then a decrease with further ageing time has been observed [35,59,61]. After these short 475 °C embrittlement treatments, an improvement of the passivation behaviour of the ferrite has been observed with no selective attack on either the ferrite or the austenite [35,59,61]. Spinodal decomposition of the ferrite has been postulated as the reason for an increase of the corrosion resistance of ferrite. Longer exposure periods, in excess of 10 h at 475 °C, results in further precipitates formed within ferrite regions and at interphase boundaries, such as R-phase. These are believed to be the reason for the reduced corrosion resistance of the ferrite [35].

The AISCC behaviour was compared to corrosion area and corrosion volume measurements in Figure 22. The extent of SCC (length, number) in the microstructures along RD increased with ageing time, with the lowest susceptibility to cracking observed, in line with the smallest corrosion attack. The microstructure bent in TD, however, showed severe cracking beneath Drop 3 after ageing for 5 h. The extent of SCC and the corrosion volume of the microstructure after bending in TD did not seem to correlate with those after bending in RD, showing that microstructure plays a key role in the extent of atmospheric corrosion and SCC.

In the as-received condition and after bending in RD, ferrite was preferentially attacked by localised corrosion. Cold deformation causes strain heterogeneity in the duplex microstructure, with austenite accommodating most plastic strains and forming strain hotspots, which have been reported to become electrochemically active sites [36]. Localised corrosion in austenite was observed under Drops 1 and 2. Localised corrosion can govern the corrosion mechanism for heavily deformed microstructures [36,60]. The austenite was more deformed after bending in RD, with strain heterogeneities observed to cause localised attack, whereas ferrite was more strained in TD.

SCC occurred primarily in the austenite, which may be considered as the phase primarily responsible for cracking. However, the propensity to SCC increased with ageing time and reached maximum susceptibility after 255 h, with numerous ferrite grains also showing fractures. Ferrite became prone to SCC with an increasing degree of 475 °C embrittlement. The crack morphology observation of ferrite would suggest chloride-induced SCC or hydrogen-assisted embrittlement/cracking, since multiple closely-spaced, fine-scale cracks arrayed in crystallographic manner are typically signs of hydrogen embrittlement [62–64]. The crack morphologies in austenite suggest that chloride-induced SCC was most likely the cause of cracking. Despite the quantitatively larger number of cracks in the microstructure bent in TD, the maximum crack length observed was only 40 µm, while longer cracks of ~100 µm were measured in microstructures bent in RD. This suggests that bending in RD is more critical than in TD regarding the corrosion and SCC susceptibility of 2205 duplex stainless steel.

6. Conclusions

In this study, the effect of process orientation and ageing at 475 °C on microstructure development and atmospheric chloride-induced stress corrosion cracking was investigated.

1. The extent of 475°C embrittlement increases with increasing ageing time and is mainly caused by microstructural changes in the ferrite. All microstructures were rendered susceptible to AISCC with exposure to 475 °C, with improved AISCC performance observed for microstructures aged for 5 or 10 h at 475 °C.
2. The austenite phase was mainly susceptible to AISCC in microstructures aged up to 50 h, whereas the ferrite became more susceptible after ageing in excess of 50 h.
3. The samples manufactured in TD had highest propensity towards AISCC. This was in line with higher strain and stress in both ferrite and austenite in TD. Bending along TD results in larger strain and stresses in the microstructure and higher propensity to corrosion and AISCC. The direction of bending deformation determines the extent and distribution of strain in the duplex microstructure.

4. XRD measurements confirmed the development of stresses in ferrite and austenite with ageing at 475 °C. The austenite showed larger stresses with exposure up to 50 h, whereas the stress in ferrite increased with ageing time.

Acknowledgments: The authors acknowledge Radioactive Waste Management, a wholly owned subsidiary of the Nuclear Decommissioning Authority (NPO004411A-EPS02), and EPSRC (EP/I036397/1) for financial support. The authors are grateful for the kind provision of Grade 2205 Duplex Stainless Steel plate by Rolled Alloys. Special thanks are addressed to Gary Harrison, School of Materials, for his support during XRD measurements.

Author Contributions: D.L.E and C.Ö. envisioned experimental planning, design, and concept of the work in the manuscript. C.Ö. and D.L.E wrote the manuscript and made discussion of the results. S.A.M.I. carried out sample preparation, heat treatment, corrosion tests, and XRD and hardness measurements. C.Ö. carried out SEM and EBSD microstructure characterisation. C.Ö. and P. R. carried out post-corrosion measurements.

Conflicts of Interest: The authors declare no conflicts of interest.

References

1. Charles, J.; Bernhardsson, S. *Duplex Stainless Steels'91—Volume 1*; Les editions de physique: Beaune, France, 1991.
2. Charles, J.; Bernhardsson, S. *Duplex Stainless Steels'91—Volume 2*; Les editions de physique: Beaune, France, 1991.
3. Nilsson, J.-O.; Chai, G. The Physical Metallurgy of Duplex Stainless Steels. In Proceedings of International Conference & Expo Duplex 2007, Grado, Italy, 18–20 June 2007.
4. Kangas, P.; Nicholls, J.M. Chloride-induced stress corrosion cracking of Duplex stainless steels. Models, test methods and experience. *Mater. Corros.* **1995**, *46*, 354–365. [CrossRef]
5. Chung, H.M. Aging and life prediction of cast duplex stainless steel components. *Int. J. Press. Vessel. Pip.* **1992**, *50*, 179–213. [CrossRef]
6. Sahu, J.K.; Krupp, U.; Ghosh, R.N.; Christ, H.J. Effect of 475 °C embrittlement on the mechanical properties of duplex stainless steel. *Mater. Sci. Eng. A* **2009**, *508*, 1–14. [CrossRef]
7. Eckstein, H.-J. *Korrosionsbeständige Stähle*; Deutscher Verlag für Grundstoffindustrie GmbH: Leipzig, Germany, 1990.
8. Chung, H.M. Spinodal decomposition of austenite in long-term-aged duplex stainless steel. In Proceedings of Fall meeting of the Metallurgical Society, Las Vegas, NV, USA, 27 February 1989.
9. Chung, H.M.; Leax, T.R. Embrittlement of laboratory and reactor aged CF3, CF8, and CF8M duplex stainless steels. *Mater. Sci. Technol.* **1990**, *6*, 249–262. [CrossRef]
10. Redjaïmia, A.; Otarola, T.; Mateo, A. Orientation Relationships between the δ-ferrite Matrix in a Duplex Stainless Steel and its Decomposition Products: The Austenite and the χ and R Frank-Kasper Phases. In Proceedings of EMC 2008 14th European Microscopy Congress, Aachen, Germany, 1–5 September 2008; pp. 479–480.
11. Redjaïmia, A.; Proult, A.; Donnadieu, P.; Morniroli, J.P. Morphology, crystallography and defects of the intermetallic χ-phase precipitated in a duplex (δ + γ) stainless steel. *J. Mater. Sci.* **2004**, *39*, 2371–2386.
12. Redjaimia, A.; Ruterana, P.; Metauer, G.; Gantois, M. Identification and characterization of a novel intermetallic compound in a Fe-22 wt. % Cr-5 wt. % Ni-3 wt. % Mo-0.03 wt. % C duplex stainless steel. *Philos. Mag. A* **1993**, *67*, 1277–1286. [CrossRef]
13. Nilsson, J.-O. Super duplex stainless steels. *Mater. Sci. Technol.* **1992**, *8*, 685–700. [CrossRef]
14. Nilsson, J.-O.; Liu, P. Aging at 400–600 °C of submerged arc welds of 22Cr-3Mo-8Ni duplex stainless steel and its effect on toughness and microstructure. *Mater. Sci. Technol.* **1991**, *7*, 853–862. [CrossRef]
15. Mateo, A.; Llanes, L.; Anglada, M.; Redjaimia, A.; Metauer, G. Characterization of the intermetallic G-phase in an AISI 329 duplex stainless steel. *J. Mater. Sci.* **1997**, *32*, 4533–4540. [CrossRef]
16. Chung, H.M.; Chopra, O.K. Characterization of Duplex Stainless Steels by TEM, SANS, and APFIM Techniques. In *Characterization of Advanced Materials*; Altergott, W., Henneke, E., Eds.; Springer US: New York, NY, USA, 1990; pp. 123–147.
17. Horvath, W.; Prantl, W.; Stroißnigg, H.; Werner, E.A. Microhardness and microstructure of austenite and ferrite in nitrogen alloyed duplex steels between 20 and 500 °C. *Mater. Sci. Eng. A* **1998**, *256*, 227–236.
18. Sahu, J.K. Effect of 475 °C Embrittlement on the Fatigue Behaviour of a Duplex Stainless Steel. Ph.D. Thesis, University of Siegen, Siegen, Germany, October 2008.

19. Tucker, J.D.; Miller, M.K.; Young, G.A. Assessment of thermal embrittlement in duplex stainless steels 2003 and 2205 for nuclear power applications. *Acta Mater.* **2015**, *87*, 15–24. [CrossRef]

20. Chandra, K.; Kain, V.; Bhutani, V.; Raja, V.S.; Tewari, R.; Dey, G.K.; Chakravartty, J.K. Low temperature thermal aging of austenitic stainless steel welds: Kinetics and effects on mechanical properties. *Mater. Sci. Eng. A* **2012**, *534*, 163–175. [CrossRef]

21. Tsai, W.-T.; Chou, S.-L. Environmentally assisted cracking behavior of duplex stainless steel in concentrated sodium chloride solution. *Corros. Sci.* **2000**, *42*, 1741–1762. [CrossRef]

22. Oldfield, J.W.; Todd, B. Room temperature stress corrosion cracking of stainless steels in indoor swimming pool atmospheres. *Br. Corros. J.* **1991**, *26*, 173–182. [CrossRef]

23. Tseng, C.-M.; Tsai, W.-T.; Liou, H.-Y. Effect of nitrogen content on the environmentally-assisted cracking susceptibility of duplex stainless steels. *Metall. Mater. Trans. A* **2003**, *34*, 95–103. [CrossRef]

24. Prosek, T.; le Gac, A.; Thierry, D.; le Manchet, S.; Lojewski, C.; Fanica, A.; Johansson, E.; Canderyd, C.; Dupoiron, F.; Snauwaert, T.; et al. Low-Temperature Stress Corrosion Cracking of Austenitic and Duplex Stainless Steels Under Chloride Deposits. *Corrosion* **2014**, *70*, 1052–1063. [CrossRef]

25. Ruel, F.; Saedlou, S.; Manchet, S.L.; Lojewski, C.; Wolski, K. The transition between Sulfide Stress Cracking and Stress Corrosion Cracking of the 2304 DSS as a function of T and pH in H_2S environment. In Proceedings of Eurocorr 2014, Pisa, Italy, 8–12 September 2014.

26. Nisbet, W.J.; Lorimer, G.W.; Newman, R.C. A transmission electron microscopy study of stress corrosion cracking in stainless steels. *Corros. Sci.* **1993**, *35*, 457–469. [CrossRef]

27. Arnold, N.; Gümpel, P.; Heitz, T.W. Chloridinduzierte Korrosion von nichtrostenden Stählen in Schwimmhallen-Atmosphären Teil 2: Einfluß von Hypochloriten. *Mater. Corros.* **1998**, *49*, 482–488. [CrossRef]

28. Arnold, N.; Gümpel, P.; Heitz, T.; Pscheidl, P. Chloridinduzierte Korrosion von Nichtrostenden Stählen in Schwimmhallen-Atmosphären Teil 1: Elektrolyt Magnesium-Chlorid (30%). *Mater. Corros.* **1997**, *48*, 679–686.

29. Takizawa, K.; Shimizu, Y.; Yoneda, E.; Tamura, I. Effects of Cold Work, Heat Treatment and Volume Fraction of Ferrite on Stress Corrosion Cracking Behavior of Duplex Stainless Steel. *Trans. Iron Steel Inst. Jpn.* **1981**, *67*, 353–361. [CrossRef]

30. Takizawa, K.; Shimizu, Y.; Yoneda, E.; Shoji, H.; Tamura, I. Effect of Cold Working and Heat Treatment on Stress Corrosion Cracking Behavior of Duplex Stainless Steel. *Trans. Iron Steel Inst. Jpn.* **1979**, *5*, 617–626.

31. Takizawa, K.; Shimizu, Y.; Higuchi, Y.; Tamura, I. Effects of Cold Work and 475 °C Aging on Corrosion Behavior of Duplex Stainless Steel. *Trans. Iron Steel Inst. Jpn.* **1984**, *70*, 904–910.

32. Kwon, H.-S.; Kim, H.-S. Investigation of stress corrosion susceptibility of duplex $(\alpha + \gamma)$ stainless steel in a hot chloride solution. *Mater. Sci. Eng. A* **1993**, *172*, 159–165. [CrossRef]

33. Leinonen, H.; Pohjanne, P.; Saukkonen, T.; Schildt, T. *Effect of Selective Dissolution on Stress Corrosion Cracking Susceptibility of Austenitic and Duplex Stainless Steels in Alkaline Solutions*; NACE International: Houston, TX, USA, 2011.

34. Engelberg, D.L.; Örnek, C. Probing propensity of grade 2205 duplex stainless steel towards atmospheric chloride-induced stress corrosion cracking. *Corros. Eng. Sci. Technol.* **2014**, *49*, 535–539. [CrossRef]

35. Örnek, C.; Engelberg, D.L. Effect of "475 °C Embrittlement" on the Corrosion Behaviour of Grade 2205 Duplex Stainless Steel Investigated Using Local Probing Techniques. *Corros. Manag.* **2013**, 9–11.

36. Örnek, C.; Engelberg, D.L. SKPFM measured Volta potential correlated with strain localisation in microstructure to understand corrosion susceptibility of cold-rolled grade 2205 duplex stainless steel. *Corros. Sci.* **2015**, *99*, 164–171. [CrossRef]

37. Örnek, C.; Engelberg, D.L. Correlative EBSD and SKPFM characterisation of microstructure development to assist determination of corrosion propensity in grade 2205 duplex stainless steel. *J. Mater. Sci.* **2015**, *51*, 1931–1948. [CrossRef]

38. Örnek, C.; Zhong, X.; Engelberg, D.L. Low-Temperature Environmentally Assisted Cracking of Grade 2205 Duplex Stainless Steel beneath a $MgCl_2$:$FeCl_3$ Salt Droplet. *Corrosion* **2015**, *72*, 384–399. [CrossRef]

39. Mino, K.; Fukuoka, C.; Yoshizawa, H. Evolution of intragranular misorientation during plastic deformation. *J. Jpn. Inst. Met.* **2000**, *64*, 50–55.

40. Tavares, S.S.M.; da Silva, M.R.; Pardal, J.M.; Abreu, H.F.G.; Gomes, A.M. Microstructural changes produced by plastic deformation in the UNS S31803 duplex stainless steel. *J. Mater. Process. Technol.* **2006**, *180*, 318–322.

41. Tavares, S.S.M.; Pardal, J.M.; da Silva, M.R.; Oliveira, C.A.S.D. Martensitic transformation induced by cold deformation of lean duplex stainless steel UNS S32304. *Mater. Res.* **2014**, *17*, 381–385. [CrossRef]

42. Guo, Y.; Hu, J.; Li, J.; Jiang, L.; Liu, T.; Wu, Y. Effect of Annealing Temperature on the Mechanical and Corrosion Behavior of a Newly Developed Novel Lean Duplex Stainless Steel. *Materials* **2014**, *7*, 6604–6619.

43. Örnek, C.; Engelberg, D.L. Kelvin Probe Force Microscopy and Atmospheric Corrosion of Cold-rolled Grade 2205 Duplex Stainless Steel. In Proceedings of Eurocorr 2014, Pisa, Italy, 8–12 September 2014; pp. 1–10.

44. Cahn, J.W. Hardening by spinodal decomposition. *Acta Metall.* **1963**, *11*, 1275–1282.

45. Hättestrand, M.; Larsson, P.; Chai, G.; Nilsson, J.-O.; Odqvist, J. Study of decomposition of ferrite in a duplex stainless steel cold worked and aged at 450–500 °C. *Mater. Sci. Eng. A* **2009**, *499*, 489–492. [CrossRef]

46. Weng, K.L.; Chen, H.R.; Yang, J.R. The low-temperature aging embrittlement in a 2205 duplex stainless steel. *Mater. Sci. Eng. A* **2004**, *379*, 119–132. [CrossRef]

47. Shoji, S.; Ohnaka, N. Effects of Relative Humidity and Kinds of Chlorides on Atmospheric Stress Corrosion Cracking of Stainless Steels at Room Temperature. *Corros. Eng.* **1989**, *38*, 92–97.

48. Tsutsumi, Y.; Nishikata, A.; Tsuru, T. Initial Stage of Pitting Corrosion of Type 304 Stainless Steel under Thin Electrolyte Layers Containing Chloride Ions. *J. Electrochem. Soc.* **2005**, *152*, B358–B363. [CrossRef]

49. Tsutsumi, Y.; Nishikata, A.; Tsuru, T. Pitting corrosion mechanism of Type 304 stainless steel under a droplet of chloride solutions. *Corros. Sci.* **2007**, *49*, 1394–1407. [CrossRef]

50. Prosek, T.; Iversen, A.; Taxen, C. *Low Temperature Stress Corrosion Cracking of Stainless Steels in the Atmosphere in Presence of Chloride Deposits*; NACE International: Houston, TX, USA, 2008; p. 17.

51. Li, S.; Hihara, L.H. Atmospheric-Corrosion Electrochemistry of NaCl Droplets on Carbon Steel. *J. Electrochem. Soc.* **2012**, *159*, C461–C468. [CrossRef]

52. Nishikata, A.; Ichihara, Y.; Hayashi, Y.; Tsuru, T. Influence of Electrolyte Layer Thickness and pH on the Initial Stage of the Atmospheric Corrosion of Iron. *J. Electrochem. Soc.* **1997**, *144*, 1244–1252. [CrossRef]

53. Symniotis, E. Galvanic Effects on the Active Dissolution of Duplex Stainless Steels. *Corrosion* **1990**, *46*, 2–12.

54. Aoki, S.; Ito, K.; Yakuwa, H.; Miyasaka, M.; Sakai, J.I. Potential Dependence of Preferential Dissolution Behavior of a Duplex Stainless Steel in Simulated Solution inside Crevice. *Zair. Kankyo Corros. Eng.* **2011**, *60*, 363–367. [CrossRef]

55. Aoki, S.; Yakuwa, H.; Mitsuhashi, K.; Sakai, J.I. Dissolution Behavior of α and γ Phases of a Duplex Stainless Steel in a Simulated Crevice Solution. *ECS Trans.* **2010**, *25*, 17–22.

56. Lee, J.-S.; Fushimi, K.; Nakanishi, T.; Hasegawa, Y.; Park, Y.-S. Corrosion behaviour of ferrite and austenite phases on super duplex stainless steel in a modified green-death solution. *Corros. Sci.* **2014**, *89*, 111–117.

57. Pettersson, R.F.A.; Flyg, J. *Electrochemical Evaluation of Pitting and Crevice Corrosion Resistance of Stainless Steels in NaCl and NaBr*; Outokumpu: Stockholm, Sweden, 2004.

58. Örnek, C.; Ahmed, A.H.; Engelberg, D.L. Effect of Microstructure on Atmospheric-Induced Corrosion of Heat-treated Grade 2205 and 2507 Duplex Stainless Steels. In Proceedings of Eurocorr 2012, Istanbul, Turkey, 8–13 September 2012; pp. 1–10.

59. Tavares, S.S.M.; Loureiro, A.; Pardal, J.M.; Montenegro, T.R.; Costa, V.C.D. Influence of heat treatments at 475 and 400 °C on the pitting corrosion resistance and sensitization of UNS S32750 and UNS S32760 superduplex stainless steels. *Mater. Corros.* **2012**, *63*, 522–526. [CrossRef]

60. Örnek, C.; Engelberg, D.L. An experimental investigation into strain and stress partitioning of duplex stainless steel using digital image correlation, X-ray diffraction and scanning Kelvin probe force microscopy. *J. Strain Anal. Eng. Des.* **2016**, *51*, 207–219. [CrossRef]

61. Örnek, C. Thermal Heat Treatments of Duplex and Super Duplex Stainless Steels in the 400 to 550 °C Range and Their Influence on the Corrosion Properties. B.Sc. Thesis, Aalen University of Applied Sciences, Aalen, Germany, July 2011. p. 136.

62. Lynch, S.P. 2-Hydrogen embrittlement (HE) phenomena and mechanisms. In *Stress Corrosion Cracking*; Raja, V.S., Shoji, T., Eds.; Woodhead Publishing: Sawston, Cambridge, UK, 2011; pp. 90–130.

63. Lynch, S.P. Hydrogen embrittlement phenomena and mechanisms. *Corros. Rev.* **2012**, *30*, 105–123. [CrossRef]

64. Lynch, S.P. 1-Mechanistic and fractographic aspects of stress-corrosion cracking (SCC). In *Stress Corrosion Cracking*; Raja, V.S., Shoji, T., Eds.; Woodhead Publishing: Sawston, Cambridge, UK, 2011; pp. 3–89.

Article

Formation and Disruption of W-Phase in High-Entropy Alloys

Sephira Riva [1],*, Chung M. Fung [2], Justin R. Searle [3], Ronald N. Clark [1], Nicholas P. Lavery [1], Stephen G. R. Brown [1] and Kirill V. Yusenko [1],*

[1] College of Engineering, Swansea University, Bay Campus, Swansea SA1 8EN, Wales, UK; 798741@swansea.ac.uk (R.N.C.); n.p.lavery@swansea.ac.uk (N.P.L.); s.g.r.brown@swansea.ac.uk (S.G.R.B.)
[2] Centre of Nanohealth, Swansea University, Singleton Park, Swansea SA2 8PP, Wales, UK; c.m.fung@swansea.ac.uk
[3] Sustainable Product Engineering Centre for Innovative Functional Industrail Coatings (SPECIFIC), Baglan Bay Innovation & Knowledge Centre, Central Avenue, Baglan, Port Talbot SA12 7AX, Wales, UK; j.r.searle@swansea.ac.uk
* Correspondence: s.riva.839245@swansea.ac.uk (S.R.); k.yusenko@swansea.ac.uk (K.V.Y.); Tel.: +44-1792-206234 (K.V.Y.)

Academic Editor: Soran Birosca
Received: 1 April 2016; Accepted: 29 April 2016; Published: 6 May 2016

Abstract: High-entropy alloys (HEAs) are single-phase systems prepared from equimolar or near-equimolar concentrations of at least five principal elements. The combination of high mixing entropy, severe lattice distortion, sluggish diffusion and cocktail effect favours the formation of simple phases—usually a *bcc* or *fcc* matrix with minor inclusions of ordered binary intermetallics. HEAs have been proposed for applications in which high temperature stability (including mechanical and chemical stability under high temperature and high mechanical impact) is required. On the other hand, the major challenge to overcome for HEAs to become commercially attractive is the achievement of lightweight alloys of extreme hardness and low brittleness. The multicomponent AlCrCuScTi alloy was prepared and characterized using powder X-ray diffraction (PXRD), scanning-electron microscope (SEM) and atomic-force microscope equipped with scanning Kelvin probe (AFM/SKP) techniques. Results show that the formation of complex multicomponent ternary intermetallic compounds upon heating plays a key role in phase evolution. The formation and degradation of W-phase, Al_2Cu_3Sc, in the AlCrCuScTi alloy plays a crucial role in its properties and stability. Analysis of as-melted and annealed alloy suggests that the W-phase is favoured kinetically, but thermodynamically unstable. The disruption of the W-phase in the alloy matrix has a positive effect on hardness (890 HV), density (4.83 g·cm^{-3}) and crack propagation. The hardness/density ratio obtained for this alloy shows a record value in comparison with ordinary heavy refractory HEAs.

Keywords: metals and alloys; phase transformations; high-entropy alloys; scandium; W-phase; Al_2Cu_3Sc

1. Introduction

Ever since their discovery, high-entropy alloys (HEAs) have become one of the major research topics in materials engineering. Outstanding mechanical and structural features, coupled with simple crystal structures, make HEAs promising as materials for high-temperature applications, in which high hardness should be coupled with high chemical and mechanical stability [1].

According to their density, melting point and hardness, HEAs can be classified as heavy/light, refractory, soft/hard materials. In many cases, refractory (with melting point above 2000 °C) HEAs are characterized by high hardness (600–900 HV): however, since their main components are Mo, Nb, Ta or Zr, their density equals or surpasses 10 g·cm^{-3} [2–4], which limits their application.

Conversely, only few alloys display micro-hardness values of 800 HV or more. Such hardness values have usually been obtained for refractory HEAs (such as MoTiVFeNiZrCoCr) or for light Al-based HEAs with moderate density (3–5 g·cm^{-3}) through mechanical alloying [5]. Nevertheless, high-temperature applications demand a sensible balance between strength and tensile ductility, and no known single-phase HEA satisfies such requirements. Perfectly tuned properties may be obtained by combining a soft matrix (*bcc*) with finely distributed intermetallics, a combination already known to improve hardness [6,7].

The design of new HEAs is driven by the principle of configurational entropy maximization, with consequent Gibb's free energy reduction. This is mostly achieved by combining at least five principal elements at 5–35 at. % concentrations. However, the relative affinities of metals, synthetic route and mechanical treatments also play a role in the formation of single-phase HEAs. Therefore, trials have often led to the development of equimolar multicomponent alloys consisting of intermetallic compounds and solid solutions of different space groups [8,9]. As a matter of fact, many HEAs originally believed to consist of a single *bcc* or *fcc* phase contain minor inclusions and intermetallic phases when investigated by Powder X-ray Diffraction (PXRD), high-resolution scanning-electron microscope (SEM) and transmission electron microscope (TEM). It may therefore be necessary to extend the definition of a HEA to include these systems [10].

While models based exclusively on ternary phase diagrams fail to predict the formation of single phase HEAs, the information extracted from simple systems can certainly prove useful in directing new synthetic approaches.

In the current study, we focus on the key role played by multicomponent intermetallic compounds in the phase formation of equimolar HEAs containing elements likely to form complex ordered binary and especially ternary phases, such as Al, Cu and Sc [11]. We present the formation and disruption of ternary Al-Cu-Sc intermetallic compounds (Al_4Cu_4Sc and Al_2Cu_3Sc) in the equimolar AlCrCuScTi alloy. We highlight the effect of the degradation of these phases on density, hardness and crack formation.

2. Materials and Methods

2.1. Synthesis and Heat Treatment

The target alloy has been prepared using induction melting. Al, Cr, Cu, and Ti were taken as fine powders or grains (Goodfellows, Huntingdon, UK) and mixed with 1 mm Sc grains; metals were used in equiatomic proportions. The total mass of the melted sample was 10 g. The powders were pressed into a pellet (11 mm in diameter and 10–15 mm in height) using a steel press mould, under a pressure equivalent to 10 t. The pellet was placed in a BN crucible (Kennametal, Newport, UK) and melted using an induction coil in a glove-box operated under Ar pressure, in order to protect metals from high-temperature oxidation. Complete melting of the sample was achieved above 1600 °C. After 5 min at the melting temperature, the sample was cooled down naturally to room temperature. The as-melted material was thermally annealed in a dynamic vacuum (10^{-2} Pa) at 1000 °C for 4 h, followed by natural cooling to room temperature. Part of the homogenized sample was fast-heated (1–2 min) to 1400 °C using an induction coil in a glove-box and quenched in high purity hydrostatic oil. The three samples are referred to in the text as "as-melted", "annealed" and "quenched".

2.2. Material Characterization

All samples were mounted in carbonised resin, polished using MetaDiTM Supreme Polycrystalline Diamond Suspension (Buehler, Esslingen am Neckar, Germany) (1 μm) and etched with a 5% solution of HNO$_3$ in ethanol.

Morphology and elemental composition were analysed using a Hitachi S-4800 Field Emission scanning-electron microscope (SEM, Hitachi, Tokyo, Japan) equipped with energy dispersive X-ray (EDX) analyser. The average elemental composition was obtained from 2.5 × 1.5 mm maps and locally.

Density was measured using flotation in water in the ATTENSION equipment (Biolin Scientific, Stockholm, Sweden). The results are an average of five measurements.

Vickers hardness was investigated on a WilsonR VH3100 Automatic Knoop/Vickers Hardness tester (Buehler, Esslingen am Neckar, Germany): 50–60 individual points under a 9.81 N (1 kg) testing load were measured to get statistically significant results. In order to investigate cracking phenomena, various forces of 2.94, 9.81, 19.62, 29.43, 49.04 and 98.10 N were applied. Lengths of indentation and cracks were measured using ×50 magnification with 0.2 μm accuracy.

Atomic-force microscope equipped with scanning Kelvin probe (AFM/SKP) measurements were performed in AC mode on the annealed alloy using a JPK NanoWizard3 Instrument (JPK, Berlin, Germany), equipped with a FM-50 Pointprobe® tip. The SKP scanning was performed at 10 nms^{-1} with 5 μms^{-1} speeds.

Powder X-ray diffraction profiles for the as melted and annealed powdered samples were collected at room temperature at the ID-09A beam-line at the ESRF (λ = 0.4145 Å, MAR 555 flat panel detector (marXperts GmbH, Norderstedt, Germany) company, city, country, beam size 10 × 15 μm²). LaB$_6$ was used as external standard for calibration. For data analysis, two-dimensional images were first integrated to one-dimensional intensity counts as a function of diffraction angle using the FIT2D software (Version 18, ESRF, Grenoble, France, 2016). Phase identification was performed using the Crystallography Open Database and the Inorganic Crystal Structure Database.

3. Results

The data related to the average composition, density and hardness of the AlCrCuScTi alloy are summarized in Table 1.

Table 1. Properties of as-melted, annealed and quenched AlCrCuScTi alloy.

AlCrCuScTi	Average Composition (±0.5 at. %) from 2.5 mm × 1.5 mm Maps					Density (±0.02 g·cm^{-3})	Hardness	
	Al	Cr	Cu	Sc	Ti		HV	MPa
Nominal	20	20	20	20	20	4.71 (estimated)	-	-
As melted	21.7	20.0	21.6	17.4	19.3	4.84	636 ± 27	6237 ± 267
Annealed	16.7	19.1	19.5	23.3	23.3	4.83	890 ± 20	8728 ± 196
Quenched	18.1	22.3	20.2	20.5	20.8	-	797 ± 10	7816 ± 98

The amount of Al in the sample is reduced during heat treatment. The high temperatures required by the synthetic route leads to the partial evaporation of Al. The alloy should therefore be considered quasi-equimolar. The density of the multicomponent homogeneous alloy (ρ_{est}) can be estimated using the following formula:

$$\rho_{est} = \frac{\sum c_i A_i}{\sum \frac{c_i A_i}{\rho_i}}$$

where c_i, A_i and ρ_i are respectively concentration, atomic weight and density of the *i*th element. Density values for pure metals were taken from [12].

Both the as-melted and thermally annealed alloys show high brittleness and can be easily ground to a fine powder. Because of the easy formation of cracks during indentation, Vickers hardness of the quenched material can only be roughly estimated. The thermal annealing results in a drastic increase of hardness and decrease in the brittleness of the alloy. To compare the brittleness of the as-melted and annealed samples, indentations with various testing forces were performed. Since the indenter is bigger than the individual domains characteristic for the sample, no significant differences in hardness values were detected among different areas of indention. Anstis *et al.* demonstrated that the crack length forming in a material during Vickers indentation can be used to characterize ductile-brittle transitions in hard and soft materials. Indentation at various forces produces pyramid-shaped traces linked to material ductility, as well as symmetrical cracks related to the sample's brittleness [13].

Plotting the logarithmic pressure load, Log$P$ *versus* the logarithm of indentation length, Log(d), and *versus* the total length of crack, Log($2c$), gives a straight line, whose slope is close to 2 and to 3/2 respectively (Figure 1). The intersection of the two lines identifies the critical load P_{crit}. For $P > P_{crit}$ brittle cracks appear while for $P < P_{crit}$ only plastic deformation is observed. This methodology gives good estimated information about the ductile-brittle transition, especially if mechanical tests cannot be performed. The as-melted alloy displays cracks at the lowest load (2.94 N), and P_{crit} can be estimated as 2.26 N. Since the material has a fairly uniform element distribution and microstructure, brittle cracks propagate in the volume over large distances. The annealed sample, on the other hand, has a more resilient grain structure due to the intermetallic phase, which inhibits cracks propagation in the matrix: the annealed alloy does not show crack formation at moderate loads (below 10 N) and above 10 N only small asymmetric cracks can be detected (Figure 3). Therefore, P_{crit} can be estimated as 10 N approximately. Higher hardness (1.5 times greater) and critical load (at least 4 times greater), combined with suppressed crack propagation, make the annealed alloy more mechanically stable in comparison with the as-melted alloy.

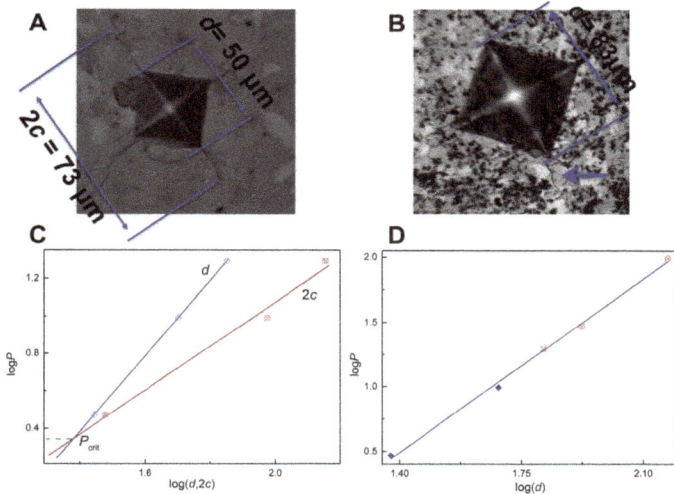

Figure 1. Optical microscope image of the indentation of the as-melted alloy with 9.81 N load (**A**) and of the annealed alloy with 29.43 N load (**B**) (the arrow shows the formation and prolongation of a single crack); (**C**) Log$P$ *vs.* Log(d) and Log($2c$) plots for as-melted alloy; (**D**) Log$P$ *vs.* Log(d) plot for annealed alloy (two points at low loadings show no cracks formation, tree data-points at high load pressure display moderate crack formation).

The as-melted, annealed and quenched AlCrCuScTi alloy microstructures are presented in Figure 2. The thermal annealing results in a relatively rapid stress release and phase equilibration. According to pyrometric data recorded during induction melting, the melting temperature for AlCrCuScTi alloy can be estimated as 1200–1250 °C. As the annealing temperature was above 4/5 of the melting temperature it allows easier diffusion of metallic atoms in the sample and further homogenisation.

The as-melted alloy presents a homogeneous polished surface and an even distribution of elements. Its microstructure is dominated by large (50 μm) spear-like Cr-rich structures surrounded by

Ti-rich areas (Figure 2G,J). Between them, Al, Sc and Cu are distributed in a relatively uniform matrix (Figure 2A,D).

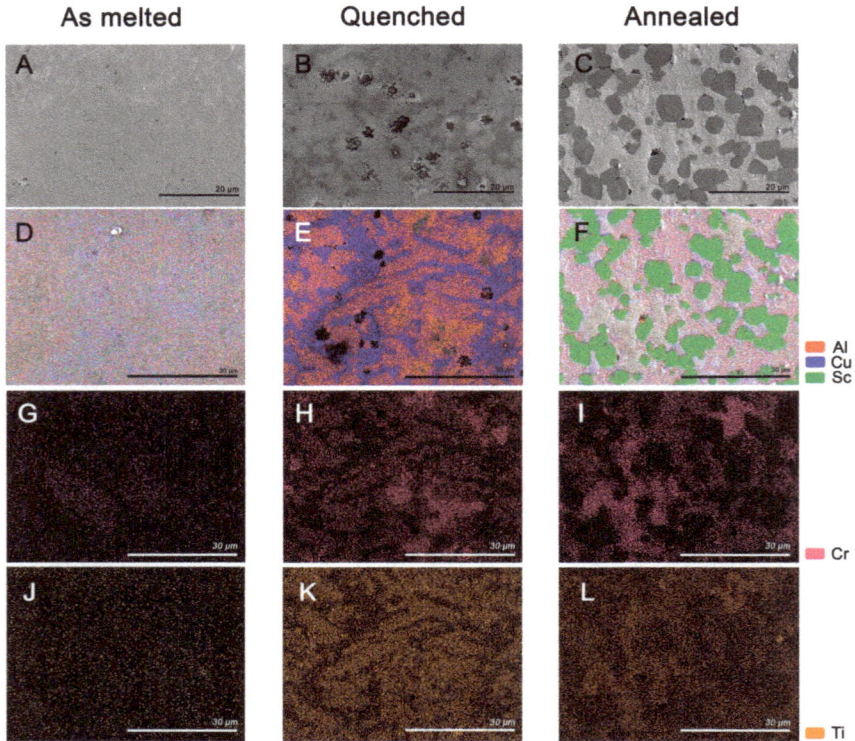

Figure 2. SEM images of as melted (**A**), quenched (**B**) and annealed (**C**) AlCrCuScTi alloy. EDX elemental composition maps of Al, Cu and Sc for as melted (**D**), quenched (**E**) and annealed (**F**) AlCrCuScTi alloy respectively. EDX elemental composition maps of Cr distribution for as melted (**G**), quenched (**H**) and annealed (**I**) AlCrCuScTi alloy respectively. EDX elemental composition maps of Ti distribution for as melted (**J**), quenched (**K**) and annealed (**L**) AlCrCuScTi alloy respectively.

After heat treatment, the alloy consists of evenly distributed hexagonal-shaped inclusions embedded in a uniform matrix (Figure 2C). However, atomic-force microscope equipped with scanning Kelvin probe (AFM-SKP) measurements performed between inclusions reveals the existence of a finer structure, a phase characterized by a different electrical potential grows along grain boundaries and over bigger grains (Figure 3). In fact, a few raised regions (of maximum height 0.7 µm) can be associated with a minor phase protruding slightly from the matrix, probably due to a higher hardness and resistance to polishing. Quite interestingly, the same areas are identified in the Volta potential map as having 150 mV lower potential values.

EDX analysis shows the dark inclusions to be Sc-rich, and surrounded by Cu-rich areas (Figure 2I,F,L). The grains consist of Al and Ti, while Cr segregates in areas far from those containing Sc.

The quenched sample has an intermediate microstructure between the as-melted and the annealed sample (Figure 2B,E,H,K). It is dominated by dendrite-like structures growing in a Cu/Ti-rich matrix Dendrites consist mostly of Al and Sc, with an inter-dendritic phase of Cr. The surface shows many defects and holes, which fill with carbon during the mounting operation.

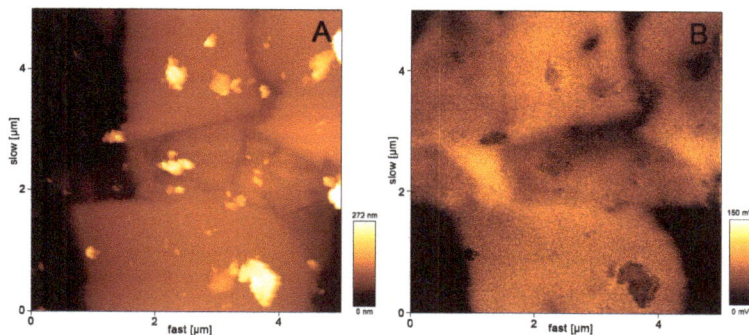

Figure 3. Thermally annealed AlCrCuScTi alloy: AFM (**A**) and SKP (**B**) images.

4. Discussion

It has been shown for a number of Al-based alloys containing minor Sc and Cu additions that a noticeable microstructural feature is the appearance of a ternary phase having a $ThMn_{12}$-type crystal structure with lattice parameters estimated as $a = 8.63$ Å, $c = 5.10$ Å [14]. Its formation and disruption after thermal annealing is responsible for the change in mechanical properties of the alloy. At the Al-rich corner of the Al-Cu-Sc ternary phase diagram, the Al-Cu-Sc W-phase forms by the combination of the Al-Cu phase (ϑ-phase, Al_2Cu) and the Al-Sc phase (Al_3Sc) [15,16]. Various compositions have been proposed for the W-phase—$Al_{5-8}Cu_{7-4}Sc$, $Al_{5.4-8}Cu_{6.6-4}Sc$, $Al_{8-x}Cu_{4+x}Sc$—but further studies are needed to determine the stability range of the phase, its homogeneity range and its detailed crystal structure.

Moreover, no studies have been performed on the Sc-rich corner of the Al-Cu-Sc phase diagram. As such, the as-melted AlCrCuScTi alloy here is the first example of W-phase formation in an over-saturated Sc alloy. The crystallization of the AlCrCuScTi alloy from a homogeneous melt can be driven by entropy with the formation of a high-entropy alloy or by the entropy of formation of binary and ternary compounds, with subsequent formation of multiphase composites. The formation of the W-phase as a main component of AlCrCuScTi alloy suggests an entropy driven pathway, which can be associated with the relatively high thermodynamic stability of Al-Cu-Sc ternary phases.

The XRD profile (Figure 4) of the as-melted alloy is dominated by cubic phases (*fcc*- and *bcc*-based alloys and partially ordered intermetallics, characterised by various cell parameters, ordering type and occupancies) and intermetallic compounds (Al_3Sc, Al_4Cu_4Sc (W-phase) and Al_2Cu_3Sc—a ternary phase formed in Sc excess). The existing phases are summarized in Table 2: despite its extremely low formation enthalpy, Sc_2O_3 is not present. The inherent multiphase complex nature of the alloy makes realistic indexing of all lines impossible.

The W-phase appears disrupted in the quenched sample (Figure 2). Al and Sc segregate in the form of dendrite-like structures from the Cu matrix. Ultimately, the thermal annealing process leads to the coalescence of Sc from these structures. Conversely, the XRD profile of the annealed alloy (Figure 4) shows traces of the pre-existing Sc-containing ternary phases, and the formation of *hcp* alloys (a combination of Sc and Ti in different ratios).

The comparison between as-melted, heat-treated and quenched alloys highlights the major role played by temperature in the formation of different intermetallics. The instability of the W-phase is related to the high content of scandium in the sample. While Sc drives the formation of the W-phase in a kinetically determined reaction, its segregation pushes the disaggregation of the very same phase when the sample approaches thermodynamic equilibrium.

The formation of ternary Al-Cu-Sc phases in an equiatomic five-component system shows an important limitation in the design of single- or nearly single-phase high-entropy alloys. The crystallization of a ternary compound with relatively high thermodynamic stability causes the formation of multiphase composites. It seems that the formation of stable ternary compounds is preferable and cannot be suppressed by configurational entropy.

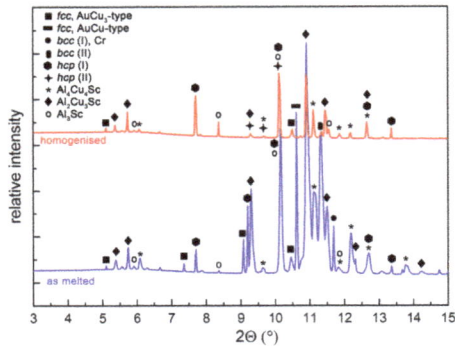

Figure 4. A comparison of PXRD profiles for the as-melted (**below**) and annealed (**above**) AlCrCuScTi alloy ($\lambda = 0.4145$ Å).

Table 2. Phases in the as melted and annealed AlCrCuScTi alloy, according to PXRD profiles.

Phase	As-Melted Alloy	Annealed Alloy
fcc AuCu$_3$-type	Yes	Traces
fcc AuCu-type	Yes	No
bcc (I) Cr	Yes	No
bcc (II)	Yes	Traces
hcp (I)	Yes	Yes
hcp (II)	No	Yes
Al$_4$Cu$_4$Sc	Yes	Traces
Al$_2$Cu$_3$Sc	Yes	Traces
Al$_3$Sc	Traces	Traces

5. Conclusions

The present study reports the formation of Al$_2$Cu$_3$Sc and Al$_4$Cu$_4$Sc W-phase in the frame of High-Entropy Alloys development. The results show that:

1. The high concentration of scandium required by equimolarity brings forth the formation of intermetallic compounds, preventing the formation of a single phase. Among them, formation of Al$_2$Cu$_3$Sc compound has been detected for the first time in a multicomponent alloy.
2. The synthesis of a single-phase Al, Cu and Sc containing HEA is not compatible with melting from powders, due to the formation of stable intermetallic compounds.
3. The product of the W-phase degradation following heat-treatment is a combination of intermetallics in a soft *hcp*-Sc/Ti alloy matrix. The thermally annealed alloy has increased hardness (890 HV) and crack resistance.

Acknowledgments: The authors gratefully acknowledge the financial support provided by the Welsh Government and Higher Education Funding Council for Wales through the Sêr Cymru National Research Network in Advanced Engineering and Materials, and the Materials Advanced Characterisation Centre (MACH1) at Swansea University. The authors thank European Synchrotron Radiation Facility (ESRF, France) for providing us measurement time and support.

Author Contributions: Sephira Riva and Kirill V. Yusenko conceived and designed the experiments; Sephira Riva and Kirill V. Yusenko performed synthetic and analytical experiments including synchrotron-based PXRD and analyzed data; Sephira Riva and Chung M. Fung performed SEM experiments; Sephira Riva, Ronald N. Clark and Justin R. Searle performed AFM experiments; Nicholas P. Lavery and Stephen G.R. Brown contributed to the experiment design and data evaluation; Nicholas P. Lavery is the director of the MACH1 centre, and he built and developed the materials synthesis equipment and methodology; all authors contributed equally writing the paper.

Conflicts of Interest: The authors declare no conflict of interest.

Abbreviations

The following abbreviations are used in this manuscript:

SEM-EDX	Scanning electrode microscope equipped with energy dispersed X-ray analyzer
AFM	Atomic force microscopy
SKP	Scanning Kelvin probe
PXRD	Powder X-ray diffraction

References

1. Samaei, A.T.; Mirsayar, M.M.; Aliha, M.R.M. Microstructure and mechanical behaviour of modern high temerature alloys. *Eng. Sol. Mech.* **2015**, *3*, 1–20. [CrossRef]
2. Pogrebnjak, A.D.; Bagdasaryan, A.A.; Yakushchenko, I.V.; Beresnev, V.M. The structure and properties of high-entropy alloys and nitride coatings based on them. *Russ. Chem. Rev.* **2014**, *83*, 1027–1061. [CrossRef]
3. Zhang, Y.; Zuo, T.T.; Tang, Z.; Gao, M.C.; Dahmen, K.A.; Liaw, P.K.; Lu, Z.P. Microstructures and properties of high-entropy alloys. *Prog. Mater. Sci.* **2014**, *61*, 1–93. [CrossRef]
4. Senkov, O.N.; Wilks, G.B.; Miracle, D.B.; Chuang, C.P.; Liaw, P.K. Refractory high-entropy alloys. *Intermetallics* **2010**, *18*, 1758–1765. [CrossRef]
5. Pradeep, K.G.; Wanderka, N.; Choi, P.; Banhart, J.; Murty, B.S.; Raabe, D. Atomic-scale compositona characterization of a nanocrystalline AlCrCuFeNiZn high-entropy alloy using atom probe tomography. *Acta Mater.* **2013**, *61*, 4696–4706. [CrossRef]
6. Fu, Z.; Chen, W.; Xiao, H.; Zhou, L.; Zhu, D.; Yang, S. Fabrication and properties of nanocrystalline $Co_{0.5}FeNiCrTi_{0.5}$ high entropy alloy by MA-SPS technique. *Mater. Des.* **2013**, *44*, 535–539. [CrossRef]
7. Dong, Y.; Lu, Y.; Kong, J.; Zhang, J.; Li, T. Microstructure and mechanical properties of multi-component $AlCrFeNiMo_x$ high-entropy alloys. *J. Alloy. Compd.* **2013**, *537*, 96–101. [CrossRef]
8. Li, L.; Huang, L.; Song, X.; Ye, F.; Lin, J.; Cheng, G. Microstructure and performance of TiAlBeSc alloys with low density. *Rare Met. Mater. Eng.* **2012**, *5*, 826–829. [CrossRef]
9. Lin, Y.C.; Cho, Y.H. Elucidating the microstructural and tribological characteristics of NiCrAlCoCu and NiCrAlCoMo multicomponent alloy cladlayers synthesized *in situ*. *Surf. Coat. Technol.* **2009**, *203*, 1694–1701. [CrossRef]
10. Santodonato, L.J.; Zhang, Y.; Feygenson, M.; Parish, C.M.; Gao, M.C.; Weber, R.J.K.; Neuefeind, J.C.; Tang, Z.; Liaw, P.K. Deviation from high-entropy configurations in the atomic distribution of a multi-principal-element alloy. *Nat. Commun.* **2015**, *6*. [CrossRef] [PubMed]
11. Riva, S.; Yusenko, K.Y.; Lavery, N.P.; Jarvis, D.J.; Brown, S.G.R. The scandium effect in multicomponent alloys. *Int. Mater. Rev.* **2016**. [CrossRef]
12. Haynes, W.M. *CRC Handbook of Chemistry and Physics: A Ready-Reference Book of Chemical and Physical Data*, 96th ed.; CRC Press: Boca Raton, FL, USA, 2015; p. 2766.
13. Anstis, G.R.; Chantikul, P.; Lawn, B.R.; Marshall, D.B. A critical evaluation of indentation techniques for measuring fracture toughness: I, Direct crack measurements. *J. Am. Ceram. Soc.* **1981**, *64*, 533–538. [CrossRef]
14. Kharakterova, M.L. Phase composition of Al-Cu-Sc alloys at temperatures of 450 and 500 °C. *Russ. Metall.* **1991**, *4*, 195–199.
15. Gazizov, M.; Teleshov, V.; Zakharov, V.; Kaibyshev, R. Solidification behaviour and the effects of homogenisation on the structure of an Al-Cu-Mg-Ag-Sc alloy. *J. Alloy. Compd.* **2011**, *509*, 9497–9507. [CrossRef]
16. Norman, A.F.; Prangnell, P.B.; McEwen, R.S. The solidification behaviour of dilute aluminium-scandium alloys. *Acta Mater.* **1998**, *46*, 5715–5732. [CrossRef]

metals

MDPI

Article

Examination of Solubility Models for the Determination of Transition Metals within Liquid Alkali Metals

Jeremy Isler and Jinsuo Zhang *

Nuclear Engineering Program, The Ohio State University, Columbus, OH 43210, USA; isler.38@osu.edu
* Correspondence: zhang.3558@osu.edu; Tel.: +1-614-292-5405; Fax: +1-614-292-3163

Academic Editor: Soran Birosca
Received: 1 February 2016; Accepted: 20 June 2016; Published: 28 June 2016

Abstract: The experimental solubility of transition metals in liquid alkali metal was compared to the modeled solubility calculated using various equations for solubility. These equations were modeled using the enthalpy calculations of the semi-empirical Miedema model and various entropy calculations. The accuracy of the predicted solubility compared to the experimental data is more dependent on which liquid alkali metal is being examined rather than the transition metal solute examined. For liquid lithium the calculated solubility by the model was generally larger than experimental values, while for liquid cesium the modeling solubility was significantly smaller than the experimental values. For liquid sodium, potassium, and rubidium the experimental solubilities were within the range calculated by this study. Few data approached the predicted temperature dependence of solubility and instead most data exhibited a less pronounced temperature dependence.

Keywords: solubility; liquid alkali; Miedema model

1. Introduction

Liquid alkali metals such as liquid sodium have been previously proposed as primary coolant in advanced nuclear reactors, with one of the Generation IV designs being a sodium-cooled fast reactor. Alkali metals are advantageous as coolants for they are liquids in the intermediate temperature range of nuclear reactors and provide excellent heat transfer properties as a coolant. For this reason the majority of the alkali metals have been considered in the past as coolants and thus have been extensively studied [1]. Furthermore, liquid sodium is applied as the liquid that fills the gap between stainless steel cladding and the metallic nuclear fuels such as U-Zr fuels in a sodium-cooled nuclear reactor system. One of the major concerns of liquid metal-cooled reactors is the material transport in liquid metal, including the corrosion products (such as transition metals) and fission products (such as rare earth metals) [2]. An integral part of this transport is the solubility limit of the transition metals and fission products in the liquid metal. This solubility is thus of interest for coolant radioactivity and material degradation, such as fuel-cladding chemical interaction [3]. Several models have been developed for the prediction of the solubility within the liquid alkali metals. The enthalpy and entropy of various transition metals in liquid alkali metals were modeled and the corresponding solubility of these transition metals was calculated using various solubility equations. These calculations were then compared to experimental data to show the accuracy of these various models.

2. Solubility Equations

The solubility of a metal, A, in a solution of liquid alkali metal, B, can be expressed as:

$$X_{\text{A in B}} = \exp\left(\frac{-\Delta G_A^{Ex}}{RT}\right) \tag{1}$$

where R is the ideal gas constant and T is the temperature in Kelvin. The molar excess Gibbs energy (ΔG_A^{Ex}) can also be expressed in terms of the molar entropy (ΔS_A^{Ex}) and enthalpy (ΔH_A^{Ex}) of mixing for infinitely dilute solution of A in liquid alkali metal B.

$$\Delta G_A^{Ex} = \Delta H_A^{Ex} - T\Delta S_A^{Ex} \tag{2}$$

The solubility equation of a metal, A, in the solution can be described in an expanded solubility equation of:

$$X_{A\ in\ B} = \exp\left(\frac{\Delta S_A^{Ex}}{R}\right)\exp\left(\frac{-\Delta H_A^{Ex}}{RT}\right) \tag{3}$$

Several approximations of Equation (3) have been proposed to model the solubility of metals in liquid alkali metal. The simplest of these formulas was suggested by de Boer et al. [4] shown in Equation (4).

$$X_{A\ in\ B} = \exp\left(\frac{-\Delta H_A^{Ex}}{RT}\right) \tag{4}$$

Lyublinski et al. [5] suggested the inclusion of the latent heat term of the solid to liquid phase of the pure metal A, $\Delta H^{s\rightarrow l}$, and the associated molar entropy term, $\Delta S^{s\rightarrow l}$ as shown in Equation (5). For this work the latent heat and associated entropy term were derived from the work of Gale and Totemeier [6].

$$X_{A\ in\ B} = \exp\left(\frac{\Delta S^{s\rightarrow l} + \Delta S_A^{Ex}}{R}\right)\exp\left(\frac{-\left(\Delta H^{s\rightarrow l} + \Delta H_A^{Ex}\right)}{RT}\right) \tag{5}$$

To model these various solubility equations, the entropy and enthalpy terms for the various systems must be calculated.

2.1. Enthalpy Calculation

The enthalpy term of the solubility equations were modeled with the Miedema model [4]. This semi-empirical model, created to predict values of enthalpy for arbitrary combinations of metals, has been suggested previously for use of calculating excess enthalpy in liquid alkali metals [4,5]. This "macroscopic-atom" model derives the enthalpy effect of the interaction at the interface between two various elements. Three major parameters are used for the calculation of enthalpy: V_A, φ, and n_{ws}. The molar volume of the solute, V_A, is considered to account for the surface area of the solute in the interaction. The work function, φ, is a parameter that is a potential that is felt by the outer electrons of the atom and is similar to electronegativity. The third parameter n_{ws}, relates to the Wigner-Seitz cell and the electron density at the boundary of the cell. All three parameters have been previously tabulated for most elements [4].

The excess enthalpy due to the interfacial enthalpy of solids is calculated by [4]:

$$\Delta H_A^{Ex} = \frac{2 \times V_A^{2/3} \times P}{\left(n_{ws\ A}^{-1/3}\right) + \left(n_{ws\ B}^{-1/3}\right)} \times \left(-(\Delta\varphi)^2 + \frac{Q}{P}(\Delta n_{ws}^{1/3})^2\right) \tag{6}$$

where $Q/P = 9.4$ and P is a constant that depends on the valence of both the atoms considered. This excess enthalpy due to the interfacial enthalpy of solids is the partial enthalpy of the solution of solid species A in liquid B.

For the application of the model an additional enthalpy term, R^*, is included to account for the interaction of a transition metal and a non-transition metal [4].

$$\Delta H_A^{Ex} = \frac{2 \times V_A^{2/3} \times P}{\left(n_{ws\ A}^{-1/3}\right) + \left(n_{ws\ B}^{-1/3}\right)} \times \left(-(\Delta\varphi)^2 + \frac{Q}{P}\left(\Delta n_{ws}^{\frac{1}{3}}\right)^2 + R^*\right) \tag{7}$$

2.2. Entropy Calculation

The entropy was modeled in two ways. The initial model of entropy was to neglect the entropy as shown in Equation (4). This modeling of the entropy is described in detail by Bakker [7]. Bakker argues that the entropy is mainly vibrational entropy, which is on the order of magnitude of the ideal gas constant, *R*. Based on this assumption the entropy term will not significantly affect the order of magnitude of the solubility.

The second approach of the entropy term was in the use of an interaction parameter. The interaction parameter relates the enthalpy and entropy as derived by Lupis [8].

$$\Delta H_A^{Ex} = \tau \Delta S_A^{Ex} \tag{8}$$

Lupis [8] reported that the value of the interaction parameter, τ, to be between 1500 and 3000 K and for most systems it is approximately 3000 K. Lyublinski et al. [5] previously used the interaction parameter of 3000 K to calculate the entropy for the solubility of transition metals in liquid lithium.

3. Results

Within this section are several figures that compare the solubility equations to previous experimental data [9–32] for transition metals in liquid alkali metals. Additional comparisons are provided in the Appendix 5 for experimental data [33–55] of transition metals in liquid Lithium (Figure A1), liquid Sodium (Figure A2), liquid Potassium (Figure A3), liquid Rubidium (Figure A4) and liquid Cesium (Figure A5).

3.1. Lithium

The calculations of the solubility for metals in liquid lithium when including the entropy were generally larger than actually experimentally measured. The solubility of tantalum in liquid lithium in Figure 1 is a typical result for transition metals in liquid lithium. As can be seen in Figure 1, the equations that calculated the entropy with an interaction term, Equations (3) and (5), overestimate the solubility of the tantalum. The solubility equation that neglects entropy, Equation (4), is approximately the correct order of magnitude but its temperature dependence appears to be too strong for most of the data. Figure 2 of the solubility of titanium in liquid lithium shows the experiment data is closer to the estimated solubility of the equations with an entropy parameter of 3000 K but is still lower than predicted.

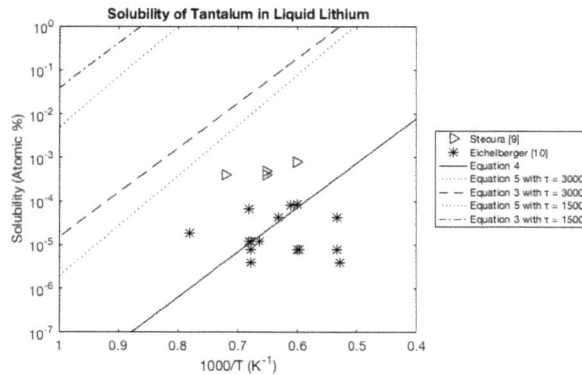

Figure 1. Solubility of tantalum in liquid lithium.

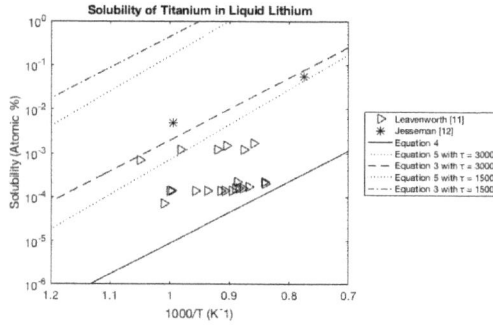

Figure 2. The solubility of titanium in liquid lithium.

While the Figures 1 and 2 represent the typical results for the liquid lithium systems there are a few cases for which the experimental data more closely resembles the solubility equations with the entropy modeled. Shown in Figures 3 and 4 are the solubility of iron and of chromium in liquid lithium. These systems have considerably more experimental data and the solubility equations that included entropy results more accurately describe the experimental data than in Figures 1 and 2. Additionally the temperature dependence of the estimated solubility for these systems agrees much more favorably with the experimental data than do the other metal solutes' solubilities in liquid lithium.

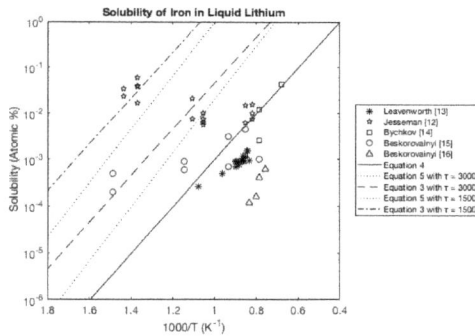

Figure 3. Solubility of iron in liquid lithium.

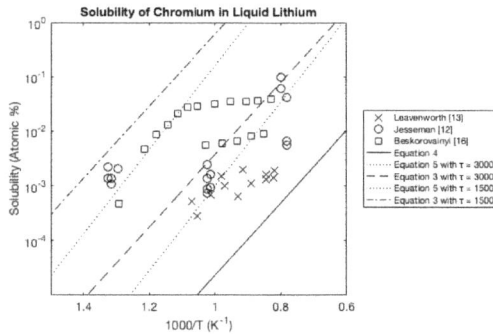

Figure 4. Solubility of chromium in liquid lithium.

3.2. Sodium

Figures 5 and 6 show the solubility of iron and chromium in liquid sodium. These figures represent the typical relation between experimental data and the solubility equations for the liquid sodium systems. As seen in these figures, the majority of experimental data is within the solubility calculated with the interaction parameter entropy. One exception is the lower temperature experimental data which has a larger solubility than predicted by the model. The experimental data exhibits a mixture of temperature dependence but the majority of the data shows a temperature dependence that is less than predicted.

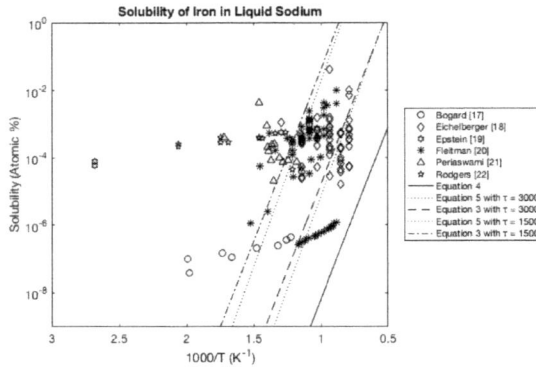

Figure 5. Solubility of iron in liquid sodium.

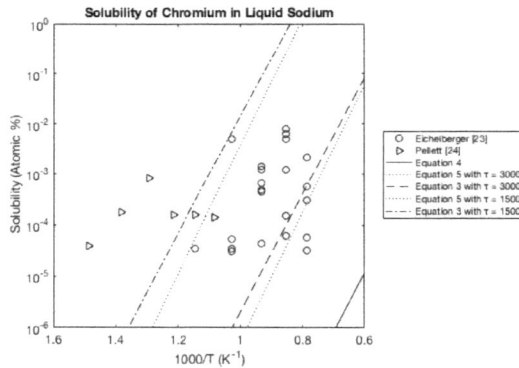

Figure 6. Solubility of chromium in liquid sodium.

3.3. Potassium

Figures 7 and 8 show the solubility of tantalum and iron in liquid potassium. Representing the typical results for the solubilities in liquid potassium, the experimental data is within the limits set by the estimations using the solubility equations with entropy determined by an interaction parameter. The estimation using an entropy parameter of 3000 K covers the solubility range of the lower temperatures and the entropy parameter of 1500 K covers the solubility range of the higher temperature. The temperature dependence of the experimental data occasionally approaches the dependence predicted by the model but the data's dependence is typically less pronounced.

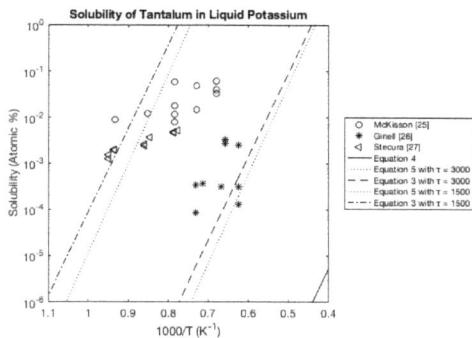

Figure 7. Solubility of tantalum in liquid potassium.

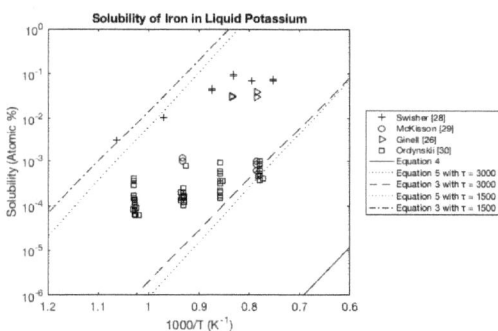

Figure 8. Solubility of iron in liquid potassium.

3.4. Rubidium

The experimental data for rubidium is sparse compared to the previous alkali liquids. The available data shows a relationship to the calculated solubility that is similar to that of the liquid sodium and potassium systems. Figures 9 and 10 show the solubility of zirconium and iron in liquid rubidium. The experimental data solubility for metals is between the calculations provided when using an entropy interaction parameter of 1500 and 3000 K.

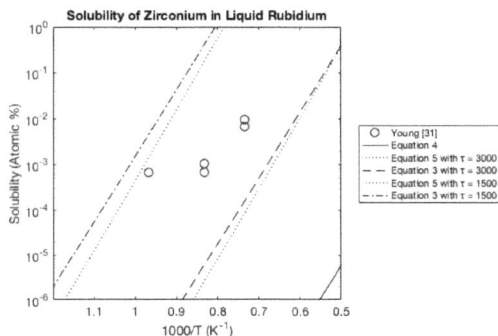

Figure 9. Solubility of zirconium in liquid rubidium.

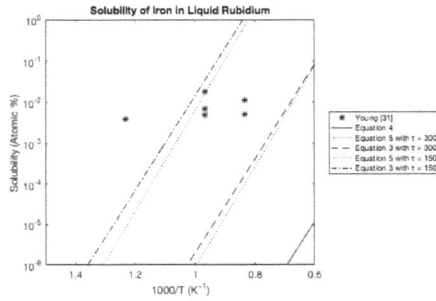

Figure 10. Solubility of iron in liquid rubidium.

3.5. Cesium

Figures 11 and 12 show of the solubility of iron and vanadium in liquid cesium. Both figures show the experimentally measured solubility being considerably larger than predicted. Also the temperature dependence of the experimental data is negligible compared to that of the model. This negligible temperature dependence is common among the various solubilities in liquid cesium.

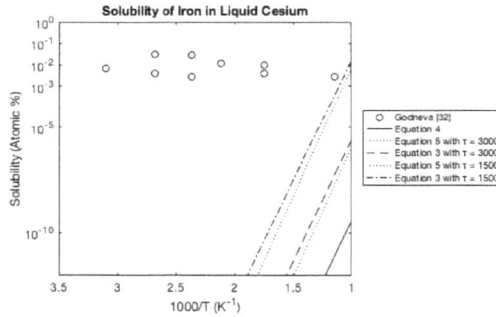

Figure 11. Solubility of iron in liquid cesium.

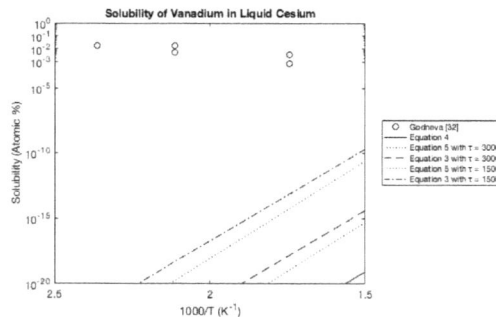

Figure 12. Solubility of vanadium in liquid cesium.

4. Discussion

The relation between experimental data and the fit with the solubility equations exhibits a stronger dependence on the liquid alkali metal studied rather than the solute. This relation is clear when considering the solubility of iron in the various liquid alkali metals. Iron is one of the most common solutes experimentally tested for solubility for all the alkali systems considered in the present study and thus allows the most comprehensive comparison between the model and the liquid alkali. The majority of experimental data for the liquid sodium, potassium, and rubidium systems are within the range predicted with the Miedema model for enthalpy and the interaction parameter for entropy. The exceptions to all three cases are some of the lower temperature experimentally measured solubilities which are larger than predicted. The measured solubility in liquid lithium is smaller than predicted with an entropy term and for cesium the measured data is larger than the prediction. These relations between each liquid alkali metal and the solubility equations hold true for the majority of transition metals.

The solute or alkali metal examined has a negligible effect on the temperature dependence of the solubility. The majority of the experimental solubility data exhibits a temperature dependence that is smaller than predicted. Due to the large scatter of data, there is some experimental data that approaches the temperature dependence of the model while other data shows no variation of solubility with temperature. The calculated temperature dependence should be seen as the maximum limit of the solubility dependence on temperature with the majority of experimental results having a less pronounced affect.

The prediction of solubility based on the enthalpy from the Miedema model has several limitations when compared to experimental data. One of the limitations is the elemental periodicity of the liquid alkali metal which appears to have a significant influence on the enthalpy term and thus the accuracy of the solubility. The intermediate sized liquid alkali metals (sodium, potassium, and rubidium) have accurately predicted solubilities when using the interaction parameter for entropy with the enthalpy calculation. However the models predicted solubility for liquid lithium is significantly larger than the experimental data and for liquid cesium the prediction is significantly smaller than the experimental data.

For several elements with multiple solubility studies performed there is a significant amount of scatter of the experimental data. This complicates the analysis of the model compared to the experimental data. This scatter has been previously attributed to solute interactions with the various container materials used and for solute interactions with other impurities in the liquid sodium [56]. The largest contribution to the scattering has been attributed to the varying oxygen concentrations in the liquid sodium for the different studies. Increased oxygen concentrations in liquid sodium have been shown to result in increased solubility measurements for many of the elements and have a larger effect at lower temperature measurements [56]. This effect is shown in many of the lower temperature ranges of the studies examined, where the measured solubility has the most severe scattering.

Neglecting the contribution of entropy on solubility, Equation (4), severely underestimates the solubility in all the liquid alkali metals except lithium. The use of the interaction parameter provides a simple relationship to calculate the solubility while still enhancing the accuracy. The plausible range of the interaction parameter of 1500 to 3000 K while helping to capture the scatter of experimental data, allows too large of a range for the calculated solubility. Lyublinski et al. [5] suggested the use of only the interaction parameter of 3000 K. This interaction parameter gives an excellent fit to some of the data and allows a narrowing of the predicted solubility. However the scatter of most of the experimental data can cause equal justification in the same system for the use of an interaction parameter of 1500 K. The accuracy in the prediction of solubility in liquid alkali metal would be greatly increased if a more definitive contribution from the entropy were determined.

The use of Equations (3) and (5) provided little difference in the models' effectiveness to predict solubility in liquid alkali metals. The enthalpy calculated by the Miedema model dominated both equations causing the results to be similar. The major variation between the equations was that Equation (5) resulted in a slightly larger temperature dependence for solubility than Equation (3). The effectiveness of the solubility prediction was not drastically altered with the inclusion of the latent heat term.

5. Conclusions

The model presented using enthalpy prediction by de Boer et al. [4] and an interaction parameter for entropy can be used for an initial estimate of the solubility of transition metals in liquid alkali metals but considerations must be made. For liquid sodium, potassium, and rubidium the use of the enthalpy from the Mideama model with an interaction parameter for entropy correctly predicted a range for the solubility. However the results of the model were much larger than experimental data for liquid lithium systems and much smaller than experimental data for liquid cesium systems. Thus the prediction was influenced more by the Miedema model's effect of periodicity on the calculation of enthalpy for liquid alkali metal than experimental data suggests.

The model's predicted temperature dependence of the solubility was larger than the majority of experimental data. While several experimental results approached the temperature dependence predicted, the majority of data suggested a less dramatic impact of temperature on solubility. The predicted temperature dependence could be used as a maximum limit on the temperature dependence on solubility with the majority of practical applications experiencing a less pronounced dependence.

The increase of the model's solubility prediction can be accomplished in several ways. The parameters for cesium and lithium should be carefully examined in an attempt to reduce the periodicity the model causes for the liquid alkalis. Another consideration is in how the entropy term is represented. The use of an interaction parameter creates a large range of solubility but this range effectively captures experimental data scatter.

Acknowledgments: This research was funded under the U.S. Department of Energy DOE-NEUP program, Project Number: 14-6482.

Author Contributions: Jeremy Isler and Jinsuo Zhang conceived and designed the experiments; Jeremy Isler performed the calculations and analyzed the data; Jinsuo Zhang contributed analysis tools; Jeremy Isler wrote the paper.

Conflicts of Interest: The authors declare no conflict of interest.

Appendix A

A.1. Lithium

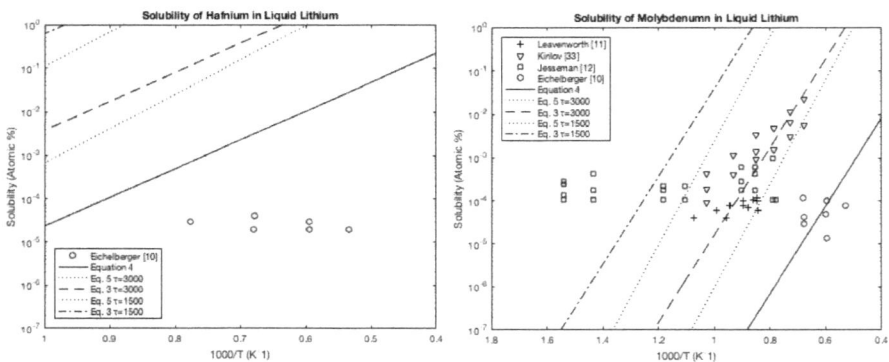

Figure A1. *Cont.*

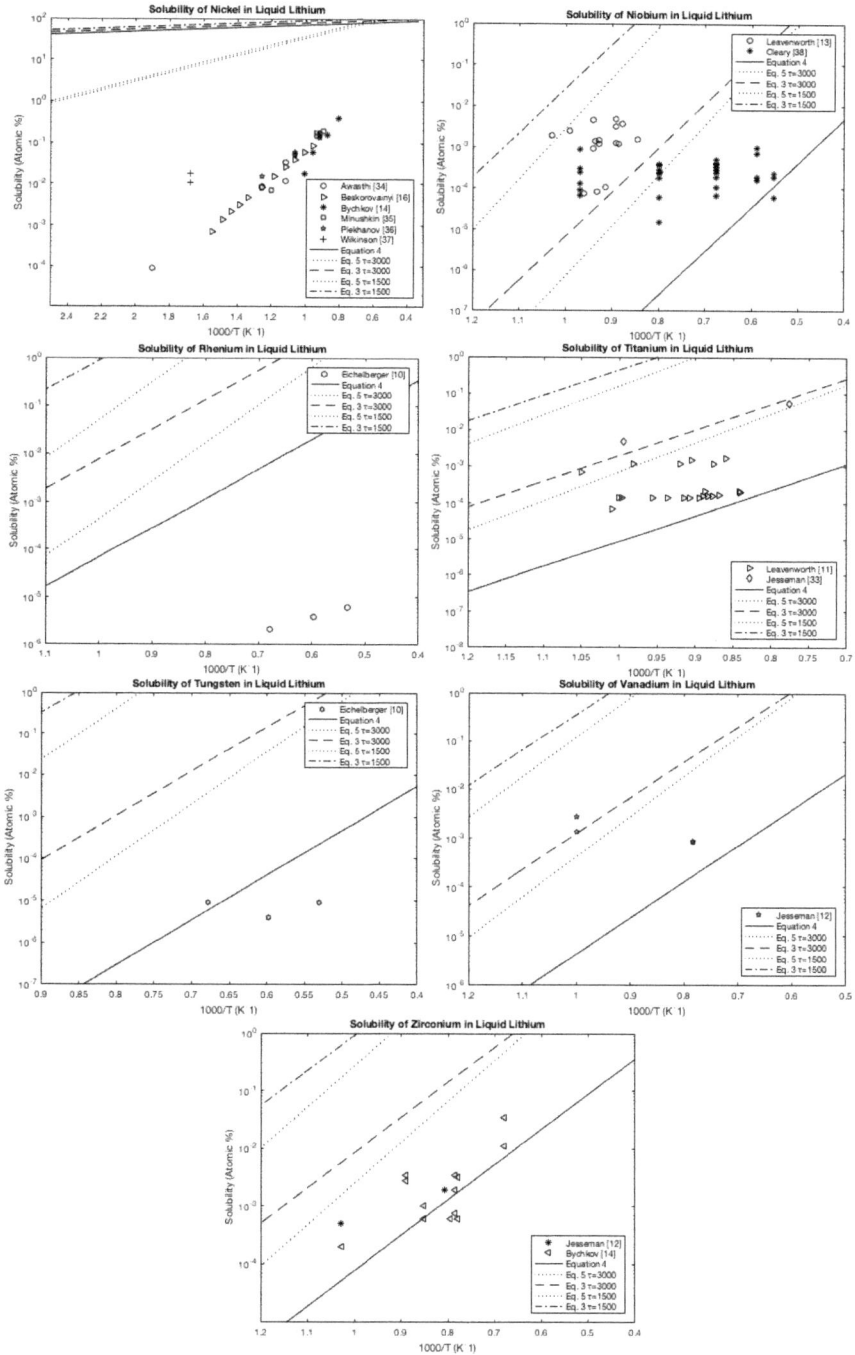

Figure A1. Solubility of additional transition metals in liquid lithium compared to the model.

A.2. Sodium

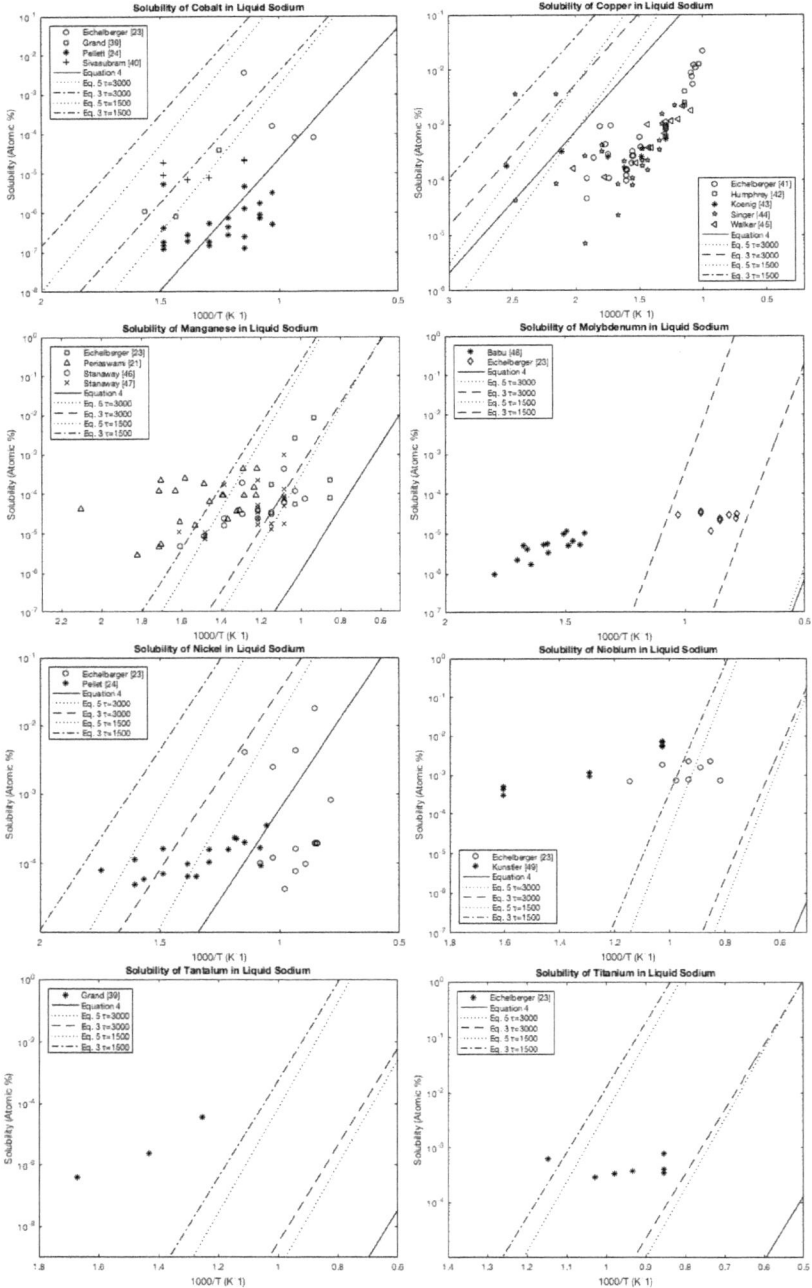

Figure A2. *Cont.*

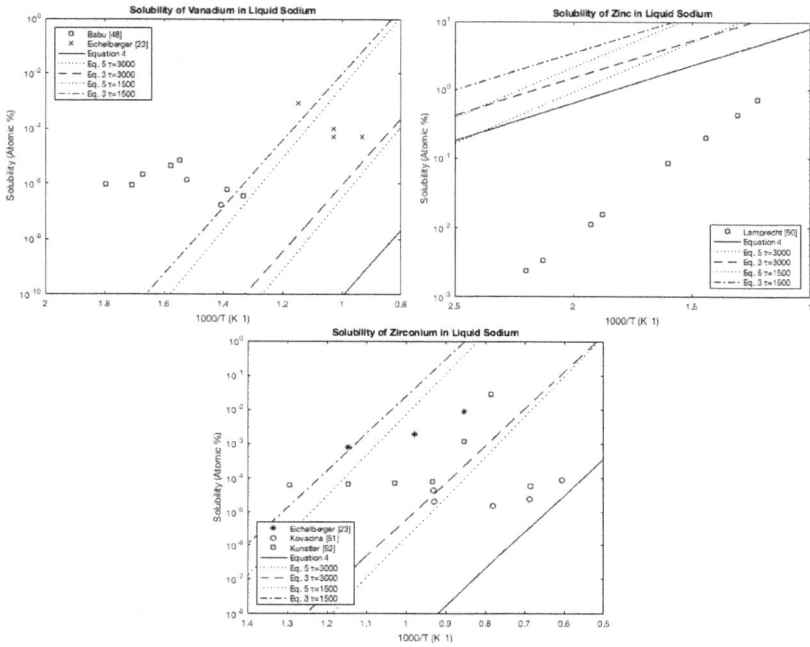

Figure A2. Solubility of additional transition metals in liquid sodium compared to the model.

A.3. Potassium

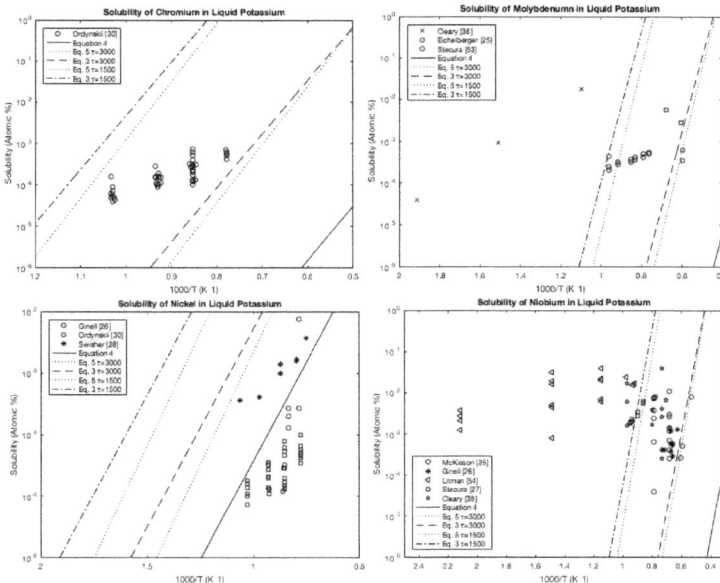

Figure A3. *Cont.*

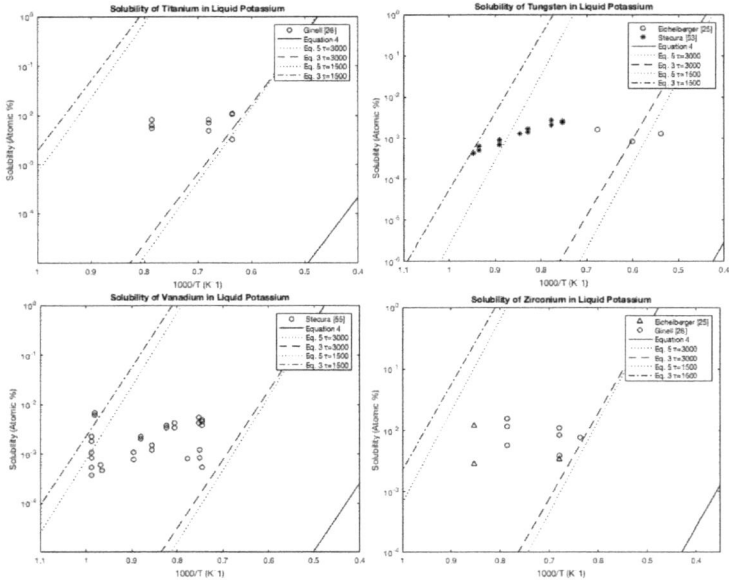

Figure A3. Solubility of additional transition metals in liquid potassium compared to the model.

A.4. Rubidium

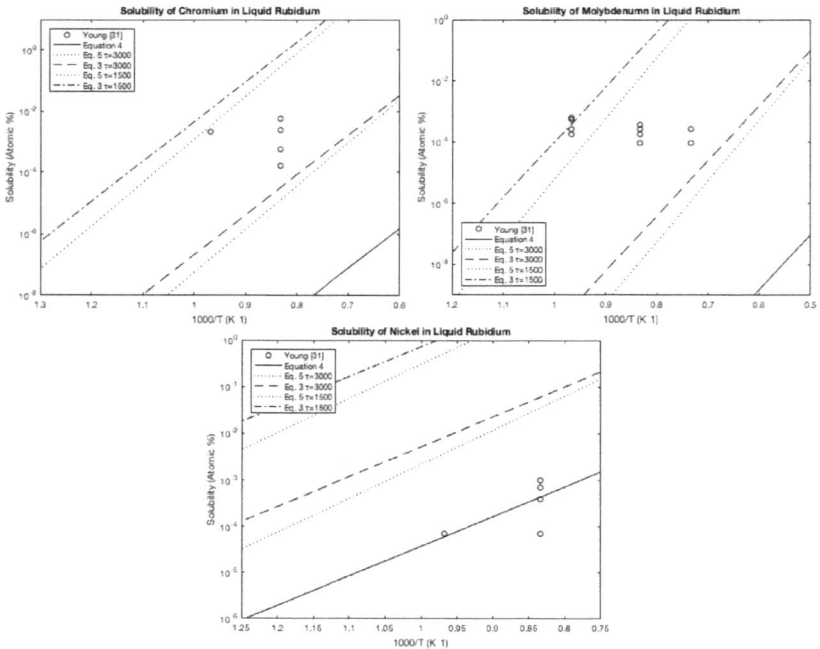

Figure A4. Solubility of additional transition metals in liquid rubidium compared to the model.

A.5. Cesium

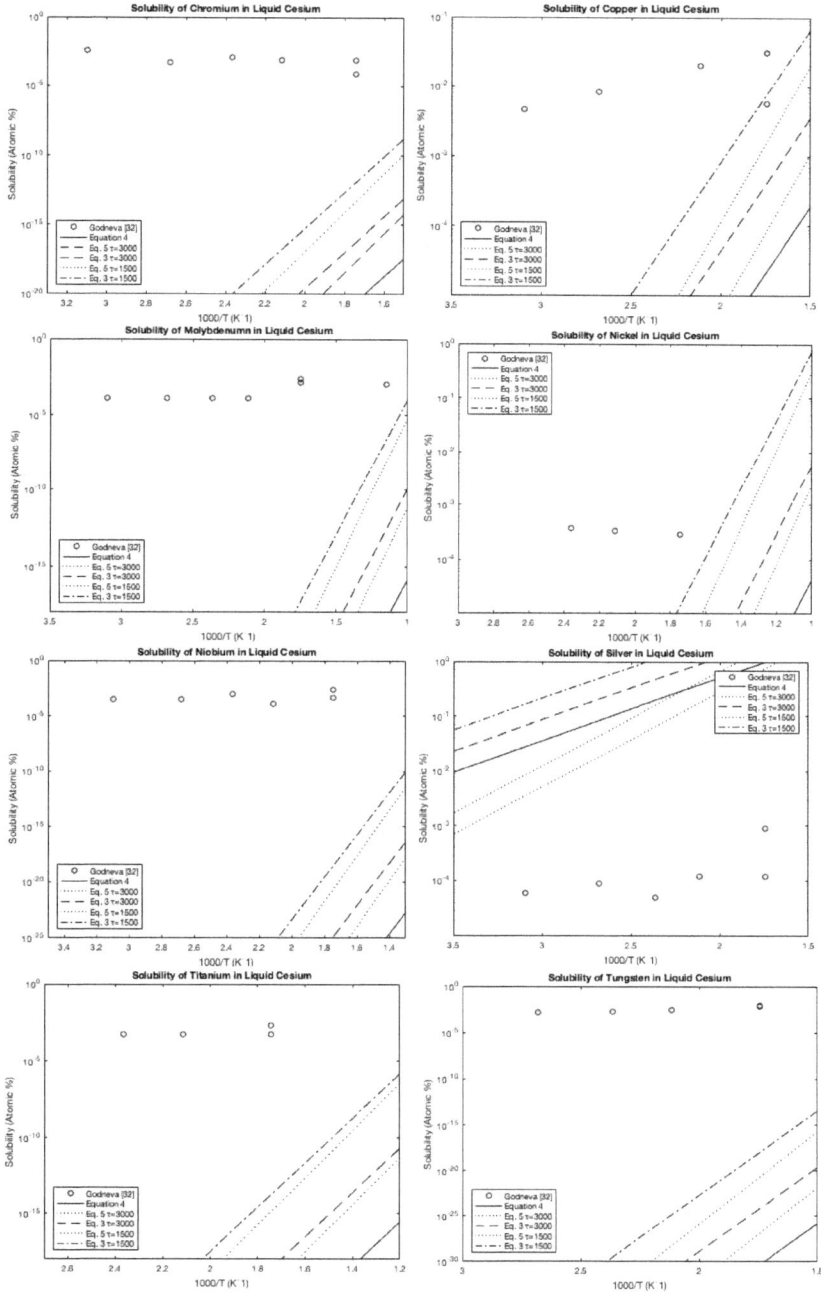

Figure A5. *Cont.*

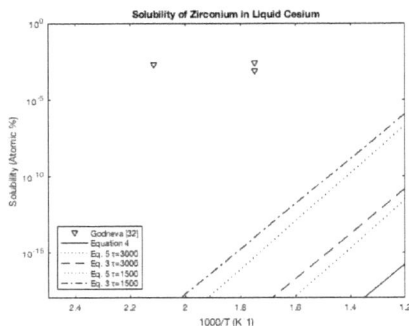

Figure A5. Solubility of additional transitional metals in liquid cesium compared to the model.

References

1. Borgstedt, H.U.; Guminski, C. *Solubility Data Series: Metals in Liquid Alkali Metals Part 1: Be to Os*; Oxford University Press: Oxford, UK, 1996.
2. IAEA. *Alkali Metal Coolants: Proceedings of the Symposium Held by the International Atomic Energy Agency in Vienna, 28 November–2 December 1966*; International Atomic Energy Agency: Wien, Austria, 1967.
3. Mariani, R.D.; Porter, D.L.; O'Holleran, T.P.; Hayes, S.L.; Kennedy, J.R. Lanthanides in metallic nuclear fuels: Their behavior and methods for their control. *J. Nucl. Mater.* **2011**, *419*, 263–271. [CrossRef]
4. De Boer, F.R.; Boom, R.; Mattens, W.C.; Miedema, A.R.; Niessen, A.K. *Cohesion in Metals: Transition Metal Alloys*; Elsevier: Amsterdam, The Netherlands, 1988.
5. Lyublinski, I.E.; Evtikhin, V.A.; Pankratov, V.Y.; Krasin, V.P. Numerical and experimental determination of metallic solubilities in liquid lithium, lithium-containing nonmetallic impurities, lead and lead-lithium eutectic. *J. Nucl. Mater.* **1995**, *224*, 288–292. [CrossRef]
6. Gale, W.F.; Totemeier, T.C. *Smithells Metals Reference Book (Eighth Edition)*; Elsevier: Amsterdam, The Netherlands, 2004.
7. Bakker, H. *Enthalpies in Alloys: Miedema's Semi-Empirical Model*; Enfield Publishing & Distribution Company: Enfield, NH, USA, 1998.
8. Lupis, C.H.P. *Chemical Thermodynamic of Materials*; Elsevier Science Ltd.: Amsterdam, The Netherlands, 1983.
9. Stecura, S. Corrosion of oxygen-doped tantalum by lithium. *Corros. Sci.* **1976**, *16*, 233–241. [CrossRef]
10. Eichelberger, R.L.; McKisson, R.L.; Johnson, B.G. *Solubility of Refractory Metals and Alloys in Potassium and in Lithium*; National Aeronautics and Space Administration: Washington, DC, USA, 1969.
11. Leavenworth, H.W.; Cleary, R.E. The solubility of Ni, Cr, Fe, Ti and Mo in liquid lithium. *Acta Metall.* **1961**, *9*, 519–520. [CrossRef]
12. Jesseman, D.S.; Roben, G.D.; Grunewald, A.L.; Fleshman, W.S.; Anderson, K.; Calkins, V.P. *Preliminary Investigation of Metallic Elements in Molten Lithium*; Technical Report NEPA-1465; Fairchild Engine and Airplane Corp.: Oak Ridge, TN, USA, 1950.
13. Leavenworth, H.; Cleary, R.E.; Bratton, W.D. *Solubility of Structural Metals in Lithium*; Technical Report PWAC-356; Pratt and Whitney Aircraft: Middletown, CT, USA, 1961.
14. Bychkov, Y.F.; Rozanov, A.N.; Yakovleva, V.B. The solubility of metals in liquid lithium. *At. Energ.* **1959**, *7*, 531–536. (In Russian)
15. Beskorovainyi, N.M.; Yakovlev, E.I. *Metallurgy & Metallography of Pure Metals*; Gordon & Breach: New York, NY, USA, 1962; pp. 189–206.
16. Beskorovainyi, N.M.; Vasilev, V.K.; Lyublinskii, I.E. Determination of the solubility of Fe, Ni, and Cr in Li by absorption X-ray spectral analysis. *Metall. Metalloved. Chist. Met.* **1980**, *14*, 135–148. (In Russian)
17. Bogard, A.D. *The Solubility of Iron in Sodium Metal, Sodium-Sodium Oxide, and Sodium-Sodium Oxide-Sodium Hydroxide*; Technical Report NRL-4131; Naval Research Lab.: Washington, DC, USA, 1953.
18. Eichelberger, R.L.; McKisson, R.L. *Studies of the Solubility of Iron in Sodium*; Technical Report Al-AEC-12834; Atomics International: Canoga Park, CA, USA, 1969.

19. Epstein, L.F. Studies on the solubility of iron in liquid sodium. *Science* **1950**, *112*, 426.
20. Fleitman, A.H.; Isaacs, H.S. The solubility and corrosion of pure iron in liquid sodium containing dissolved oxygen. In Proceedings of 1970 Metals Conference, Cleveland, OH, USA, 1970.
21. Periaswami, G.; Ganesan, V.; Babu, S.R.; Mathews, C.K. Solubility of manganese and iron in sodium. In *Material Behavior and Physical Chemistry in Liquid Metal Systems*; Borgstedt, H.U., Ed.; Plenum Press: New York, NY, USA, 1982; pp. 411–420.
22. Rodgers, S.J.; Mausteller, J.W.; Batutis, E.F. *Iron and Nickel Concentrations in Sodium*; Technical Report NP-5241; Mine Safety Appliances Co.: Cranberry Township, PA, USA, 1954.
23. Eichelberger, R.L.; McKisson, R.L. *Solubility Studies of Cr, Co, Mn, Mo, Ni, Nb, Ti, V, and Zr in Liquid Sodium*; Technical Report AL-AEC-12955; Atomics International: Canoga Park, CA, USA, 1970.
24. Pellett, C.R.; Thompson, R. Measurement of transition metal solubilities in liquid sodium; cobalt, nickel, and chromium. In *Liquid Metal Engineering and Technology: Proceedings of the Third International Conference Held in Oxford on 9–13 April 1984*; British Nuclear Energy Society: London, UK, 1984; Volume 3, pp. 43–48.
25. McKisson, R.L.; Eichelberger, R.L.; Dahleen, R.C.; Scarborough, J.M.; Argue, G.R. *Solubility Studies of Ultra Pure Alkali Metals*; Technical Report NASA-CR-610; North American Aviation, Inc.: Canoga Park, CA, USA, 1966.
26. Ginell, G.S.; Teitel, R.J. Determination of solubility of several transition metals in molten potassium. *Trans. Am. Nucl. Soc.* **1965**, *8*, 393–394.
27. Stecura, S. *Apparent Solubilities of Commercially Pure and Oxygen-Doped Tantalum and Niobium in Liquid Potassium*; Technical Report NASA-TN-D-5875; National Aeronautics and Space Administration: Cleveland, OH, USA, 1970.
28. Swisher, J.H. *Solubility of Iron, Nickel, and Cobalt in Liquid Potassium and Effect of Oxygen Gettering Agents on Iron Solubility*; Technical Report NASA-TN-D-2734; NASA Lewis Research Center: Cleveland, OH, USA, 1965.
29. McKisson, R.L.; Eichelberger, R.L.; Dahleen, R.C.; Scarborough, J.M.; Argue, G.R. *The Solubility of Cu, Mo, Nb, Fe, Ni, and Cr in High Purity Sodium*; Technical Report CONF-650411; United States Atomic Energy Commission: Washington, DC, USA, 1965.
30. Ordynskii, A.M.; Popov, R.G.; Raikova, G.P.; Samsonov, N.V.; Tarbov, A.A. Solubility of components of stainless steel in liquid potassium. *Teplofiz. Vys. Temp.* **1981**, *19*, 1192–1197. (In Russian)
31. Young, P.F.; Arabian, R.W. *Determination of Temperature Coefficient of Solubility of Various Metals in Rubidium and the Corrosive Effects of Rubidium on Various Alloys at Temperatures from 1000 to 2000 °F*; Technical Report AGN-8063; Aerojet-General Nucleonics: San Ramon, CA, USA, 1963.
32. Godneva, M.M.; Sedelnikova, N.D.; Geizler, E.S. Solubilities and Change of Weight of Some Metals in Caesium. *Zh. Prikl. Khim.* **1974**, *47*, 2177–2180. (In Russian)
33. Kirillov, V.B.; Krasin, V.P.; Lyublinskij, I.E.; Kuzin, A.N. Effect of nitrogen and oxygen impurities on dissolution and metal mass transfer in lithium and sodium melts. *Zh. Fiz. Khim.* **1988**, *62*, 3191–3195.
34. Awasthi, S.P.; Borgstedt, H.U.; Frees, G. Solubility of metals in sodium and lithium. In *Liquid Metal Engineering and Technology: Proceedings of the Third International Conference Held in Oxford on 9–13 April 1984*; British Nuclear Energy Society: London, UK, 1984; Volume 1, pp. 265–269.
35. Minushkin, B. *Solution Rates and Equilibrium Solubility of Nickel and Iron in Liquid Lithium*; Technical Report NDA-2141-1; United Nuclear Corp.: White Plains, NY, USA, 1961.
36. Plekhanov, G.A.; Fedortsov-Lutikov, G.P.; Glushko, Y.V. The effect of non-metallic impurities on the solubility of steel components in liquid lithium. *At. Energ.* **1978**, *45*, 143–145. [CrossRef]
37. Wilkinson, W.D.; Yaggee, F.L. *Attack on Metals by Lithium*; Technical Report ANL-4990; Argonne National Lab.: Lemont, IL, USA, 1950.
38. Cleary, R.E.; Blecherman, S.S.; Corliss, J.E. *Solubility of Refractory Metals in Lithium and Potassium*; Technical Report TIM-850; Pratt and Whitney Aircraft: Middletown, CT, USA, 1965.
39. Grand, J.A.; Baus, R.A.; Bogard, A.D.; Williams, D.D.; Lockhart, L.B.; Miller, R.R. The solubility of tantalum and cobalt in sodium by activation analysis. *J. Phys. Chem.* **1959**, *63*, 1192–1194.
40. Sivasubramanian, K.; Mitragotri, D.S.; Bhat, N.P. *Measurement of Solubility of Cobalt in Liquid Sodium*; Bhabha Atomic Research Centre: Bombay, India, 1995; pp. 262–263.
41. Eichelberger, R.L.; McKisson, R.L. *Solubility of Copper in Sodium*; Technical Report Al-AEC-12671; Atomics International: Canoga Park, CA, USA, 1968.

42. Humphreys, J.R. *Interdivision Document K-3-774*; Los Alamos Scientific Laboratory: Los Alamos, NM, USA, 1958.

43. Koenig, R.F. *Corrosion of Zirconium and Its Alloys in Liquid Metals*; Technical Report KAPL-982; Knolls Atomic Power Lab.: Niskayuna, NY, USA, 1953.

44. Singer, R.M.; Weeks, J.R. *On the Solubilities of Cu, Ni, and Fe in Liquid Sodium*; Technical Report BNL-1307; Brookhaven National Lab.: Upton, NY, USA, 1969.

45. Walker, R.A.; Pratt, J.N. The solubility of copper in liquid sodium. *J. Nucl. Mater.* **1969**, *32*, 340–345. [CrossRef]

46. Stanaway, W.P.; Thompson, R. Solubility of Metals, Iron and Manganese in Sodium. In Proceedings of the Second International Conference on Liquid Metal Technology in Energy Production, Richland, WA, USA, 20–24 April 1980; Dahlke, J.M., Ed.; American Nuclear Society: Washington, DC, USA, 1980.

47. Stanaway, W.P.; Thompson, R. The solubility of transition metals, Mn and Co in liquid sodium. In *Material Behavior and Physical Chemistry in Liquid Metal Systems*; Borgstedt, H.U., Ed.; Plenum Press: New York, NY, USA, 1982; pp. 421–427.

48. Babu, S.R.; Periaswami, C.; Geetha, R.; Mahalingam, T.R. Solubility of molybdenum and vanadium in liquid sodium. In *Liquid Metal Engineering and Technology: Proceedings of the Third International Conference Held in Oxford on 9–13 April 1984*; British Nuclear Energy Society: London, UK, 1984; Volume 1, pp. 271–275.

49. Kunstler, K.; Baum, H. *Deposition of Niobium in the System Sodium-Steel*; Technical Report ZFK-340; German Democratic Republic: Dresden, Germany, 1977; pp. 47–48. (In German)

50. Lamprecht, G.J.; Crowther, P. Solubility of metals in liquid sodium: The systems sodium-silver, sodium-zinc, and sodium-cerium. *Trans. AIME* **1968**, *242*, 2169–2171.

51. Kovacina, T.A.; Miller, R.R. *The Solubility of Columbian-1% Zirconium in Sodium by Activation Analysis*; Technical Report NRL-6051; Naval Research Lab.: Washington, DC, USA, 1964.

52. Kunstler, K.; Heyne, H. Solubility of zirconium in liquid sodium. In *Material Behavior and Physical Chemistry in Liquid Metal Systems*; Borgstedt, H.U., Ed.; Plenum Press: New York, NY, USA, 1995; pp. 311–319.

53. Stecura, S. *Solubilities of Molybdenum and Tungsten in Liquid Potassium*; Technical Report NASA-TN-D-5504; NASA Lewis Research Center: Cleveland, OH, USA, 1969.

54. Litman, A.P. *The Effect of Oxygen on the Corrosion of Niobium by Liquid Potassium*; Technical Report ORNL-3751; Oak Ridge National Lab.: Oak Ridge, TN, USA, 1965.

55. Stecura, S. Solubilities of molybdenum, tungsten, vanadium, titanium, and zirconium in liquid potassium. In *Corrosion by Liquid Metals*; Draley, J.E., Ed.; Springer: New York, NY, USA, 1970; pp. 601–611.

56. Awasthi, S.P.; Borgstedt, H.U. An Assessment of solubility of some transition metals (Fe, Ni, Mn, and Cr) in liquid sodium. *J. Nucl. Mater.* **1983**, *116*, 103–111. [CrossRef]

metals

MDPI

Article

Grain Boundary Assemblies in Dynamically-Recrystallized Austenitic Stainless Steel

Marina Tikhonova *, Pavel Dolzhenko, Rustam Kaibyshev and Andrey Belyakov

Laboratory of Mechanical Properties of Nanostructured Materials and Superalloys, Belgorod State University, Pobeda 85, Belgorod 308015, Russia; dolzhenko.p@yandex.ru (P.D.); rustam_kaibyshev@bsu.edu.ru (R.K.); belyakov@bsu.edu.ru (A.B.)
* Correspondence: tikhonova@bsu.edu.ru; Tel.: +7-4722-585457

Academic Editor: Soran Birosca
Received: 26 September 2016; Accepted: 2 November 2016; Published: 7 November 2016

Abstract: The grain boundary misorientation distributions associated with the development of dynamic recrystallization were studied in a high-nitrogen austenitic stainless steel subjected to hot working. Under conditions of discontinuous dynamic recrystallization, the relationships between the grain or subgrain sizes and flow stresses can be expressed by power law functions with different grain/subgrain size exponents of about -0.76 (for grain size) or -1.0 (for subgrain size). Therefore, the mean grain size being much larger than the subgrain size under conditions of low flow stress gradually approaches the size of the subgrains with an increase in the flow stress. These dependencies lead to the fraction of high-angle boundaries being a function of the flow stress. Namely, the fraction of ordinary high-angle boundaries in dynamically-recrystallized structures decreases with a decrease in the flow stress. On the other hand, the fraction of special boundaries, which are associated with annealing twins, progressively increases with a decrease of the flow stress.

Keywords: dynamic recrystallization; austenitic stainless steel; grain boundary engineering; hot working

---

1. Introduction

Dynamic recrystallization (DRX) is a very effective tool to obtain a desirable microstructure in various metallic materials [1–4]. The DRX microstructures depend sensitively on the deformation conditions, i.e., temperature (T) and strain rate ($\dot{\varepsilon}$), which are commonly represented by a temperature-compensated strain rate, the Zener-Hollomon parameter ($Z = \dot{\varepsilon} \exp Q/RT$, where Q and R are the activation energy and the universal gas constant, respectively) [5,6]. One of the most important structural parameters is grain size, which significantly affects properties of metals and alloys; especially, their mechanical/deformation behavior [7]. The grain size in DRX microstructures can be controlled by deformation conditions, e.g., it increases with an increase in the deformation temperature and/or a decrease in the strain rate [8,9]. The flow stress during hot deformation of various metals and alloys also depends on the deformation conditions and can be expressed by a power law function of Z. Therefore, a power law generally holds between the flow stress and the DRX grain size with a grain size exponent of about -0.7 for hot working conditions [10–12].

In contrast to the size of DRX grains, which has been a subject of numerous investigations along with the fraction and distribution of DRX grains [2], the grain boundary character distributions that develop during DRX have not been studied in sufficient details. The DRX mechanism that operates under hot working conditions involves a local bulging of a grain boundary portion leading to the development of a DRX nucleus. Then, the DRX nucleus grows out, consuming the work-hardened surroundings until it impinges with other growing DRX grains, or the driving force for the DRX grain growth drops below a critical value due to the increase in the deformation stored energy in the growing DRX grain during deformation. Therefore, there are several grain types in the DRX microstructure, including DRX nuclei, growing DRX grains, and work-hardened grains. The latter ones involve a number of low-angle dislocation subboundaries. Moreover, annealing twins may appear in the growing DRX grains as a result of a growth accident of a migration boundary in metals and alloys with low stacking fault energy (SFE) [12,13], since the growth of DRX grains is associated with the migration of their boundaries. The fraction and density of twin-related boundaries, which are characterized by special coincident site lattices ($\Sigma 3^n$ CSL), have been frequently studied for static recrystallization and grain growth [14–16]. It has been suggested that both the fraction and density of $\Sigma 3^n$ CSL boundaries depend on a relative change in the grain size and, thus, can be expressed by functions of the size ratio of an instant grain (after recrystallization or grain growth) to the initial one (just before recrystallization/grain growth) [17–19]. A similar relationship between the fraction of $\Sigma 3^n$ CSL boundaries and grain size ratio has been obtained for continuous post-dynamic recrystallization of an austenitic stainless steel [20]. On the other hand, the development of $\Sigma 3^n$ CSL boundaries in low-SFE materials during conventional discontinuous DRX under hot working conditions has not been clarified.

The aim of the present paper is to study the grain boundary assemblies that develop during discontinuous DRX in an austenitic stainless steel. It places particular emphasis on quantitative relationships between the grain boundary character distributions and hot working conditions. A high-nitrogen chromium-nickel stainless steel was selected in the present study as a representative of advanced structural steels designated for crucial engineering applications sustainable against loading under low-temperature conditions.

2. Materials and Methods

A high-nitrogen austenitic stainless steel, 0.025%C–22%Cr–10.2%Ni–0.36%N–6.2%Mn–0.34%Si–1.9%Mo–0.003%S–0.005%P and the balance Fe (all in wt %), was cast at the Central Research Institute for Machine-Building Technology, Moscow, Russia. The steel was solution annealed and hot rolled at a temperature of 1373 K. The average grain size was approximately 420 μm. Rectangular samples with initial dimensions of 10 mm × 12 mm × 15 mm were machined for compression tests, which were carried out using an Instron 300LX testing machine (Instron Ltd., Norwood, MA, USA) equipped with a three-sectioned high-temperature furnace (Instron Ltd., Norwood, MA, USA). A powder of boron nitride was used as a lubricant to minimize the friction between the specimen and anvil. The specimens were compressed at temperatures of 1073 K to 1348 K, and strain rates of 10^{-4} s^{-1} to 10^{-2} s^{-1}, and then immediately quenched by water jet while the deformation was ceased (the quench delay was 1–2 s).

The structural investigations were carried out on the sample sections parallel to the compression axis using a Nova Nano SEM 450 scanning electron microscope (FEI Corporation, Hillsboro, OR, USA) equipped with an electron backscatter diffraction (EBSD) analyzer (Oxford Instruments, Oxfordshire, UK) incorporating an orientation imaging microscopy (OIM) system (Oxford Instruments, Oxfordshire, UK). The OIM images were subjected to clean-up procedures, setting a minimal confidence index of 0.1. The grain size (D) was measured as an average distance between ordinary high-angle boundaries (HABs), i.e., those with misorientations of $\theta \geq 15°$, excepting the $\Sigma 3^n$ CSL boundaries, along and crosswise to the compression axis. In the case of uncompleted DRX, the DRX grains were distinguished from non-recrystallized portions by means of kernel average misorientation, assuming that the kernel average misorientation within DRX portions did not exceed $1°$. The subgrain size (d) was measured as an average distance between subboundaries with misorientations of $1° < \theta < 15°$.

3. Results

3.1. Stress-Strain Curves

Figure 1 shows a series of true stress-strain curves obtained during hot isothermal compressions at different temperatures and strain rates. Most of the curves exhibit typical DRX behavior with a peak flow stress followed by a steady state deformation behavior [3,21]. Note here that an apparent increase in the flow stress at large strains is associated with a growing contribution from contact friction stresses. The flow stresses vary with the deformation temperature and strain rate. The flow stresses decrease with an increase in the deformation temperature (Figure 1a) or with a decrease in the strain rate (Figure 1b). The peak flow stress increases by 30%–40% with a ten-fold increase in the strain rate, if the temperature is a constant. A decrease in the deformation temperature results in the peak flow stress occurring at larger strain that is much similar to other studies on DRX at hot working conditions [1,2,5,22]. This suggests, therefore, that the DRX kinetics slow down as the deformation temperature decreases. At relatively low deformation temperatures below 1173 K, the steady state deformation behavior can hardly be distinguished through the flow curves in Figure 1a.

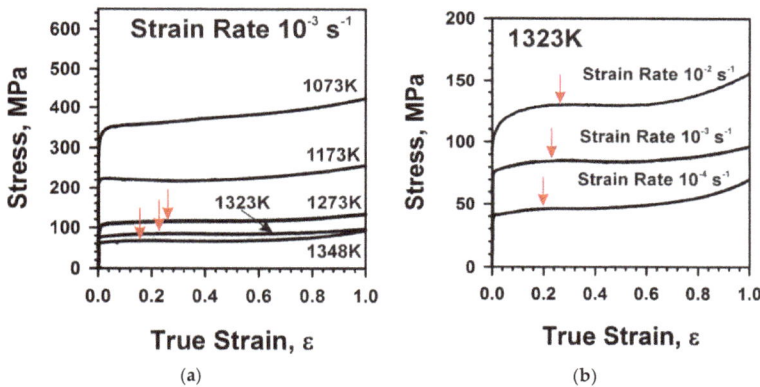

Figure 1. The true stress-true strain curves of the high nitrogen austenitic stainless steel under various deformation conditions, i.e., at strain rate of 10^{-3} s^{-1} (a) and at a temperature of 1323 K. (b) The red arrows indicate the peak stress associated with DRX.

3.2. Deformation Microstructures

Typical deformation microstructures that developed during hot isothermal compressions at temperatures of 1073–1323 K are shown in Figure 2. The deformation microstructures are characterized by pancake-shaped original grains, which are separated from each other by thin chains of ultrafine DRX grains, after compression at 1073 K (Figure 2). The fraction of DRX grains increases with the increase in the deformation temperature. The coarse remnants of the original grains are surrounded by numerous DRX grains in the sample compressed at 1173 K (Figure 2). The development of DRX grains readily takes place along the original grain boundaries [1,3,10]. Therefore, so-called necklace microstructures are evolved in the partially-recrystallized samples in Figure 2. It is clearly seen in Figure 2 that uniform fine-grained microstructures are completely developed in the samples processed at temperatures of $T \geq 1273$ K. As it could be expected, these fine-grained DRX microstructures involve various boundaries, including rather large fractions of low-angle subboundaries and twin-related $\Sigma 3^n$ CSL boundaries (some of them are indicated by arrows in Figure 2) besides ordinary high-angle grain boundaries.

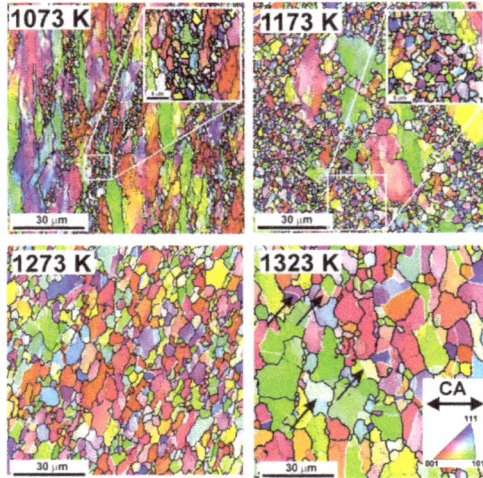

Figure 2. Typical microstructures that evolved in the high-nitrogen austenitic stainless steel during compression to a strain of 1.2 at 1073–1323 K and a strain rate of 10^{-3} s^{-1}. The thin and thick lines correspond to the subgrain and grain boundaries, respectively. The white line corresponds to CSL $\Sigma 3^n$ boundaries. The colors reflect the inverse pole figures for the compression axis (CA). The arrows indicate some typical twins by growth accident.

The presence of large fractions of non-recrystallized portions in the DRX necklace microstructures evolved at 1073–1173 K results in bimodal grain size distributions (Figure 3). The fraction of large non-recrystallized grains comprises 0.74 at 1073 K. Correspondingly, the fraction of DRX grains at this temperature is 0.26. An increase in the deformation temperature leads to an increase in the DRX fraction. The peak of the grain size distribution against the large grain sizes, which corresponds to non-recrystallized remnants of the original grains, decreases and spreads out towards smaller sizes. At 1173 K, the fraction of recrystallized portions increases to 0.5.

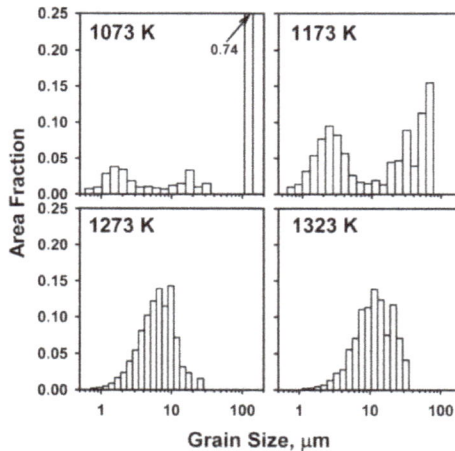

Figure 3. Grain size distributions that evolved in the high-nitrogen austenitic stainless steel during compression to a strain of 1.2 at the indicated temperatures and a strain rate of 10^{-3} s^{-1}.

Fully-recrystallized microstructures develop in the samples compressed at 1273–1348 K. The corresponding grain size distributions are characterized by a single peak against grain sizes ranging from 1–2 μm to 10–40 μm, depending on the deformation temperature (Figure 3).

3.3. DRX Grains and Their Boundaries

Let us consider the DRX grain evolution during compressions at different temperatures and strain rates in more detail. The temperature and strain rate effects on the average grain and subgrain sizes are shown in Figure 4. Note here that the non-recrystallized portions were omitted while considering the DRX grains/subgrains and their boundaries in the partially-recrystallized microstructures. The grain/subgrain sizes exhibit apparently weak temperature and strain rate dependencies at relatively low temperatures. The average DRX grain and subgrain sizes fall in almost the same range of 1–2 μm during compressions at 1073–1173 K and 10^{-4} to 10^{-2} s^{-1}. In contrast, the mean DRX grain size rapidly increases with an increase in the temperature, especially, at the low strain rate of 10^{-4} s^{-1} (Figure 4a). On the other hand, the rate of increase of DRX subgrain size with an increase in the deformation temperature is much lower as compared with that for DRX grain size, irrespective of the strain rate (Figure 4b).

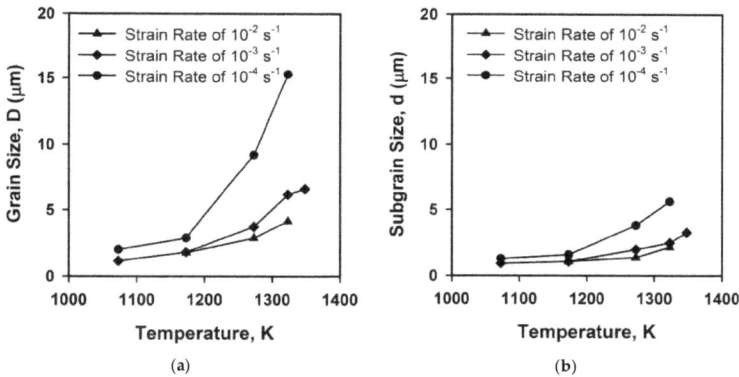

Figure 4. DRX grain (**a**) and subgrain (**b**) size that evolved in the high-nitrogen austenitic stainless steel during compression at different temperatures and strain rates.

The distributions of the DRX grain boundary misorientations that developed in the samples subjected to compressions at a strain rate of 10^{-3} s^{-1} are shown in Figure 5. These developed at different temperatures, and the DRX misorientation distributions are quite similar in appearance and, qualitatively, can be represented as a superposition of three characteristic misorientations. Namely, low-angle misorientations typical of deformation subboundaries, random misorientations (Mackenzie [23]) with a broad peak at 45°, and a sharp peak at 60° corresponding to annealing twins. The changes in the misorientation distributions after compression at different temperatures are mainly associated with the changes in the fractions of low-angle subboundaries and 60° twin boundaries. The peak below 15°, corresponding to low-angle boundaries, sharpens and rises with an increase in the deformation temperature. The twin-related peak against 60° demonstrates the same behavior. After compression at 1073 K, the fraction of 60° misorientations is about 0.05. Increasing the deformation temperature promotes the annealing twin formation. As a result, the fraction of 60° misorientations exceeds 0.15 in the sample compressed at 1323 K. The relative changes in the fractions of 60° misorientations are clearly correlated with the relative changes in the grain/subgrain sizes. The smallest fraction of 60° misorientations corresponds to the sample compressed at 1073 K, in which the grain and subgrains sizes are almost the same.

It has been shown that the steady-state size of DRX grains can be expressed by a power-law function of the flow stress [10,11,24,25]. The relationships between the flow stresses normalized by shear modulus and the sizes of DRX grains and subgrains developed in the present steel during compressions at various temperatures and strain rates are presented in Figure 6. The flow stresses can be related to the DRX grain size through a power-law function with a grain size exponent of −0.76. Such a grain size exponent is typical of discontinuous DRX under hot working conditions [26,27]. On the other hand, the subgrain size is expressed by a different power law function of the flow stress with a size exponent of −1. Therefore, the size difference between DRX grains and subgrains diminishes with increasing flow stress.

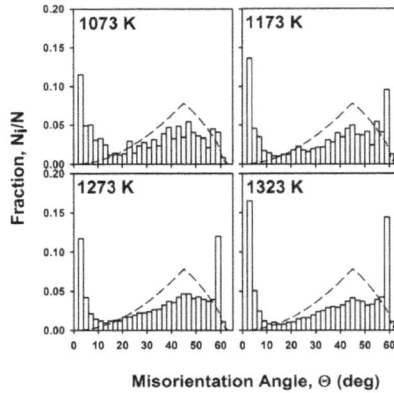

Figure 5. Distributions of the DRX grain boundary misorientations developed in the high-nitrogen austenitic stainless steel during compression to a strain of 1.2 at the indicated temperatures and a strain rate of 10^{-3} s^{-1}. The dashed lines indicate random (Mackenzie) distribution.

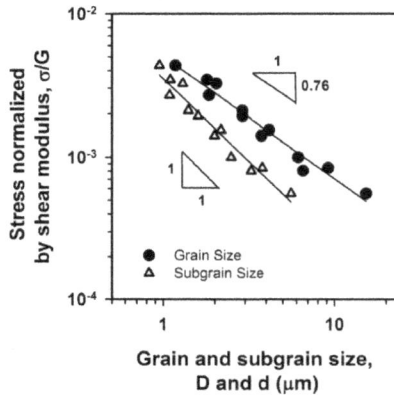

Figure 6. Relationship between the flow stress normalized by the shear modulus (σ/G) and the DRX grain size (D) and subgrain size (d) for the high-nitrogen austenitic stainless steel.

The flow stress and the DRX grain/subgrain sizes are sensitively dependent on the processing conditions [11,28]. Figure 7 shows the relationship between the DRX grain and subgrain sizes and the deformation conditions, which are represented by the temperature-compensated strain rate, Z. The latter was estimated with an activation energy of 280 kJ·mol^{-1} [29]. Increasing Z, generally, leads to a decrease of both the grain and subgrain sizes. In the range of $Z < 10^{11}$ s^{-1}, corresponding to hot working conditions, the microstructural evolution during deformation of chromium-nickel austenitic stainless steels has been suggested as being associated with the development of discontinuous DRX, when the DRX grain and subgrain sizes exhibit rather strong temperature and/or strain rate dependencies [11]. The DRX grain and subgrain sizes that develop in the present steel can be expressed as $D{\sim}Z^{-0.27}$ and $d{\sim}Z^{-0.2}$, respectively. These dependencies predict a certain change in the fraction of ordinary grain boundaries in the DRX microstructures with a change in deformation conditions.

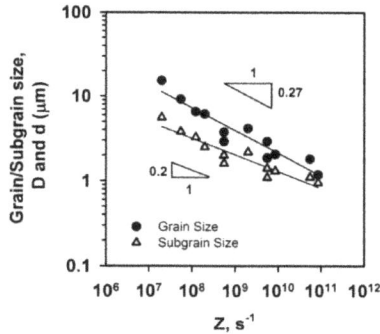

Figure 7. Dependences of the DRX grain/subgrain sizes on the temperature-compensated strain rate (Z) for the high nitrogen austenitic stainless steel.

The fractions of ordinary high-angle grain boundaries and twin-related $\Sigma 3^n$ CSL boundaries in the DRX microstructures that develop in the present steel during hot working are shown in Figure 8. The fraction of ordinary grain boundaries decreases from approx. 0.7 to 0.5 with a decrease in Z from 10^{11} s^{-1} to 10^7 s^{-1}. That is to say, the fraction of high-angle grain boundaries in the DRX microstructures should decrease with an increase in deformation temperature and/or a decrease in strain rate. Such behavior is directly related to the variations of the DRX grain/subgrain sizes with deformation conditions. The difference between the DRX grain size and subgrain size markedly increases with a decrease in Z (Figure 7) or flow stress (Figure 6), then, the fraction of grain boundaries decreases as a ratio of d/D. On the other hand, a decrease in Z (i.e., an increase in deformation temperature and/or a decrease in strain rate) results in an increase in the fraction of $\Sigma 3^n$ CSL boundaries (Figure 8). Thus, the larger DRX grain size corresponds to the larger fraction of twin-related boundaries.

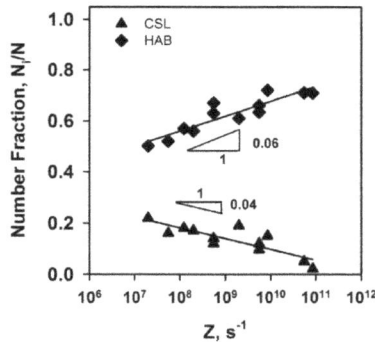

Figure 8. Dependences of the fractions of HAB and CSL boundaries on the temperature-compensated strain rate (Z) for the high-nitrogen austenitic stainless steel.

4. Discussion

Annealing twins have been suggested to appear by growth accidents of a migrating boundary during recrystallization and grain growth [13,18]. Considering an increase in the grain size from D_0 to D the following relationships between the grain size ratio (D/D_0) and the fraction of twin-related $\Sigma 3^n$ CSL boundaries (F_{CSL}) were obtained [20].

$$F_{CSL} = \frac{N_{CSL_0} + K \ln \frac{D}{D_0}}{N_{CSL_0} + K \ln \frac{D}{D_0} + 1} \tag{1}$$

$$N_{CSL_0} = \left(\frac{1}{F_{CSL_0}} - 1 \right)^{-1} \tag{2}$$

where N_{CSL_0} and F_{CSL_0} are the number and fraction of $\Sigma 3^n$ CSL boundaries in the initial microstructure just before the grain growth, respectively, and the numerical factor of $K \sim 0.5$ depends on the probability of twin formation by a growth accident.

According to the common representation of the discontinuous DRX mechanism, the grain boundary bulging creates a DRX nucleus, which is actually a subgrain that locates under the bulge and almost free of lattice dislocation. Then, this DRX nucleus rapidly grows out and becomes a new grain replacing other pre-existing work-hardened grains [2,3]. Therefore, the size of deformation subgrains can be roughly considered as the starting grain size (D_0), while considering the grain coarsening during discontinuous DRX. Then, the fraction of $\Sigma 3^n$ CSL boundaries can be evaluated by Equation (1), taking $D_0 = d$ for the present DRX microstructures that developed under various deformation conditions.

Figure 9 shows the relationship between F_{CSL} and D/d that were obtained for austenitic stainless steels in the present (filled symbols) and previous (open symbols) studies [20,30] as well as that calculated by Equation (1) with $K = 0.5$ (solid line). It should be noted that previous studies dealt with continuous post-dynamic recrystallization after large strain deformation by multiple warm forgings at different temperatures, whereas the present study considers conventional DRX microstructures.

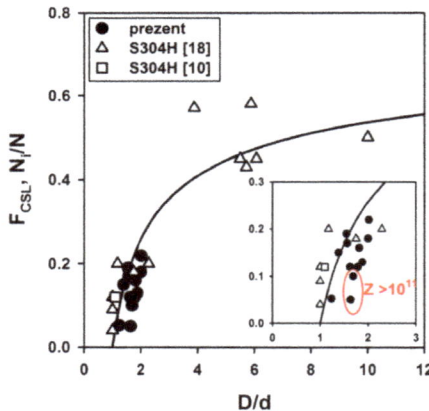

Figure 9. Relationship between the fraction of CSL boundaries (F_{CSL}) and the grain/subgrain size ratio (D/d).

Nevertheless, all data points in Figure 9 locate quite near the position predicted by Equation (1), suggesting that the fraction of $\Sigma 3^n$ CSL boundaries can be expressed by a unique function of the grain size change irrespective of the grain coarsening mechanisms. Some data points to the right of the curve (s. insert in Figure 9) were obtained under conditions of $Z > 10^{11}$ s^{-1}, when the contribution of discontinuous DRX to the new grain development becomes small, i.e., new grains develop without remarkable grain growth [11].

Let us consider the effect of deformation conditions from cold to hot working on the grain boundary assemblies evolved after sufficiently large strains, i.e., under conditions of apparent steady state deformation. The change in the fraction of ordinary high-angle boundaries (F_{HAB}) in Figure 9 clearly correlates with the change in subgrain/grain size ratio (d/D) in Figures 7 and 8. The hot deformation conditions selected in the present study correspond to the range of discontinuous DRX. Therefore, the fraction of ordinary grain boundaries in discontinuous DRX microstructures decreases with a decrease in Z (increase in temperature and/or decrease in strain rate). In a previous paper, a three-stage relationship between flow stress and dynamic grain size has been suggested for a wide temperature interval [31,32]. Three power law functions with different grain size exponents of about -0.7, -0.3, and -1.0 have been reported for discontinuous DRX during hot working, continuous DRX during warm working, and grain refinement during cold working. On the other hand, the subgrain size can be related to the flow stress with a size exponent of -1.0 in the whole temperature range. Such dependences suggest a three-stage variation of boundary characteristics on the processing conditions, as schematically shown in Figure 10.

A decrease in the deformation temperature corresponds to the sequential change from hot working to warm working and then to cold working (the regions of I to II to III in Figure 10). The fraction of ordinary high-angle grain boundaries increases as the deformation temperature decreases within the conditions of hot working (region I). Correspondingly, slow down grain growth with decreasing temperature results in a gradual decrease in the fraction of $\Sigma 3^n$ CSL boundaries. The transition to warm working is accompanied by a change in the grain size dependence on deformation conditions/flow stress. The grain size that evolves at sufficiently large strains rapidly approaches the subgrain size, i.e., $D/d \rightarrow 1$, as the deformation temperature decreases. Therefore, the fraction of high-angle grain boundaries comes close to 1, while F_{CSL} becomes negligibly small (region II in Figure 10). A further decrease in the deformation temperature to cold working leads to the cessation of the grain growth. The ultrafine grains that develop under conditions of severe plastic deformation are entirely bounded by high-angle boundaries [33,34], leading to $F_{HAB} \sim 1$. The $\Sigma 3^n$ CSL boundaries may appear during severe plastic deformation as result of deformation twinning, but this is not an annealing phenomenon.

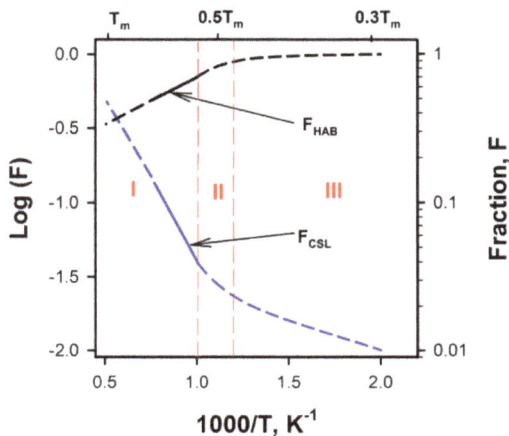

Figure 10. Variations of the fractions of HAB (F_{HAB}) and CSL (F_{CSL}) boundaries in austenitic stainless steels with deformation conditions.

5. Conclusions

The deformation microstructures that developed in a high-nitrogen austenitic stainless steel during hot working accompanied by dynamic recrystallization (DRX) were studied. The main results can be summarized as follows:

1. The relationship between the DRX grain/subgrain size and flow stress can be expressed by power-law functions with different size exponents of -0.76 (grains) and -1.0 (subgrains), which is typical of discontinuous DRX under hot working conditions. Correspondingly, power-law functions with exponents of -0.27 and -0.2 are held for the effect of temperature-compensated strain rate (Z) on the grain and subgrain sizes, respectively.

2. The fraction of ordinary high-angle boundaries in DRX microstructures increases from approx. 0.5 to 0.7, while the fraction of twin-related $\Sigma 3^n$ CSL boundaries decreases from 0.25 to 0.05 as Z increases from 10^7 s^{-1} to 10^{11} s^{-1}. Both the fraction of ordinary grain boundaries and the fraction of $\Sigma 3^n$ CSL boundaries depend on the grain and subgrain sizes, i.e., D and d, and can be expressed by functions of the ratio of D/d.

Acknowledgments: The authors acknowledge with gratitude the financial support received from the Ministry of Education and Science, Russia, project No. 14.575.21.0070 (ID No. RFMEFI57514X0070), and the technical support from the personnel of the Joint Research Centre, Belgorod State University.

Author Contributions: The present paper is a result of fruitful collaboration of all co-authors. R. Kaibyshev designed the research theme; A. Belyakov and M. Tikhonova selected the experimental procedure; P. Dolzhenko carried out the experimental study. Then, all co-authors discussed and analyzed the obtained results.

Conflicts of Interest: The authors declare no conflict of interest.

References

1. Sakai, T.; Jonas, J.J. Dynamic recrystallization: Mechanical and microstructural considerations. *Acta Metall.* **1984**, *32*, 189–209. [CrossRef]
2. Sakai, T.; Belyakov, A.; Kaibyshev, R.; Miura, H.; Jonas, J.J. Dynamic and post-dynamic recrystallization under hot, cold and severe plastic deformation conditions. *Prog. Mater. Sci.* **2014**, *60*, 130–207. [CrossRef]
3. Humphreys, F.J.; Hatherly, M. *Recrystallization and Related Annealing Phenomena*; Elsevier Science: New York, NY, USA, 2004.
4. Huang, K.; Logé, R.E. A review of dynamic recrystallization phenomena in metallic materials. *Mater. Des.* **2016**, *111*, 548–574. [CrossRef]
5. McQueen, H.J.; Jonas, J.J. Recovery and Recrystallization during High Temperature Deformation. In *Treatise on Materials Science and Technology*; Arsenault, R.J., Ed.; Academic Press: New York, NY, USA, 1975; Volume 6, pp. 394–493.
6. Maki, T.; Akasaka, K.; Okuno, K.; Tamura, I. Dnamic recrystallization of austenite in 18-8 stainless steel and 18 Ni maraging steel. *Trans. Iron Steel Inst. Jpn.* **1982**, *22*, 253–261. [CrossRef]
7. Belyakov, A.; Sakai, T.; Miura, H.; Kaibyshev, R. Grain refinement under multiple warm deformation in 304 type austenitic stainless steel. *ISIJ Int.* **1999**, *39*, 592–599. [CrossRef]
8. Sakai, T. Dynamic recrystallization microstrustures under hot working conditions. *J. Mater. Process. Technol.* **1995**, *53*, 349–361. [CrossRef]
9. Belyakov, A.; Tikhonova, M.; Yanushkevich, Z.; Kaibyshev, R. Regularities of grain refinement in an austenitic stainless steel during multiple warm working. *Mater. Sci. Forum* **2013**, *753*, 411–416. [CrossRef]
10. Dudova, N.; Belyakov, A.; Sakai, T.; Kaibyshev, R. Dynamic recrystallization mechanisms operating in a Ni-20%Cr alloy under hot-to-warm working. *Acta Mater.* **2010**, *58*, 3624–3632. [CrossRef]
11. Tikhonova, M.; Belyakov, A.; Kaibyshev, R. Strain-induced grain evolution in an austenitic stainless steel under warm multiple forging. *Mater. Sci. Eng. A* **2013**, *564*, 413–422. [CrossRef]
12. Wang, X.; Brunger, E.; Gottstein, G. The role of twinning during dynamic recrystallization in alloy 800 H. *Scr. Mater.* **2002**, *46*, 875–880. [CrossRef]
13. Mahajan, S. Critique of mechanisms of formation of deformation, annealing and growth twins: Face-centered cubic metals and alloys. *Scr. Mater.* **2013**, *68*, 95–99. [CrossRef]

14. Watanabe, T. Approach to Grain Boundary Design for Strong and Ductile Polycrystals. *Res. Mech.* **1984**, *11*, 47–84.

15. Randle, V. Twinning-related grain boundary engineering. *Acta Mater.* **2004**, *52*, 4067–4081. [CrossRef]

16. Fang, X.; Liu, Z.; Tikhonova, M.; Belyakov, A.; Wang, W. Evolution of texture and development of $\Sigma 3^n$ grain clusters in 316 austenitic stainless steel during thermal mechanical processing. *J. Mater. Sci.* **2013**, *48*, 997–1004. [CrossRef]

17. Cahoon, J.R.; Qiangyong, L.I.; Richards, N.L. Microstructural and processing factors influencing the formation of annealing twins. *Mater. Sci. Eng. A* **2009**, *526*, 56–61. [CrossRef]

18. Pande, C.S.; Imam, M.A.; Rath, B.B. Study of Annealing Twins in FCC Metals and Alloys. *Metall. Trans. A* **1990**, *21*, 2891–2896. [CrossRef]

19. Pande, C.S.; Imam, M.A. Grain growth and twin formation in boron-doped nickel polycrystals. *Mater. Sci. Eng. A* **2009**, *512*, 82–86. [CrossRef]

20. Tikhonova, M.; Kuzminova, Y.; Fang, X.; Wang, W.; Kaibyshev, R.; Belyakov, A. $\Sigma 3$ CSL boundary distributions in an austenitic stainless steel subjected to multidirectional forging followed by annealing. *Philos. Mag.* **2014**, *94*, 4181–4196. [CrossRef]

21. Jonas, J.J.; Sellars, C.M.; Tegart, W.J. Strength and structure under hot-working conditions. *Metall. Rev.* **1969**, *130*, 1–33. [CrossRef]

22. Belyakov, A.; Miura, H.; Sakai, T. Dynamic recrystallization under warm deformation of a 304 type austenitic stainless steel. *Mater. Sci. Eng. A* **1998**, *255*, 139–147. [CrossRef]

23. Mackenzie, J.K. Second Paper on the Statistics Associated with the Random Disorientation of Cubes. *Biometrika* **1958**, *45*, 229–240. [CrossRef]

24. Derby, B. The Dependence of Grain Size on Stress during Dynamic Recrystallization. *Acta Metall. Mater.* **1991**, *39*, 955–962. [CrossRef]

25. Belyakov, A.; Gao, W.; Miura, H.; Sakai, T. Strain induced grain evolution in polycrystalline copper during warm deformation. *Metall. Mater. Trans. A* **1998**, *29*, 2957–2965. [CrossRef]

26. Beladi, H.; Cizek, P.; Hodgson, P.D. On the Characteristics of Substructure Development through Dynamic Recrystallization. *Metall. Mater. Trans. A* **2009**, *40*, 1175–1189. [CrossRef]

27. Cizek, P. The microstructure evolution and softening processes during high-temperature deformation of a 21Cr-10Ni-3Mo duplex stainless steel. *Acta Mater.* **2016**, *106*, 129–143. [CrossRef]

28. Abson, D.J.; Jonas, J.J. Substructure strengthening in zirconium and zirconium-tin alloys. *J. Nucl. Mater.* **1972**, *42*, 73–85. [CrossRef]

29. Frost, H.J.; Ashby, M.F. *Deformation Echanism Maps*; Pergamon press: Oxford, UK, 1982.

30. Tikhonova, M.; Kaibyshev, R.; Fang, X.; Wang, W.; Belyakov, A. Grain boundary assembles developed in an austenitic stainless steel during large strain warm working. *Mater. Charact.* **2012**, *70*, 14–20. [CrossRef]

31. Tikhonova, M.; Enikeev, N.; Valiev, R.Z.; Belyakov, A.; Kaibyshev, R. Submicrocrystalline Austenitic Stainless Steel Processed by Cold or Warm High Pressure Torsion. *Mater. Sci. Forum* **2016**, *838–839*, 398–403. [CrossRef]

32. Belyakov, A.; Zherebtsov, S.; Salishchev, G. Three-stage relationship between flow stress and dynamic grain size in titanium in a wide temperature interval. *Mater. Sci. Eng. A* **2015**, *628*, 104–109. [CrossRef]

33. Humphreys, F.J.; Prangnell, P.B.; Bowen, J.R.; Gholinia, A.; Harris, C. Developing Stable Fine-Grain Microstructures by Large Strain Deformation. *Philos. Trans. R. Soc. Lond. A* **1999**, *357*, 1663–1681. [CrossRef]

34. Valiev, R.Z.; Langdon, T.G. Principles of equal-channel angular pressing as a processing tool for grain refinement. *Prog. Mater. Sci.* **2006**, *51*, 881–981. [CrossRef]

Article

The Role of the Bainitic Packet in Control of Impact Toughness in a Simulated CGHAZ of X90 Pipeline Steel

Bin Guo [1,3], Lei Fan [1,2], Qian Wang [1,2], Zhibin Fu [1,2], Qingfeng Wang [1,2,*] and Fucheng Zhang [1,2]

1   Laboratory of Metastable Materials Science and Technology, Yanshan University, Qinhuangdao 066004, China; guobinwg@126.com (B.G.); lwc84268092wjx@163.com (L.F.); wq986086441@139.com (Q.W.); fzbknc@163.com (Z.F.); zfc@ysu.edu.cn (F.Z.)
2   National Engineering Research Center for Equipment and Technology of Cold Strip Rolling, Yanshan University, Qinhuangdao 066004, China
3   Research and Development Center of WISCO, Wuhan 430080, China
*   Correspondence: wqf67@ysu.edu.cn; Tel.: +86-335-2039-067

Academic Editor: Soran Birosca
Received: 3 September 2016; Accepted: 20 October 2016; Published: 27 October 2016

Abstract: X90 pipeline steel was processed with the simulated coarse grain heat affect zone (CGHAZ) thermal cycle with heat input varying from 30 kJ/cm to 60 kJ/cm, the microstructures were investigated by means of optical microscope (OM), scanning electron microscope (SEM), electron backscattering diffraction (EBSD), and transmission electron microscope (TEM), and the impact properties were evaluated from the welding thermal cycle treated samples. The results indicate that the microstructure is primarily composed of lath bainite. When decreasing the heat input, both bainite packet and block are significantly refined, and the toughness has an increased tendency due to the grain refinement. The fracture surfaces all present cleavage fracture for the samples with different heat inputs. Moreover, the average cleavage facet size for the CGHAZ is nearly equal to the average bainite packet size, and the bainitic packet boundary can strongly impede the crack propagation, indicating that the bainitic packet is the most effective unit in control of impact toughness in the simulated CGHAZ of X90 pipeline steel.

Keywords: coarse grain heat affected zone (CGHAZ); impact toughness; bainitic packet; EBSD; cleavage facet

1. Introduction

High-grade pipeline steels have been developed for many years to meet the requirements of the high-pressure transportation for crude oil and nature gas [1]. With the further application of pipeline steels for long distance transportation and in harsh environments, the combination of excellent toughness and weldability becomes more and more important, in addition to high strength [2]. However, the balance of high strength and good toughness in steels can be disturbed by the thermal cycles experienced during welding, resulting in the poor toughness of local areas [3]. Particularly, the toughness in the coarse grain heat affected zone (CGHAZ) of high grade pipeline steels deteriorates severely after the welding thermal cycle [4]. In order to evaluate and improve the toughness of the welding joint [5,6], it is necessary to show the evolution of microstructure and toughness in the CGHAZ with varying welding heat inputs, as well as the effective microstructure unit in control of toughness.

Many studies have been done to investigate the relationship between the microstructure evolution and the resultant mechanical properties in CGHAZ [7–9]. The microstructure of high grade pipeline steel after the welding process generally includes lath bainite, granular bainite, and the secondary phase martensite/austenite (M/A) constituent, which mainly disperses at the bainite ferrite boundaries [10,11]. For lath bainite or martensite, the austenite grain is divided into packets (bundled by group of blocks with the same habit plane), and each packet is further subdivided into blocks (bundled by laths with the same orientation) [12–14]. Early works by Lee et al. [15] and Kim et al. [16] indicated that prior austenite is the effective microstructure unit in control of the toughness of bainite steels, and crack can propagate across the packet boundaries without changing the direction. On the contrary, studies by Wang et al. [14] and Rancel et al. [17] showed that packet boundaries can strongly hinder fracture propagation and can act as an effective microstructure unit in control of the toughness for cleavage fracture. Zhang et al. [18] further confirmed that block is the minimum structure unit in control of the toughness of steels. In recent works by Yang et al. [19], they attributed the excellent toughness in the CGHAZ to the optimized density of the high angle boundary (crystallographic packet boundary) with a misorientation angle greater than 15°, as well as the refinement of the brittle M/A constituent.

The aim of this work was to investigate the relationship between the microstructure and the impact toughness in a simulated CGHAZ of X90 pipeline steel with varied heat inputs. The microstructure of the specimens with different heat inputs was characterized by optical microscopy (OM), scan electron microscopy (SEM), electron back scattering diffraction (EBSD), and transmission electron microscopy (TEM) to better understand the microstructure evolution in the welding thermal cycle and to clarify the microstructure unit in control of the impact toughness in a simulated CGHAZ of X90 pipeline steel.

2. Materials and Methods

The steel used in this study was X90 commercial pipeline steel with a thickness of 14.7 mm (chemical composition is listed in Table 1). It should be noted that 0.09 wt. % Nb was added to the steel because it may additionally provide a strong pinning effect on the moving grain boundary during the welding thermal cycle with the peak temperature of 1350 °C [20]. The microstructure of the pipe steel is shown in Figure 1, which consists of granular bainite and M/A islands.

Table 1. Chemical compositions of the test steel (wt. %).

C	Si	Mn	P	S	Cr	Cu	Mo	Ni	Nb	V	Ti	Al
0.06	0.30	1.90	0.005	0.002	0.36	0.26	0.19	0.14	0.09	0.02	0.02	0.03

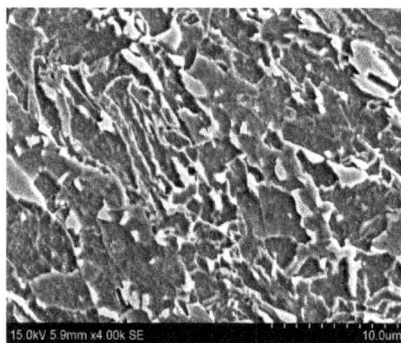

Figure 1. SEM (scanning electron microscope) image of X90 steel with multi-phase: granular bainite (black matrix) + M/A islands (white secondary phase). M/A—martensite/austenite.

The cubic specimens with a size of $11 \times 11 \times 80$ mm^3 for the simulated welding thermal cycle experiment were cut from the steel pipe along the longitudinal direction. The single-pass welding thermal cycles (Figure 2) were conducted in a Gleeble-3500 system to simulate the CGHAZ at heat inputs of 30, 40, 50, and 60 kJ/cm with a mean heating rate of 100 °C/s and a peak temperature of 1350 °C for 1 s. Four specimens were simulated for each heat input—one for microstructure observation and three for Charpy impact tests.

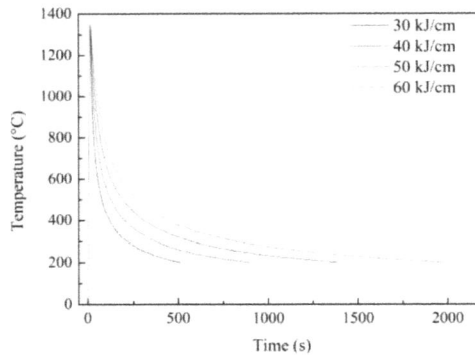

Figure 2. Graph of the simulated thermal cycle on the Gleeble for the steel with different heat inputs.

After the simulated thermal cycle, specimens for metallographic observation were cut on the plane perpendicular to the axis of the simulated specimens and near the monitoring thermocouple. After being polished and etched by a 4% nital solution, microstructures were observed via optical microscopy OM (Axiover-200MAT, Göttingen, Germany) and scanning electron microscopy SEM (Hitachi S-4800, Hitachi, Japan). At least 10 micrographs were taken for each specimen to determine the average prior austenite grain (PAG) size and bainite packet size. Quantitative measurements of the volume percent of M/A phase were made on SEM micrographs with a point count method. For each condition, at least 10 fields at 1000× magnification were measured with a grid of 400 points, giving a typical standard deviation of ±0.5 vol. %. Thin foils for transmission electron microscopy were prepared using the twin-jet method and observed with a high-resolution transmission electron microscope TEM (JEM-2010, Tokyo, Japan). Electron back scattering diffraction EBSD (TSL-EBSD, Tokyo, Japan) examination was performed on a SEM (Hitachi S-3400N, Hitachi, Japan) equipped with a TSL-EBSD analysis system with a step size of 0.2 μm and applied to estimate the average bainite block width and grain boundary misorientation distribution. The highlight line method (point to point approach) was used to study the misorientation distribution between the neighboring prior austenite boundaries, and the bainite packet and block boundaries.

The standard Charpy V-Notch (CVN) specimens (size: 10 × 10 × 55 mm³) were prepared from the thermal cycle treated samples and tested at −20 °C on a Tinius Olsen impact tester according to ASTM E23 (Norwood, MA, USA). After the Charpy impact tests, fracture surfaces of the specimens were observed by SEM, and the cleavage facet size was calculated as the equivalent circle diameter related to the individual cleavage facet area. The average size was statistically measured by averaging at least 1000 facets for each heat input. The specimens cut on the plane perpendicular to the fracture surface were observed by SEM to characterize the secondary crack propagation after being etched by a 4% nital solution.

3. Results

3.1. Microstructures of the Simulated CGHAZ

Optical micrographs (OM) of the specimens varying with welding heat input are shown in Figure 3. The transformed microstructures of the specimens with different welding heat inputs are predominately lath bainite and a relatively small amount of granular bainite. With the decreasing welding heat input, the lath bainite increases at the expense of granular bainite. In addition, the prior austenite grain is gradually refined (Figure 3), and the average size decreases from 68.6 to 30.4 μm (Table 2).

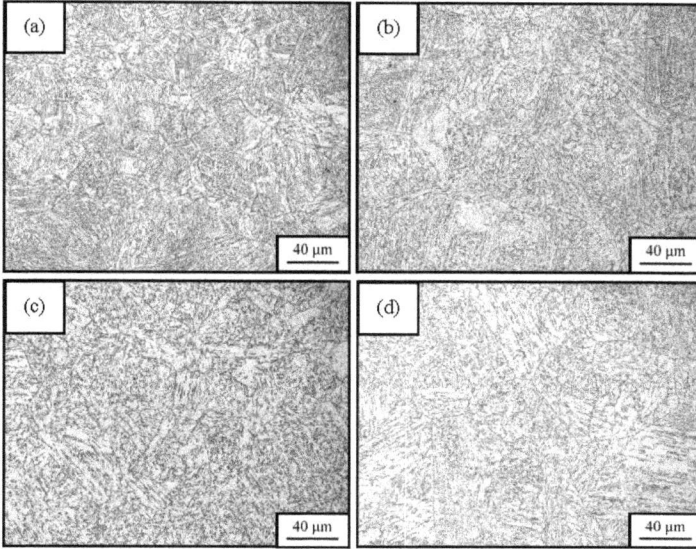

Figure 3. The optical micrographs of the simulated thermal cycle treated specimens with different heat inputs: (**a**) 30 kJ/cm; (**b**) 40 kJ/cm; (**c**) 50 kJ/cm; (**d**) 60 kJ/cm.

Table 2. Results of the microstructure quantification.

Heat Input (kJ/cm)	Block Width (μm)	Packet Size (μm)	PAG (μm)	Cleavage Facet Size (μm)	$f_{M/A}$ (%)
60	16.2	40.8	68.6	42.3	6.1
50	13.1	29.7	50.4	30.3	5.9
40	12.5	26.1	40.6	25.1	5.3
30	8.6	18.9	30.4	20.5	3.2

SEM micrographs of the specimens varying with welding heat input are shown in Figure 4. It is revealed that the prior austenite grain was divided into several bainite packets, and a minority of M/A constituent disperses at the bainite ferrite boundaries. With a decreasing welding heat input, the bainite packet are refined with an average size of bainite packet decreasing from 40.8 to 18.9 μm, and the amount of the M/A constituent, $f_{M/A}$, decreases from 6.1% to 3.2%, as shown in Figure 4 and Table 2.

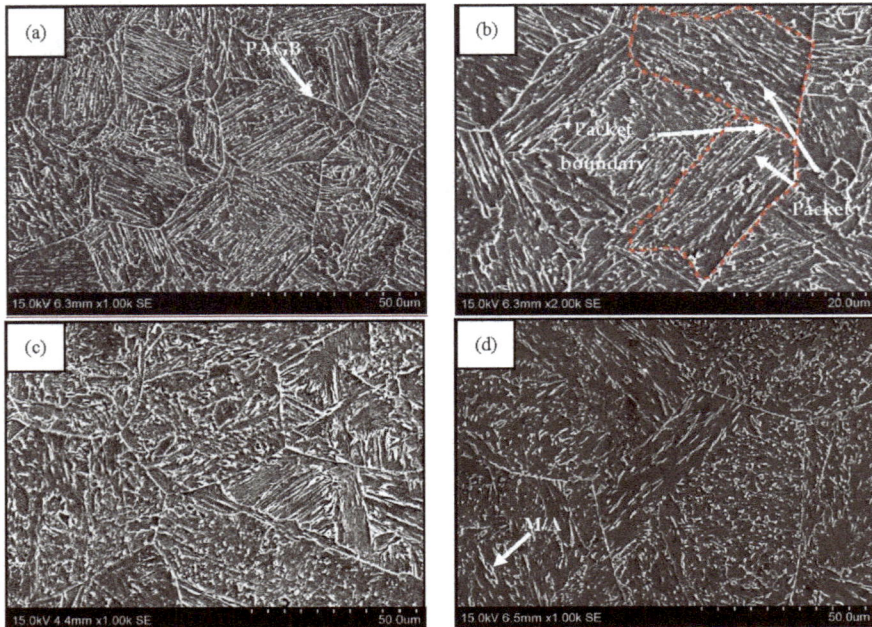

Figure 4. SEM images of the specimens with different heat inputs: (**a,b**) 30 kJ/cm; (**c**) 40 kJ/cm; (**d**) 60 kJ/cm. PAGB—prior austenite grain boundary and M/A—martensite/austenite.

The block boundary cannot be revealed clearly by means of conventional OM and SEM until the occurrence of the EBSD technique. Recently, the block structure was successfully indicated with the assistance of EBSD for SEM and a definition of grain misorientation higher than 15° as the block boundary [14]. The EBSD misorientation maps of the specimens varying with welding heat input are shown in Figure 5. The bainitic block is indicated by arrows in Figure 5a,c, and the average block widths at different welding heat inputs are shown in Table 2. It is indicated that the average block width decreases from 16.2 to 8.6 μm as the welding heat input decreases from 60 to 30 kJ/cm. Moreover, the prior austenite grain and bainitic packet boundaries can also be recognized in the EBSD misorientation maps. The misorientation distributions between the neighboring prior austenite grain boundary, the bainite packet boundary, and the block boundary are measured by the highlight line method (Figure 5b), shown in Figure 5d–f, respectively. It is indicated that the prior austenite grain boundary (Line 1) and bainitic packet boundary (Line 2) have high misorientation angles near 55° and 48°, respectively, while the misorientation angle of the bainite block boundary (Line 3) is only near 23°.

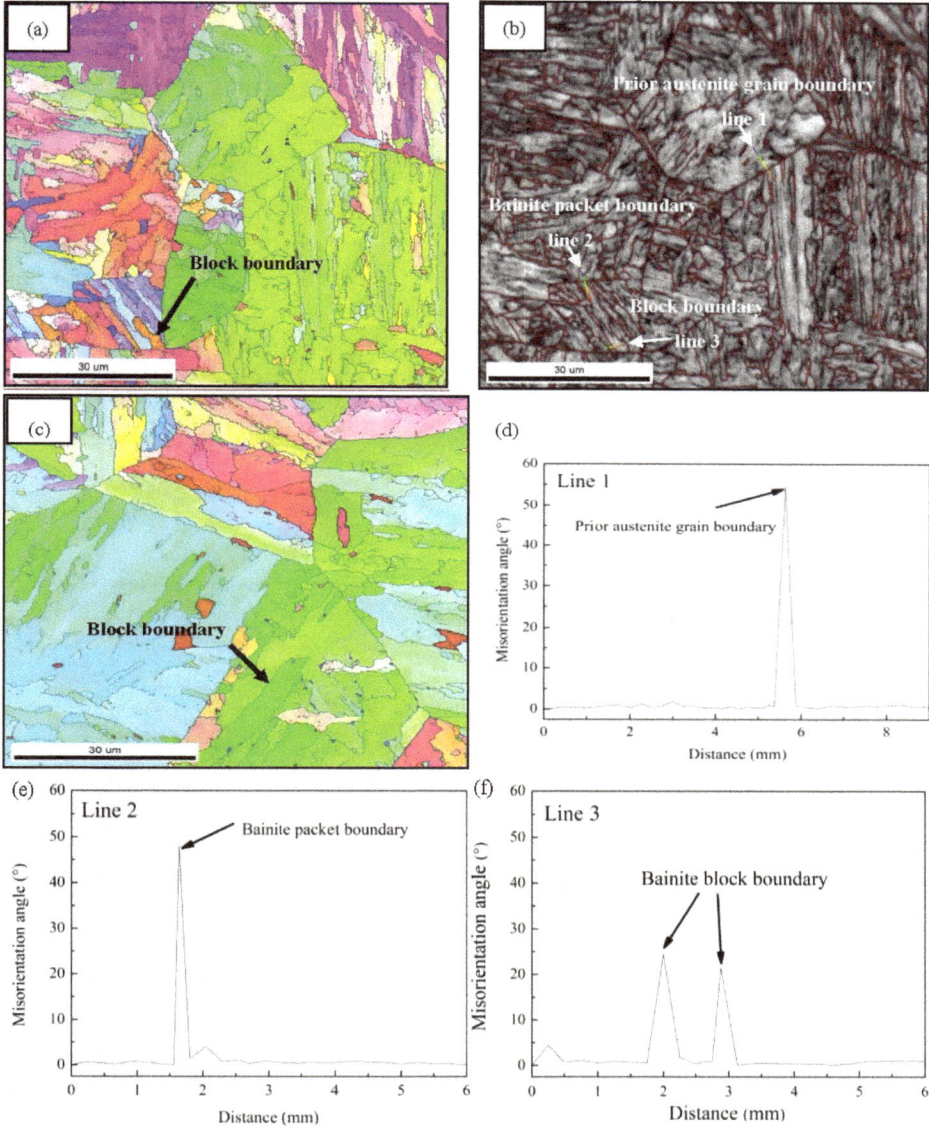

Figure 5. EBSD (electron backscattering diffraction) misorientation maps of X90 pipeline steel processed with different heat inputs: (**a**,**b**) 30 kJ/cm; (**c**) 60 kJ/cm. The black (**a**,**c**) and the red (**b**) grain boundary all represent high angle boundary with misorientation angles greater than 15°. Line 1 (**d**), line 2 (**e**), and line 3 (**f**) reveal the misorientation distributions between the neighboring prior austenite grain boundary, the bainite packet boundary, and the block boundary, respectively.

Representative bright field TEM micrographs of the specimens with heat inputs of 30 kJ/cm and 60 kJ/cm are presented in Figures 6 and 7, respectively. In general, the microstructure is predominantly composed of bainite ferrite laths with high density of dislocations based on TEM observations. It is revealed that bainite ferrite laths in the sample with lower heat input (Figure 6a) are much finer than in the sample with higher heat input (Figure 7a). Moreover, a higher density of dislocations (Figure 6b) are observed in the sample with lower heat input.

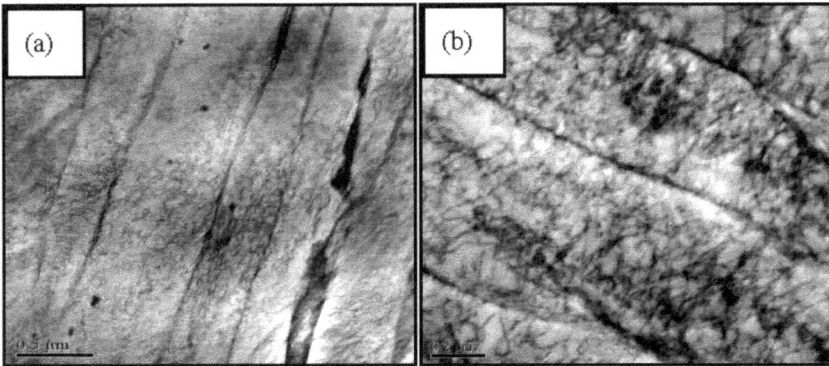

Figure 6. The TEM (transmission electron microscope) micrographs of the typical microstructure morphology in the specimen with heat input of 30 kJ/cm: (**a**) fine bainite ferrite lath; (**b**) dislocation distribution.

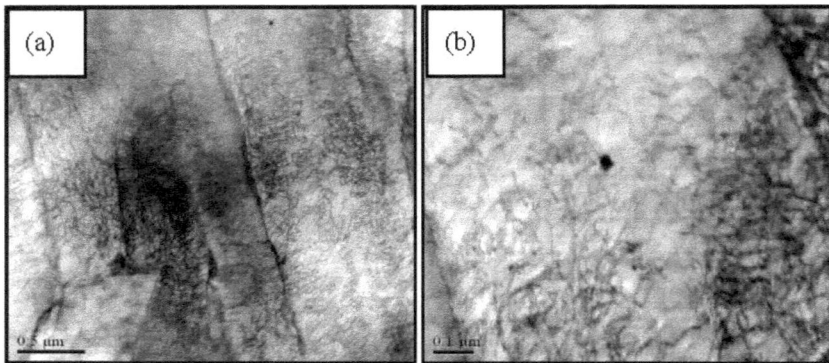

Figure 7. TEM micrographs of the typical microstructure morphology in the specimen with heat input of 60 kJ/cm: (**a**) broadening bainite ferrite lath; (**b**) dislocation distribution.

3.2. Charpy Impact Properties of the Simulated CGHAZ

The low-temperature ($-20\,^\circ$C) Charpy impact energies of specimens varying with welding heat input are shown in Figure 8. It is indicated that, with the heat input decreasing from 60 to 40 kJ/cm, the impact energy remains at a low level due to a relatively high amount of M/A constituents acting (Table 2) as the brittle phase [21,22]. However, it still slightly increases as a result of microstructure refinement (Figures 3–5). When further decreasing the heat input to 30 kJ/cm, the impact energy is significantly improved simultaneously by even finer microstructures (Figures 3–5) and a significant decrease in the amount of M/A constituents (Table 2). The impact energy stands for the resistance energy to fracture, and a higher impact energy indicates that the materials are more resistant to fracture, leading to the harder inducement and propagation of cracks [23].

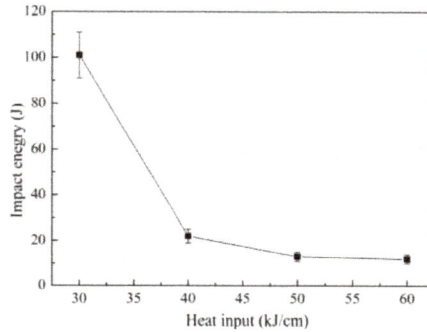

Figure 8. Charpy impact energies of the specimens with different heat inputs.

The morphologies of the fracture surfaces of the Charpy impact specimens (tested at −20 °C) varying with heat input are shown in Figure 9. All fracture surfaces present cleavage fracture, and the average cleavage facet sizes of the specimens with different heat inputs are listed in Table 2. When the heat input is 60 kJ/cm, the cleavage facet size is 42.3 μm, calculated as the equivalent circle diameter related to the cleavage facet area. With the heat input decreased to 30 kJ/cm, the cleavage facet size remarkably decreases to 20.5 μm. Moreover, despite the cleavage facet, secondary cracks can also be observed in the SEM fractographs. It is revealed that the secondary crack in the sample with lower heat input (Figure 9a) is much shorter than that in the sample with higher heat input (Figure 9b), which indicates that the microstructure in the specimen with lower heat input is more resistant to crack propagation than the one with higher heat input.

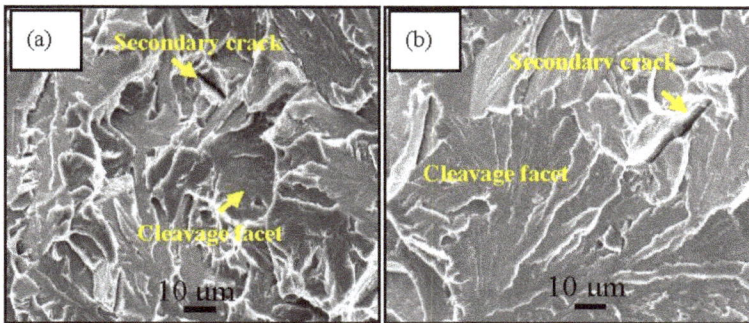

Figure 9. SEM images of fracture surfaces of the specimens with different heat inputs: (**a**) 30 kJ/cm; (**b**) 60 kJ/cm.

In order to investigate crack propagation, the secondary cracks underneath the fracture surface of the impact samples were observed, as shown in Figure 10. Figure 10a reveals that the secondary crack firstly deviates at the prior austenite grain boundary (identified by the white arrow in Figure 10a), then propagates through the substructures (blocks) in the packet, and finally deflects at the boundary between the two packets (identified by the yellow arrow in Figure 10a). Figure 10b shows that the secondary cracks propagate through the block boundaries (identified by the white arrow in Figure 10b) directly without deviation, and then are arrested at the boundary between the two packets (identified by the yellow arrow in Figure 10b). Hence, it is indicated that the block boundary (a misorientation angle of 23°) could not impede the propagation of the cracks, while the packet boundary (a misorientation angle of 48°) and the prior austenite grain boundary (a misorientation angle of 55°) could effectively deflect the crack propagation direction or arrest the cracks in the test steel.

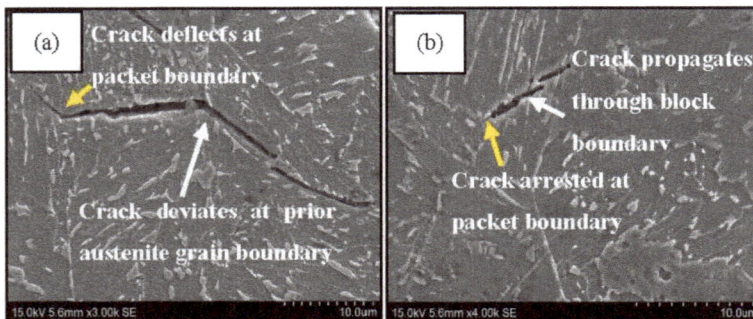

Figure 10. The secondary cracks underneath the fracture surface of impact samples: (**a**) cracks deviate at the prior austenite grain boundary and the bainite packet boundary; (**b**) cracks propagate through the bainite block boundary and are arrested at the packet boundary.

4. Discussion

4.1. Microstructure Evolution during the Simulated Welding Thermal Cycle

The evolution of the final microstructure can be explained as follows. Coarse austenite grain will be formed when the specimen is reheated to 1350 °C. During the continuous cooling process, lath bainite ferrite nucleates, austenite (γ)/ferrite (α) interface moves forwardly, and supersaturated carbon in ferrite diffuses to the γ/α interface [24]. The untransformed austenite thus becomes carbon-enriched due to the partitioning of carbon. With the lowering of the transforming temperature, the remaining austenite transforms into lath bainite ferrite continuously, and the retained austenite is further carbon-enriched and fully stabilized until the transformation becomes thermodynamically impossible [25]. In the following cooling course, part of the carbon-enriched austenite could transform to martensite, and the retained austenite would coexist with the martensite.

Lath bainite is known as the typical transformation product in the CGHAZ [7]. With the decreasing welding heat input, i.e., the increasing cooling rate, prior austenite grains are significantly refined (Table 2) together with the increasing austenite grain boundaries density due to the shorter staying time in the high temperature zone (austenite zone). Moreover, the transformation start temperature decreases accordingly [26]. All these will result in a higher thermodynamic driving force for lath bainite formation [27], an increase in dislocation density in lath bainite [28], and a refinement of lath bainite ferrite, which is related to an increase in the nucleation rate and a decrease in grain growth [29].

4.2. Effect of Grain Boundary on Impact Toughness in the Simulated CGHAZ

Grain boundary is well-known as the microstructure unit controlling the toughness of steels [14,30,31]. It owns high potential energy to impede crack propagation, and the increase of grain boundary density, namely the refinement of grain, can significantly improve the toughness of steels. Particularly, there are three kinds of grain boundaries in the CGHAZ of X90 pipeline steel: the prior austenite grain boundary, the bainite packet boundary, and the bainite block boundary. Moreover, in this study, the fracture mode for the specimens at all heat inputs is cleavage fracture; thus, the average cleavage facet size (Table 2) was measured and analyzed to better evaluate the effectiveness of different kinds of microstructure unit in control of impact toughness in the CGHAZ of X90 pipeline steel.

Figure 11a presents the linear relationship between the average cleavage facet size and average prior austenite grain size, and the fitting function is as follows:

$$d_c = 2.2 + 0.57d_a \tag{1}$$

where d_c is the average cleavage facet size, d_a is the average prior austenite grain size, and the correlation coefficient of this linear fit is 0.99.

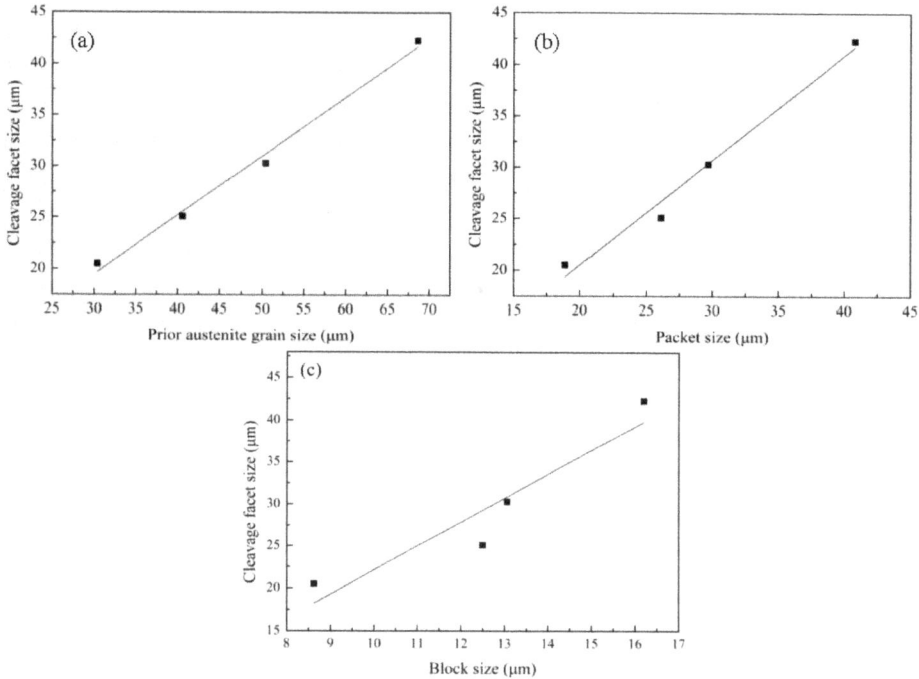

Figure 11. Cleavage facet size as a function of (**a**) prior austenite grain size; (**b**) bainite packet size; and (**c**) block width.

The average cleavage facet size of the experimental steel processed with different heat inputs is in a good linear relation to the average prior austenite grain size (Figure 11a). Moreover, it is observed that the secondary crack deviates at the prior austenite grain boundary (identified by white arrow in Figure 10a), which indicates the prior austenite grain boundary could effectively impede the crack propagation. However, the average prior austenite grain size is nearly twice as much as the average cleavage facet size, suggesting that the prior austenite grain may not be the most effective microstructure unit in control of impact toughness in the CGHAZ of X90 pipeline steel.

Figure 11b presents the linear relationship between the average cleavage facet size and average bainite packet size, and the fitting function is as follows:

$$d_c = 0.1 + 1.02d_p \tag{2}$$

where d_c is the average cleavage facet size, d_p is the average bainite packet size, and the correlation coefficient of this linear fit is 0.96.

The average cleavage facet size of the experimental steel processed with different heat inputs is in a good linear relation to the average bainite packet size (Figure 11b). Moreover, it is observed that the secondary crack deviates or can be arrested at the bainite packet boundary (identified by yellow arrow in Figure 10a,b, respectively), which indicates the bainite packet boundary could effectively impede the crack propagation or arrest the cracks. It should be further noted that the average cleavage facet size is nearly equal to the average bainite packet size, indicating that the bainite packet ought to be the most effective microstructure unit in control of impact toughness in the CGHAZ of X90 pipeline steel.

Figure 11c presents the linear relationship between the average cleavage facet size and average bainite block width, and the fitting function is as follows:

$$d_c = -6.4 + 2.85d_b \tag{3}$$

where d_c is the average cleavage facet size, d_b is the average bainite block width, and the correlation coefficient of this linear fit is 0.79.

Despite the relatively good linear fitting result, it is observed that the secondary crack propagates through the bainite block boundary (identified by white arrow in Figure 10b), which indicates the bainite block boundary could not effectively impede the crack propagation. Moreover, the average cleavage facet size is nearly triple as much as the average bainite block width, suggesting that the bainite block is not an effective microstructure unit in control of impact toughness in the CGHAZ of X90 pipeline steel.

Hence, it is indicated that the bainite packet is the most effective microstructure unit in control of impact toughness in the CGHAZ of X90 pipeline steel. Moreover, based on the mechanics model of cleavage fracture, the fracture stress σ_f can be expressed as [14,32]:

$$\sigma_f = \left[\frac{4E\gamma_P}{(1 - v^2)d} \right]^{1/2} \tag{4}$$

where E, γ_P, and v are all constants, which represents Young's modulus, the plastic deformation energy, and the Poisson's ratio, respectively, and d is the effective grain size, which corresponds to bainite packet size (d_p) in this study.

The critical fracture stress represents the maximum external load the structures can undertake to keep materials from being fractured. The increasing fracture stress means that the materials becomes more resistant to fracture [14,31]. In this study, d_p reduces significantly (Table 2) when decreasing the welding heat input, leading to the notable increase in the critical fracture stress. Therefore, with the decreasing welding heat input, the increasing impact toughness (Figure 8) is achieved as a result of the grain refinement (mainly packet refinement in this study) and the consequent increase of the critical fracture stress.

Several studies have been carried out to deal with the relationship between the impact toughness and the effective microstructure unit for lath bainite or lath martensite steels [14,17,19,33]. The studies by Wang et al. [14] and Rancel et al. [16] indicated that the packet boundaries can strongly hinder fracture propagation and can act as an effective microstructure unit for cleavage fracture. Moreover, in recent work by Yang et al. [18,34], they revealed that high angle boundary (including packet boundary) remarkably contributes to a good toughness of the CGHAZ in high-grade pipeline steel. The present results are in good agreement with the previous works above on the relationship between the impact toughness and the effective microstructure unit.

5. Conclusions

The microstructures and the impact toughness of X90 pipeline steel after a simulated thermal cycle with different heat inputs was investigated in this work. The major conclusions are summarized as follows:

1   The microstructure of the simulated CGHAZ is predominantly composed of lath bainite with different welding heat inputs. With the decreasing heat input, prior austenite grain, bainite packet, and block are all obviously refined.

2   With the decreasing heat input, the impact toughness significantly improves. The fracture surfaces all present cleavage fracture for the samples with different heat inputs, and the average cleavage facet size decreases when the heat input is decreased.

3   The average bainite packet size for the CGHAZ is nearly equal to the average cleavage facet size, and the bainite packet boundary can strongly impede the crack propagation, indicating that the bainitic packet is the most effective unit in control of impact toughness in the simulated CGHAZ of X90 pipeline steel.

Acknowledgments: This work was financially supported by the Research and Development Center of Wuhan Iron and Steel (Group) Company of China under the Contract No. 2014Z05, and by the National Natural Science Foundation of China (Grant No. 51471142 and No. 5167010961).

Author Contributions: Qingfeng Wang and Bin Guo conceived and designed the experiments; Lei Fan performed the experiments; Fucheng Zhang and Qian Wang analyzed the data; Zhibin Fu contributed analysis tools; Qingfeng Wang and Bin Guo wrote the paper.

Conflicts of Interest: The authors declare no conflict of interest.

References

1. Shin, S.Y.; Hwang, B.; Lee, S.; Kim, N.J.; Ahn, S.S. Correlation of microstructure and charpy impact properties in API X70 and X80 line-pipe steels. *Mater. Sci. Eng. A* **2007**, *458*, 281–289. [CrossRef]

2. Zhao, W.; Wang, W.; Chen, S.; Qu, J. Effect of simulated welding thermal cycle on microstructure and mechanical properties of X90 pipeline steel. *Mater. Sci. Eng. A* **2011**, *528*, 7417–7422. [CrossRef]

3. Davis, C.L.; King, J.E. Cleavage initiation in the intercritically reheated coarse-grained heat-affected zone: Part I. Fractographic evidence. *Metall. Mater. Trans. A* **1994**, *25*, 563–573. [CrossRef]

4. Hu, J.; Du, L.X.; Wang, J.J.; Gao, C.R. Effect of welding heat input on microstructures and toughness in simulated CGHAZ of V-N high strength steel. *Mater. Sci. Eng. A* **2013**, *577*, 161–168. [CrossRef]

5. Wei, R.; Wu, K.M.; Gao, Z. Effect of Heat Input on Impact Toughness of Coarse-Grained Heat-Affected Zone of a Nb-Ti Microalloyed Pipeline Steel. *Adv. Mater. Res.* **2012**, *538*, 2026–2031.

6. Li, R.; Zuo, X.; Hu, Y.; Wang, Z.; Hu, D. Microstructure and properties of pipeline steel with a ferrite/martensite dual-phase microstructure. *Mater. Charact.* **2011**, *62*, 801–806. [CrossRef]

7. Lambert-Perlade, A.; Gourgues, A.F.; Pineau, A. Austenite to bainite phase transformation in the heat-affected zone of a high strength low alloy steel. *Acta Mater.* **2004**, *52*, 2337–2348. [CrossRef]

8. Guo, A.; Misra, R.D.K.; Liu, J.; Chen, L.; He, X.; Jansto, S.J. An analysis of the microstructure of the heat-affected zone of an ultra-low carbon and niobium-bearing acicular ferrite steel using EBSD and its relationship to mechanical properties. *Mater. Sci. Eng. A* **2010**, *527*, 6440–6448. [CrossRef]

9. You, Y.; Shang, C.; Chen, L.; Subramanian, S. Investigation on the crystallography of the transformation products of reverted austenite in intercritically reheated coarse grained heat affected zone. *Mater. Des.* **2013**, *43*, 485–491. [CrossRef]

10. Sung, H.K.; Shin, S.Y.; Cha, W.; Oh, K.; Lee, S.; Kim, N.J. Effects of acicular ferrite on charpy impact properties in heat affected zones of oxide-containing API X80 linepipe steels. *Mater. Sci. Eng. A* **2011**, *528*, 3350–3357. [CrossRef]

11. Zhu, Z.X.; Marimuthu, M.; Kuzmikova, L.; Li, H.J.; Barbaro, F. Influence of Ti/N ratio on simulated CGHAZ microstructure and toughness in X70 steels. *Sci. Technol. Weld. Join.* **2013**, *18*, 45–51. [CrossRef]

12. Morito, S.; Tanaka, H.; Konishi, R.; Furuhara, T.; Maki, T. The morphology and crystallography of lath martensite in Fe-C alloys. *Acta Mater.* **2003**, *51*, 1789–1799. [CrossRef]

13. Mateo, C.G.; Rivas, L.M.; Caballero, F.G.; Milbourn, D.; Sourmail, T. Vanadium effect on a medium carbon forging steel. *Metals* **2016**, *6*, 130. [CrossRef]

14. Hu, H.; Xu, G.; Zhou, M.; Yuan, Q. Effect of Mo content on microstructure and property of low-carbon bainitic steels. *Metals* **2016**, *6*, 173. [CrossRef]

15. Lee, S.; Kim, B.C.; Lee, D.Y. Fracture mechanism in coarse grained HAZ of HSLA steel welds. *Scr. Metall.* **1989**, *23*, 995–1000. [CrossRef]
16. Kim, Y.M.; Kim, S.K.; Lim, Y.J.; Kim, N.J. Effect of microstructure on the yield ratio and low temperature toughness of linepipe steels. *ISIJ Int.* **2002**, *42*, 1571–1577. [CrossRef]
17. Rancel, L.; Gómez, M.; Medina, S.F.; Gutierrez, I. Measurement of bainite packet size and its influence on cleavage fracture in a medium carbon bainitic steel. *Mater. Sci. Eng. A* **2011**, *530*, 21–27. [CrossRef]
18. Zhang, C.; Wang, Q.; Ren, J.; Li, R.; Wang, M.; Zhang, F. Effect of martensitic morphology on mechanical properties of an as-quenched and tempered 25CrMo48V steel. *Mater. Sci. Eng. A* **2012**, *534*, 339–346. [CrossRef]
19. You, Y.; Shang, C.; Nie, W.; Subramanian, S. Investigation on the microstructure and toughness of coarse grained heat affected zone in X-100 multi-phase pipeline steel with high Nb content. *Mater. Sci. Eng. A* **2012**, *558*, 692–701. [CrossRef]
20. Barbaro, F.J.; Zhu, Z.; Kuzmikova, L.; Li, H.; Gray, J.M. Towards improved steel alloy designs for control of weld heat affected zone properties. In Proceedings of the BaoSteel Academic Conference, Shanghai, China, 4–6 June 2013.
21. Li, C.; Wang, Y.; Han, T.; Han, B.; Li, L. Microstructure and toughness of coarse grain heat-affected zone of domestic X70 pipeline steel during in-service welding. *J. Mater. Sci.* **2011**, *46*, 727–733. [CrossRef]
22. Moeinifar, S.; Kokabi, A.H.; Hosseini, H.R.M. Influence of peak temperature during simulation and real thermal cycles on microstructure and fracture properties of the reheated zones. *Mater. Des.* **2010**, *31*, 2948–2955. [CrossRef]
23. Shanmugam, S.; Ramisetti, N.K.; Misra, R.D.K.; Hartmann, J.; Jansto, S.G. Microstructure and high strength-toughness combination of a new 700 MPa Nb-microalloyed pipeline steel. *Mater. Sci. Eng. A* **2008**, *478*, 26–37. [CrossRef]
24. Zhao, M.-C.; Yang, K.; Xiao, F.-R.; Shan, Y.-Y. Continuous cooling transformation of undeformed and deformed low carbon pipeline steels. *Mater. Sci. Eng. A* **2003**, *355*, 126–136. [CrossRef]
25. Wang, S.-C.; Yang, J.-R. Effects of chemical composition, rolling and cooling conditions on the amount of martensite/austenite (M/A) constituent formation in low carbon bainitic steels. *Mater. Sci. Eng. A* **1992**, *154*, 43–49. [CrossRef]
26. Yakubtsov, I.A.; Poruks, P.; Boyd, J.D. Microstructure and mechanical properties of bainitic low carbon high strength plate steels. *Mater. Sci. Eng. A* **2008**, *480*, 109–116. [CrossRef]
27. Furuhara, T.; Kawata, H.; Morito, S.; Maki, T. Crystallography of upper bainite in Fe-Ni-C alloys. *Mater. Sci. Eng. A* **2006**, *431*, 228–236. [CrossRef]
28. Bhadeshia, H.K.D.H. *Bainite in Steels*; The Institute of Materials: London, UK, 1992; pp. 1–8.
29. Olasolo, M.; Uranga, P.; Rodriguez-Ibabe, J.M.; López, B. Effect of austenite microstructure and cooling rate on transformation characteristics in a low carbon Nb-V microalloyed steel. *Mater. Sci. Eng. A* **2011**, *528*, 2559–2569. [CrossRef]
30. Naylor, J.P.; Krahe, P.R. Cleavage planes in lath type bainite and martensite. *Metall. Trans. A* **1975**, *6*, 594–598. [CrossRef]
31. Lan, L.; Qiu, C.; Zhao, D.; Gao, X.; Du, L. Microstructural characteristics and toughness of the simulated coarse grained heat affected zone of high strength low carbon bainitic steel. *Mater. Sci. Eng. A* **2011**, *529*, 192–200. [CrossRef]
32. Daigne, J.; Guttmann, M.; Naylor, J.P. The influence of lath boundaries and carbide distribution on the yield strength of 0.4% C tempered martensitic steels. *Mater. Sci. Eng.* **1982**, *56*, 1–10. [CrossRef]
33. Yasuto, F.; Yu-ichi, K. Study on critical CTOD property in heat affected zone of C-Mn microalloyed steel. *Trans. Jpn. Weld. Soc.* **1992**, *32*, 65–72.
34. Hidenori, T.; Yutaro, S.; Atsushi, T.; Yu-ichi, K.; Koji, M.; Yusaku, T. Visualization and analysis of variant grouping in continuously cooled low-carbon steel welds. *Metall. Mater. Trans. A* **2014**, *45*, 3554–3559.

metals

Article

Microstructure and Mechanical Properties of J55ERW Steel Pipe Processed by On-Line Spray Water Cooling

Zejun Chen [1,2,*], Xin Chen [2] and Tianpeng Zhou [2]

[1]   State Key Laboratory of Mechanical Transmission, College of Materials Science and Engineering, Chongqing University, Chongqing 400044, China

[2]   College of Materials Science and Engineering, Chongqing University, Chongqing 400044, China; cxcxmail@163.com (X.C.); ztp1229@163.com (T.Z.)

*   Correspondence: zjchen@cqu.edu.cn; Tel.: +86-23-6511-1547

Academic Editor: Soran Birosca

Received: 17 August 2016; Accepted: 4 April 2017; Published: 23 April 2017

Abstract: An on-line spray water cooling (OSWC) process for manufacturing electric resistance welded (ERW) steel pipes is presented to enhance their mechanical properties and performances. This technique reduces the processing needed for the ERW pipe and overcomes the weakness of the conventional manufacturing technique. Industrial tests for J55 ERW steel pipe were carried out to validate the effectiveness of the OSWC process. The microstructure and mechanical properties of the J55 ERW steel pipe processed by the OSWC technology were investigated. The optimized OSWC technical parameters are presented based on the mechanical properties and impact the performance of steel pipes. The industrial tests show that the OSWC process can be used to efficiently control the microstructure, enhance mechanical properties, and improve production flexibility of steel pipes. The comprehensive mechanical properties of steel pipes processed by the OSWC are superior to those of other published J55 grade steels.

Keywords: electric resistance welding (ERW); steel pipe; on-line spray water cooling (OSWC); microstructure; mechanical properties

---

1. Introduction

In view of the ever-increasing pipeline and operating pressure, the development of high strength steels makes a significant contribution to pipeline project cost reduction. Increasing the strength of pipeline steel allows for the thickness of the pipeline walls to be significantly reduced, with a consequent reduction in weight and cost. The high strength, in combination with high toughness and formability, are important requirements for the pipeline steels [1]. Many efforts have been made to improve the strength and performance of pipeline steels. The most effective method to improve the welding performance of pipeline steels is microalloying [2–7]. For example, adding Ti [2,3], adding Nb [4–6], adding Mo and Cr [7], etc.

Generally, there are four methods to produce steel pipes: fusion welding, electric resistance welding (ERW), seamless hot rolling, and double submerged arc welding (DSAW). The manufacturing procedures of ERW steel pipes begin with a coiled plate of steel of appropriate thickness and specific width to form a pipe that conforms to particular specifications. Steel ribbon is pulled through a series of rollers that gradually form it into a cylindrical tube. As the edges of the cylindrical plate come together, an electric charge is applied at proper points to heat the edges so they can be welded together [8]. The conventional ERW manufacturing procedures of steel pipes and tubes are shown in Figure 1.

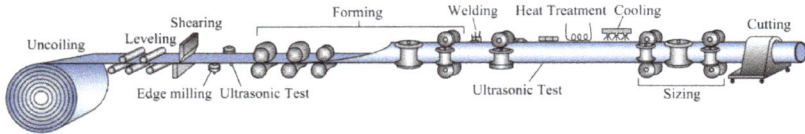

Figure 1. Conventional electric resistance welding (ERW) manufacturing procedures of the steel pipes and tubes.

ERW steel pipe production is high-speed and comparatively economical, because most of the process can be automated. ERW steel pipes have uniform wall thickness and outer dimensions, and they can be made with a wide range of specifications. Owing to these advantages, the production and application of ERW steel pipes have risen steadily in recent years. The performance requirements of steel pipes have continuously increased with the development of the oil and gas industry. There is strong demand to develop steel pipe with both high strength and excellent formability.

High-performance ERW steel pipes can substitute for seamless steel pipes in some cases, which can significantly reduce the engineering costs. However, it is difficult to obtain good performance by using the conventional ERW process. The reason is that ERW steel pipes are manufactured by cold-roll forming of steel bands, and the ductility of steel pipes is inevitably inferior to that of the steel band due to the work hardening of cold roll forming. Furthermore, the quench hardening caused by rapid cooling after welding has the same effect on the mechanical properties of the steel pipe in the weld joint [9]. Another weakness of the conventional ERW process is that the production flexibility of steel pipes with different specifications is limited by certain production lines. The main bottleneck to increasing the productivity of ERW steel pipe is the speed of the straight seam welding, which is much more notable when manufacturing small-diameter steel pipes. A manufacturing technique that involves reducing diameter of large ERW steel pipe is more efficient and reasonable, and could easily increase the productivity of small-diameter ERW steel pipes.

To improve the mechanical properties, the Kawasaki Steel Corporation developed a tube product called HISTORY (high-speed tube welding and optimum reducing technology) pipe [10]. The HISTORY pipe not only has high strength, excellent formability, and uniformity of the seam hardness due to the realization of ultrafine microstructure and tiny, dispersed, and spheroid cementite in the stretch reducer, but also features high productivity and flexibility of production by applying a stretch-reducing process to pipe-making [11,12].

The thermo-mechanical control process (TMCP) is very important to the microstructural control of high-performance products [13]. Recently, a new generation of thermo-mechanical controlled process (NG-TMCP) for steel production was developed [14], and it has been applied to all kinds of steel products (e.g., line pipe steels [15]). The NG-TMCP takes the super on-line accelerated cooling (Super-OLAC) and the precise control of cooling routes by a heat-treatment on-line process (HOP) as core technologies. The NG-TMCP consists of an advanced accelerated cooling device with the purpose of reaching the highest cooling rates, as well as induction heating equipment for HOP [16]. The steel products can obtain high strength by transformation strengthening, high toughness by refinement of the transformed microstructure, and reduce the alloying elements [17].

In the conventional manufacturing technology of ERW steel pipes, off-line heat treatment is an essential procedure to enhance the mechanical properties, especially the impact performance of the weld joint. This process reduces the productivity and increases the costs of ERW steel pipe. A high-efficiency manufacturing technology for ERW steel pipe is presented based on a rapid on-line spray water cooling (OSWC) process in conjunction with the pipe reducing process. The manufacturing technology based on OSWC process can enhance the mechanical properties, improve the productivity and the flexibility of specifications, and decrease the cost of production. The microstructure and mechanical properties of ERW steel pipes produced by the OSWC process were investigated, and industrial tests were carried out to validate the technique.

2. Materials and Industrial Tests

2.1. Materials and Specifications

J55-grade steel pipes were investigated in this paper. In API Specification 5CT, ISO 11960 [18], the composition of J55 grade steel pipes is only specified in regard to the maximum sulfur (S) and phosphorus (P) content. The content of carbon (C) can be changed within a large range. In this study, the dimensions of J55 ERW steel pipe are $\Phi139.7$ mm $\times$ 7.72 mm $\times$ 600 mm, and the chemical compositions are shown in Table 1.

Table 1. Chemical compositions of J55 ERW steel pipe.

C	Si	Mn	P	S	V + Nb + Ti
$\leq$0.21	$\leq$0.30	$\leq$1.40	$\leq$0.025	$\leq$0.015	$\leq$0.15%

The requirements for the mechanical properties of J55 grade steel in API Specification 5CT and ISO 11960 are shown in Table 2. The data are regarded as criteria to determine the effectiveness of the OSWC process.

Table 2. Mechanical properties of J55 steel in API Specification 5CT/ISO 11960 [18].

Properties	Yield Strength (MPa)	Tensile Strength (MPa)	Elongation (%)	Impact Energy (J)
J55 grade steel	379–552	>517	>24	>20 (0 °C)

2.2. Industrial Manufacturing Process

The new ERW manufacturing procedures are shown in Figure 2. The reducing process and on-line spray water cooling process were added, and the detailed route can be described as follows: steel strip → slitting strip → cold roll forming → high-frequency welding → induction heating → reducing and sizing → on-line spray water cooling process → finishing → ultrasonic testing → cutting. The main bottleneck to increasing productivity in ERW steel pipe manufacturing is the weld speed of small-diameter ERW steel pipe. To enhance the production efficiency, large-diameter steel pipe (e.g., $\Phi193.7$ mm) is welded using ERW. Then, the pipe is heated beyond the austenitizing temperature by electric induction heating, and then the diameters and wall thickness are reduced to various dimensions (e.g., $\Phi60.3$ mm to $\Phi177.8$ mm). The reducing process improves the flexibility of ERW steel pipe production. After applying this procedure, the OSWC process was immediately performed for hot deformed steel pipes. The aim of high-temperature thermo-mechanical treatment is the on-line control of the microstructure and mechanical properties of the pipe. The rapid OSWC process can efficiently refine the grain, improve the strength and formability, reduce the differences in mechanical properties between welded joints and the steel band substrate, improve the uniformity of the microstructure along the circumference, and reduce the adverse effects of the weld joint.

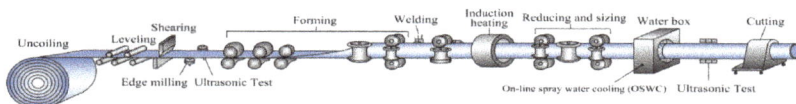

Figure 2. The new ERW manufacturing procedures of the steel pipes and tubes processed by on-line spray water cooling (OSWC).

In the OSWC system, the arrangement of nozzles is very important for the cooling effect and uniformity. There are two typical arrangements of nozzles: a centripetal arrangement along the circumference, and an arrangement tangent to a concentric circle with radius r. The radius of this circle is less than the radius (R) of the steel pipe ($r < R$). The centripetal arrangement of nozzles results in severe lateral spatter of the water. Large quantities of cooling water flow into the steel pipe, leading to a rapid temperature drop at the bottom of the inside wall of the steel pipe. The non-uniform temperature distribution results in non-uniformity of the microstructure and mechanical properties around the circumference of the pipe.

In contrast, the tangential arrangement of nozzles can greatly reduce the lateral spatter of cooling water and improve the temperature uniformity of the steel pipe around the circumference. The number of spray water boxes was determined by the specifications and performance requirements of ERW steel pipes. The arrangement of spray water boxes and the operating state of spraying water are shown in Figure 3.

(a)   (b)

Figure 3. Arrangement of water boxes and its operating state (a) arrangement of spray water boxes; (b) operating state of spraying water.

The aim of industrial tests is to validate the OSWC process for improving the mechanical properties of J55 ERW steel pipe. The detailed OSWC parameters of the industrial test are shown in Table 3.

Table 3. OSWC parameters of industrial test for ERW steel pipes.

No.	Original Steel Pipe	Specification of Reducing	Motion Velocity of Steel Pipe (m/s)	Cooling Parameters				
	Outer Diameter × Wall Thickness (mm × mm)			Heating Temperatur (°C)	Cooling Temperatur (°C)	Rebound Temperatur (°C)	Water Pressure (kPa)	Flux (m³/h)
P1#				air cooling				
P2#	193.7 × 7.3	139.7 × 7.72	1.08	980 ± 10	800 ± 15	670	65	124
P3#				980 ± 10	800 ± 15	660	100	169
P4#				980 ± 10	800 ± 15	655	150	217
P5#				980 ± 10	800 ± 15	645	200	258

The cooling parameters of OSWC process (water pressure, flux, and motion velocity of steel pipe etc.) have important effects on the cooling rate of steel pipe. The different cooling paths determine the microstructure evolution of steel pipe. The convective heat transfer coefficient between water and steel pipe is an important parameter for investigating and predicting temperature behavior during the OSWC process. An inverse heat conduction methodology can be used to obtain the mathematical model of convective heat transfer for the annular spray water cooling process. The heat transfer model can be used to determine the temperature history of steel pipe during the OSWC process. The detailed description for analyzing the cooling rate based on the cooling parameters of process can be found in [19,20].

The mechanical properties of ERW steel pipes processed by different spray water cooling schemes were measured using an autograph tensile testing machine with a maximum load of 600 kN. The dimensions of a typical tensile test specimen are shown in Figure 4.

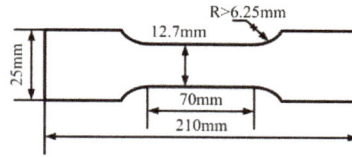

Figure 4. Dimensions of the tensile specimen.

3. Industrial Testing Results

In the new manufacturing process, the OSWC process is a critical procedure for improving the mechanical properties of ERW steel pipe. The microstructure of steel pipe will transform again into austenite due to the intermediate induction heating. To obtain excellent mechanical properties, the microstructure evolution of hot deformed steel pipe must be efficiently controlled according to its continuous cooling transformation (CCT) diagram. The CCT diagram of the J55 steel used in this industrial testis shown in Figure 5, and it is the foundation for analyzing the microstructure evolution and mechanical properties during the controlled cooling process of the steel pipe.

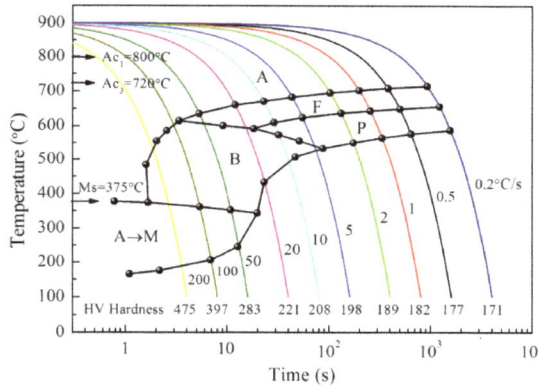

Figure 5. Continuous Cooling Transformation (CCT) diagram of J55 steel.

3.1. Mechanical Properties of Steel Pipes

The mechanical properties of steel pipe obtained by industrial production tests are shown in Table 4.

Table 4. Mechanical properties of steel pipe obtained in industrial tests. YS: yield strength; UTS: ultimate tensile strength.

No.	Longitudinal Direction of Steel Pipe			UTS of Weld Joint (MPa)
	YS (MPa)	UTS (MPa)	Elongation (%)	
P1#	410	640	32.7	615
P2#	480	685	32.9	670
P3#	500	700	32.9	685
P4#	500	705	33.0	685
P5#	510	700	33.1	685

Table 4 shows that the strength of steel pipe after undergoing OSWC is greater than that obtained with air cooling. The OSWC process can improve the strength of the ERW steel pipe, and the elongations are basically kept constant. The strength of the weld joint is lower than in other parts of the pipe.

The hardness data of pipes processed using different spray water cooling schemes are shown in Figure 6. Figure 6a shows that the hardness of the steel pipe heat-treated by scheme P4# is higher than that obtained with other schemes. There are similar trends of hardness for the base metal, weld joint, and heat-affected zones. Figure 6b shows that the hardness of the weld joints is lower than that of the base metal and the heat-affected zone for all spray water cooling processes. The hardness of the steel pipes that undergo the OSWC process is higher than that obtained with the air cooling process, regardless of the area of steel pipe. This was the case in the base metal, weld joint, and heat-affected zone. The differences in hardness between the base metal and weld joint were also increased.

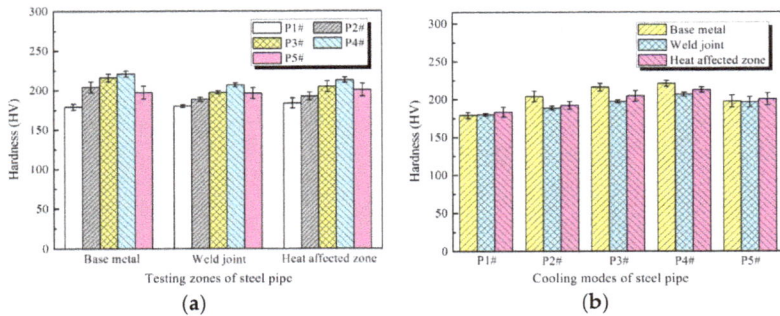

Figure 6. Hardness distribution of steel pipes (**a**) comparison of different processes; and (**b**) comparison of different zones.

Figure 7 shows the impact energy of steel pipe for the base metal and weld joint. It can be seen that the weld joint is the weak point of the ERW steel pipe, and the impact properties of the weld joint are obviously lower than those of the base metal. However, the OSWC process can improve the impact property of the steel pipe, including the base metal and weld joint. The industrial tests show that the impact property of the pipe treated by OSWC scheme P4# is superior to that obtained with other processes.

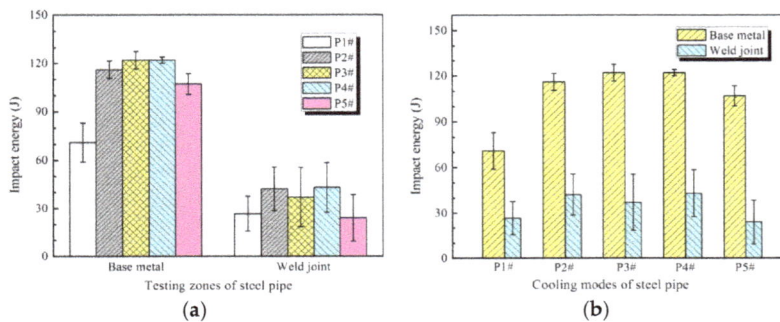

Figure 7. Impact energy of steel pipes: (**a**) comparison of different processes; and (**b**) comparison of different zones.

3.2. Microstructure of Steel Pipes

Figure 8 shows the optical microscope (OM) images of weld joints subjected to different spray water cooling processes.

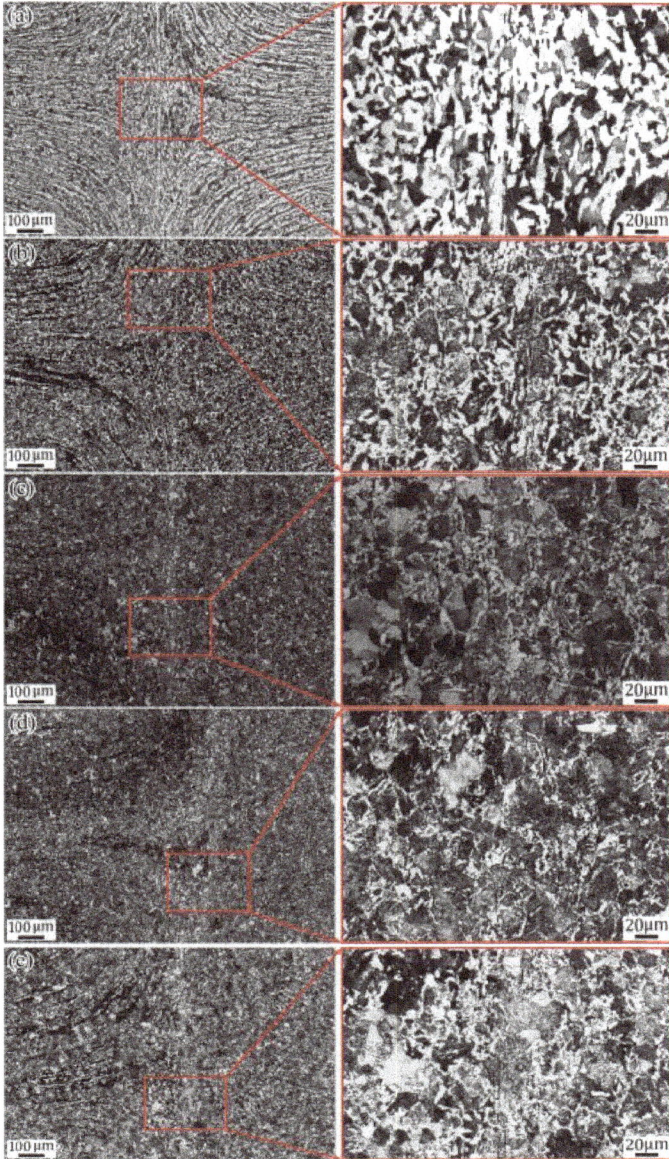

Figure 8. Optical microscope (OM) images of weld joint of steel pipes (**a**) P1#; (**b**) P2#; (**c**) P3#; (**d**) P4#; and (**e**) P5#.

In Figure 8, the welding traces of ERW steel pipes can be easily distinguished in optical microscope images of the weld joint. The microstructure of the steel pipe at the weld joint is mainly composed of ferrite and pearlite, and the grain sizes obtained by the OSWC process are finer than those obtained by the air cooling process. The welding traces of air-cooled steel pipe are more evident than those obtained by the spray water cooling. The difference in microstructure between the base metal and weld joint leads to a significant reduction in the impact performance of the steel pipe. Compared with the air cooling process, the microstructure streamlines of the welding joint are weakened after undergoing the spray cooling process. Therefore, the impact properties of steel pipes heat-treated by the OSWC process are better than that obtained by air cooling.

Figure 9 shows the microstructure and compositions of the outer range of the welding joint of steel pipe subjected to different spray water cooling processes. The percentages of each component for different spray water cooling conditions are shown in Table 5.

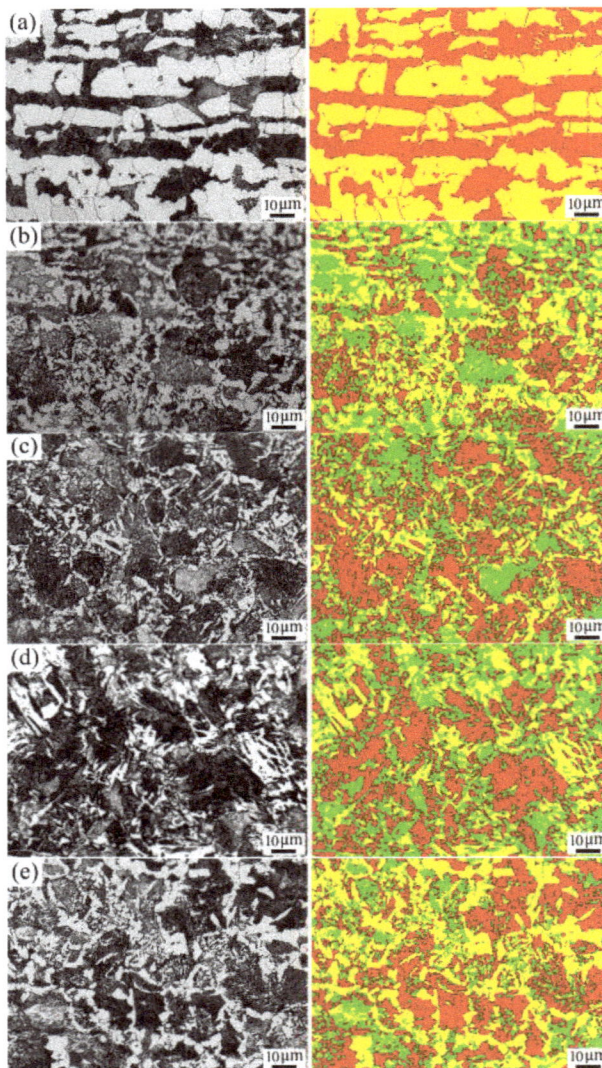

Figure 9. Microstructure and compositions of base metal of steel pipes (**a**) P1#; (**b**) P2#; (**c**) P3#; (**d**) P4#; and (**e**) P5#.

Table 5. Percentages of microstructural constituents for different spray water cooling conditions.

Percentage (%)	P1#	P2#	P3#	P4#	P5#
Ferrite	43	31	18	20	34
Pearlite	57	34	50	45	44
Composite structure	-	35	32	35	22

Figure 9 and Table 5 show that the microstructure of steel pipe heat-treated by air cooling is mainly composed of ferrite and pearlite. The average grain size of ferrite and pearlite is about 9 μm, and the percentages of ferrite and pearlite are 43% and 57%, respectively. For the steel pipes heat-treated by the OSWC process, the microstructure is mainly composed of ferrite, pearlite, and composite structures of these components. Compared with the air cooling process, the OSWC process reduces the percentage of ferrite and pearlite, and the composite structures of ferrite and pearlite occupy a considerable proportion. In addition, there are obvious zonal microstructure for the ERW steel pipe heat-treated by air cooling. The zonal characteristics of ferrite and pearlite decrease the mechanical properties to a certain extent. The spray water cooling process can reduce the zonal distribution of ferrite and pearlite, and improve the uniformity of the microstructure of the steel pipe. This is beneficial for improving the anisotropy and the mechanical properties of ERW steel pipes.

4. Results and Discussion

The poor service conditions propose a high demand on the mechanical properties of high-grade oil steel pipe. Table 2 shows the API standard of mechanical properties for J55-grade steel pipe. The ERW technology for producing high performance steel pipe is a complicated and systematic manufacturing process. We must make efforts to strictly control the chemical compositions and control the thermo-mechanical process to improve the mechanical properties and performance of steel pipes.

The chemical compositions of steel pipe are the foundation to obtaining excellent properties, including strength, plasticity, weldability, and toughness. Table 6 shows the reported chemical compositions of J55-grade steels [18,21–23].

Table 6. Chemical compositions of J55-grade steels (wt %).

J55 Grade Steels	C	Si	Mn	P	S	Cr	Ni	Cu	Nb	V	Ti
				No More Than							
this paper	0.21	0.30	1.40	0.025	0.015	-	-	-	sum of Nb, V, Ti ≤ 0.15		
[18]	-	-	-	0.030	0.030	-	-	-	-	-	-
[21]	0.12	0.30	1.27	0.008	0.004	0.04	0.10	0.06	0.002	0.081	0.012
[21]	0.07	0.19	1.23	0.015	0.002	-	-	0.001	0.02	0.002	0.01
[22]	0.15–0.22	0.15–0.22	1.2–1.4	0.02	0.010	0.40	0.30	0.40	-	-	0.003
[23]	0.28	0.2	0.99	0.016	0.018	0.09	0.06	0.11	-	-	-

In the J55-grade steels, each element plays a role in determining the possible performance. The C content of J55-grade steels tends to be below 0.2%. The C content can affect the content of pearlite and improve the strength. However, with increased of C content, the relatively high amount of carbon zones plays a role in cavity and nucleation core formation owing to segregation, which leads to reduced plasticity and toughness and generates a significant difference between the longitudinal direction and transverse direction [21]. Therefore, the carbon content must be decreased to ensure the plasticity and toughness of steel pipe.

Silicon (Si) can improve the yield strength, but it reduces the toughness. The Si content needs the right balance between strength and toughness within the range of 0.2% to 0.30% [21]. The role of manganese (Mn) is mainly to reduce the transformation temperature from austenite to pearlite and ferrite and to refine the ferrite grain size. The addition of Mn within the range of 1% to 1.6% can improve the strength and toughness by the solution strengthening effect [21].

S and P are harmful elements for steels, and easily generate MnS inclusions. They not only weaken the effect of Mn, but also reduce the impact property of steels in the transverse direction. Low S and P levels (<0.01%) not only ensure good weldability and the impact toughness at sub-zero temperature, but also improve the corrosion resistance of the steels [24]. The addition of niobium (Nb) and vanadium (V) can refine the grains and strengthen the effect of precipitation. The addition of titanium (Ti) can improve the crystallization temperature of austenite, promote the refinement of grains, and improve the strength and toughness by precipitation strengthening. Moreover, the research of shows that the effect of combined additions of Nb, V, and Ti on the mechanical properties is far greater than their effects individually [25].

The addition of copper (Cu) can improve the effect of precipitation strengthening of Nb and V and the corrosion resistance. The addition of nickel (Ni) mainly improves the toughness of ferrite and refines the grains [21]. The addition of chromium (Cr) can improve the strength and corrosion resistance, but it reduces the plasticity and toughness of steels. Based on the published information in Table 6, the main addition elements of J55 grade steel are C, Si, Mn, Nb, V, and Ti. The harmful elements S and P should be effectively controlled to be as low as possible to ensure excellent properties.

For a given chemical composition, the mechanical properties of steels are mainly determined by its microstructure and manufacturing process. In the conventional manufacturing technology of ERW steel pipe, a key process is to execute the heat treatment after welding. One purpose is to eliminate the residual welding stress, and another is to optimize the mechanical properties of steel pipes. The austenite is rapidly cooled to the transformation temperature zone during the controlled cooling process, which results in the refinement of ferrite. This is because the higher cooling rate can decrease the Ar$_3$ temperature, increase the nucleation, and restrain the growth of grains after phase transformation [25].

The controlled OSWC process is a key procedure for ensuring the mechanical properties of ERW steel pipes. It can control the temperature history of the pipes and further affect the microstructure evolution and mechanical properties. Figures 6 and 7, and Table 4 show that the OSWC process can improve the mechanical properties of steel pipes in comparison with air cooling. To illustrate the effectiveness of the OSWC process, we compared the mechanical properties of steel pipe produced by the OSWC process with other published results. In industrial tests, the comprehensive mechanical properties of steel pipe produced by the P4# process are better than the others. The strength and elongation of J55-grade steels reported in the literature are shown in Figures 10 and 11, respectively.

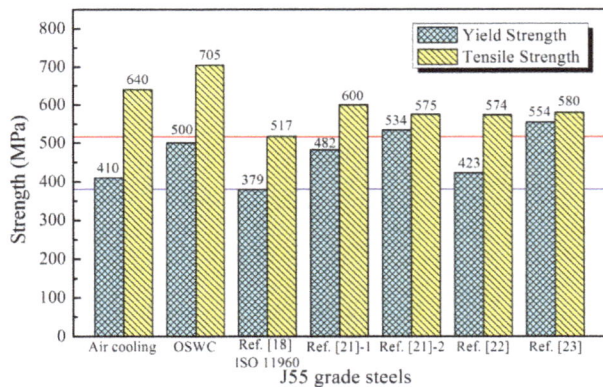

Figure 10. Strengths of J55-grade steels reported in the literature.

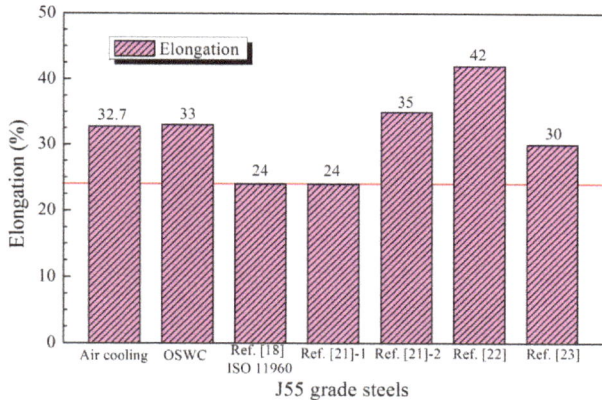

Figure 11. Elongations of different J55-grade steels reported in the literature.

Figures 10 and 11 show that the reported mechanical properties of steels are qualified in comparison with the API standard [18]. The steel pipe produced by the OSWC process obtains the best tensile strength among these J55-grade steels. The yield strength is 121 MPa higher than the API standard, and the elongation is 9% higher. High yield strength was also obtained in another study [23], but the tensile strength was not very high, which led to a yield ratio of 0.96. There were also poor margins of plasticity and toughness. Therefore, the ERW steel pipe produced by the proposed OSWC process has better comprehensive mechanical properties than other J55-grade steels.

For the J55-grade steels, the microstructure is mainly composed of ferrite, pearlite, and their composite structures. The characteristics of the microstructure include grain size, content, shape, and distribution of ferrite and pearlite, as well as the inclusion type, size, shape, and distribution. These parameters determine the mechanical properties of the steels. Figure 8 shows that the microstructure of the weld joints is obviously different from those of other ERW steel pipe. Although the microstructural streamlines of the welding joint have faded only a little, they do not disappear after undergoing the OSWC process. The grains sizes are refined significantly by the OSWC process. Grain refinement is beneficial for improving the properties of the weld joint and base metal. It improves both the effect of fine grain boundary strengthening and the low-temperature impact toughness of the steels [24]. Table 4 and Figures 7–9 show that the enhancements of strength and impact toughness are caused by the grain refinement. However, the mechanical properties of the weld joint are still poorer than that of the steel pipe substrate, especially for the impact toughness. Constantly improving the impact performance of the weld joint is still an important research issue for ERW steel pipes.

Similar to the weld joint, the microstructure of the outer range of the welding joint is also refined after undergoing the OSWC process. Moreover, the compositions of the microstructure of steel pipes produced by the OSWC process are different from that of air-cooled steel pipe. For a steel pipe produced by air cooling, the zonal microstructure is very evident in comparison with the OSWC-processed steel pipes. The extent of zonal structure decreases with the transformation temperature due to the relative difference of ferrite nucleation rates [26]. In other words, the OSWC process can decrease the zonal structure of ferrite and pearlite because the rapid cooling reduces the transformation temperature of steels.

The microstructure terms for ferrite and pearlite were used as reference [27] descriptions. Polygonal ferrite, acicular ferrite, and granular ferrite are denoted as PF, AF, and GF, respectively. The OSWC-process reduces the percentage of ferrite and pearlite, and they form composite structures. The composite microstructure can lead to a lower yield ratio, very high strength, and good ductility through the combination of different phases [28]. This is why the yield ratio of steel pipe in this paper is superior to other published results, as Figure 10 shows. Based on the CCT diagram of steel in Figure 5, we can analyze the microstructure of J55 steel subjected to different OSWC processes. In industrial tests, the ERW steel pipes were deformed from Φ193.7 mm × 7.3 mm to Φ139.7 mm × 7.72 mm at high temperature before performing the OSWC process. The hot deformation leads to the CCT diagram moving toward the top left corner and promotes the AF transformation [29,30]. The deformation of austenite accelerates the transformation of ferrite and pearlite more markedly when the undercooling is lower in the earlier stages of transformation. The effect is increased with the increase of deformation. Furthermore, it has been reported that the ferrite mainly nucleates at the boundaries of the austenite grains and defects [31]. The defects generated by deformation in the austenite increase the amount and rate of nucleation along the grain boundaries and refine the grains of ferrite and pearlite.

The effective refinement mainly happened in the austenite decomposition following the stretch-reduction-diameter process, owing to the formation of intragranular ferrite [32]. The strength and toughness of the steels can be easily adjusted through the fraction of different types of ferrites in the matrix by optimizing the parameters of the OSWC process [19,33]. Figure 9 shows that the microstructure of steel subjected to the OSWC process is composed of AF, PF, pearlite, and their composite structures based on the CCT diagram. The transformation conditions of AF and PF are related to the cooling rate. The low cooling rate easily promotes the predominant formation of a mixture of PF and pearlite.

In addition, the controlled cooling temperature during the OSWC process has an important influence on the precipitation behaviors of Nb, V, and Ti contained in J55-grade steel. A higher cooling temperature results in the reduction of dissolved vanadium predominantly by complex precipitation epitaxially growing on Ti and V carbonitride. A lower controlled cooling temperature leads to the precipitation of a great deal of V(C, N) primarily by nucleation and greatly increases the undissolved amount [34]. The precipitation type, size, shape, and distribution have an important influence on the mechanical properties of steels. Therefore, in order to obtain excellent properties, the precipitation characteristics of Nb, V, and Ti in austenite must be reasonably controlled during the OSWC process and phase transformation process of J55 ERW steel pipe.

For the same chemical compositions, the mechanical properties are mainly determined by the OSWC process and microstructure evolution. In the proposed manufacturing process, the microstructure of steel transforms into austenite due to the intermediate induction heating. If the on-line thermo-mechanical controlled cooling process can be precisely executed, the microstructure of steel pipe will be well refined, and the steel pipes will have high strength and excellent formability. The main strengthening methods of J55-grade steels are fine-grain strengthening and precipitation strengthening. The key points of industrial production are to control the precipitation behaviors of micro-alloy carbonitrides, the shape and size of deformed austenite, and its phase transformation by using the OSWC process. The undercooling and deformation also have an important impact on the microstructure of ferrite, which should be fully considered during the OSWC process.

5. Conclusions

In view of the shortcomings of the conventional manufacturing technology for ERW steel pipes, a new high-efficiency manufacturing technology based on the on-line spray water cooling (OSWC) process was developed. Industrial tests show that the new manufacturing process can control the microstructure, enhance the mechanical properties of ERW steel pipes and improve the production flexibility for different specifications. The main results were as follows:

(1) The mechanical properties are determined by the microstructure which is controlled by the parameters of the OSWC process.
(2) Industrial tests show that the OSWC process can efficiently improve and control the microstructure and mechanical properties of ERW steel pipe. The comprehensive mechanical properties of steel pipe produced by the OSWC process are superior to those of other published

J55-grade steels. The OSWC process can be applied and promoted similar on-line heat treatment of other steel products.

(3) The parameters of OSWC process have effect on the cooling rates of steel pipe. The microstructure of OSWC processed steel pipes are mainly composed of ferrite, pearlite, and composite structures of these components based on different OSWC process parameters. The Nb, V, and Ti carbonitrides, which are precipitated during the OSWC process, are helpful in enhancing the mechanical properties.

The pipeline steels often serve under extreme conditions; before application we must investigate whether the properties can satisfy the engineering requirements. The main point of this paper is to present an on-line spray water cooling (OSWC) process for manufacturing electric resistance welding (ERW) steel pipes. The mechanical properties are mainly focused on the YS, UTS, elongation and impact toughness. The impact toughness of the weld joint still needs to be further improved. The tearing resistance of the material, the defect tolerability of the ERW seam weld, and the weldability of the OSWC-processed material should be taken into account in further research. Furthermore, we need to further investigate the effects of precipitation type, size, shape, and distribution on the mechanical properties of steel pipe during the OSWC process.

Acknowledgments: The authors would like to acknowledge the project is supported by the Fundamental Research Funds for the Central Universities (No. CDJZR14135504), and supported by National Natural Science Foundation of China (No. 51421001).

Author Contributions: Zejun Chen conceived and designed the experiments; Tianpeng Zhou and Xin Chen performed the experiments; Zejun Chen and Xin Chen analyzed the data and contributed materials and tools and Zejun Chen wrote the paper.

Conflicts of Interest: The founding sponsors had no role in the design of the study; in the collection, analyses, or interpretation of data; in the writing of the manuscript, and in the decision to publish the results.

Abbreviations

The following abbreviations are used in this article:

ERW	electric resistance welding
OSWC	on-line spray water cooling
DSAW	double submerged arc welding
HISTORY	high-speed tube welding and optimum reducing technology
TMCP	thermo-mechanical controlled process
NG-TMCP	new generation of thermo-mechanical controlled process
Super-OLAC	super on-line accelerated cooling
HOP	heat-treatment on-line process
CCT	continuous cooling transformation
OM	optical microscope
YS	yield strength
UTS	ultimate tensile strength
PF	polygonal ferrite
AF	acicular ferrite
GF	granular ferrite

References

1. Pedrosa Vilaroucocastro, I.R.; de Castro, R.S.; Yadava, Y.P.; Ferreira, R.A.S. Study of phase transformations in API 5L X80 steel in order to increase its fracture toughness. *Mater. Res.* **2013**, *16*, 489–496. [CrossRef]
2. Zhu, Z.X.; Han, J.; Li, H.J. Effect of alloy design on improving toughness for X70 steel during welding. *Mater. Des.* **2015**, *88*, 1326–1333. [CrossRef]
3. Zhu, Z.X.; Kuzmikova, L.; Marimuthu, M.; Li, H.J.; Barbaro, F. Role of Ti and N in line pipe steel welds. *Sci. Technol. Weld. Join.* **2013**, *18*, 1–10. [CrossRef]

4.	Zhu, Z.X.; Marimuthu, M.; Kuzmikova, L.; Li, H.J.; Barbaro, F.; Zheng, L.; Bai, M.Z.; Jones, C. Influence of Ti/N ratio on simulated CGHAZ microstructure and toughness in X70 steels. *Sci. Technol. Weld. Join.* **2013**, *18*, 45–51. [CrossRef]

5.	Zhu, Z.X.; Han, J.; Li, H.J.; Lu, C. High temperature processed high Nb X80 steel with excellent heat-affected zone toughness. *Mater. Lett.* **2016**, *163*, 171–174. [CrossRef]

6.	Zhu, Z.X.; Han, J.; Li, H.J. Influence of heat input on microstructure and toughness properties in simulated CGHAZ of X80 steel manufactured using high-temperature processing. *Metall. Mater. Trans. A* **2015**, *46*, 5467–5475. [CrossRef]

7.	Chen, X.W.; Qiao, G.Y.; Han, X.L.; Wang, X.; Xiao, F.R.; Liao, B. Effects of Mo, Cr and Nb on microstructure and mechanical properties of heat affected zone for Nb-bearing X80 pipeline steels. *Mater. Des.* **2014**, *53*, 888–901. [CrossRef]

8.	Buckland, B. An introduction into the production and specification of steel pipe. *Piledriver* **2005**, *Q1*, 20–24.

9.	Toyooka, T.; Yorifuji, A.; Itadani, M.; Kawabata, Y.; Nishimori, M.; Koyama, Y.; Kodaka, M. Development of ultra-fine grain steel tube with high strength and excellent formability. In Proceedings of the Seoul 2000 FISITA World Automotive Congress, Seoul, Korea, 12–15 June 2000; pp. 1–6.

10.	Koyama, Y.; Toyooka, T. High strength and high elongation tubular products—HISTORY steel tube—With good bendability. *Kaw. Steel Tech. Rep.* **2000**, *43*, 55–57.

11.	Toyooka, T.; Itadani, M.; Yorifuji, A. Development of manufacturing process "HISTORY" for producing innovative high frequency welded steel tubes with excellent properties. *Kaw. Steel Tech. Rep.* **2002**, *47*, 1–8.

12.	Katsumi, M.; Yutaka, N. Manufacturing processes and products of steel pipes and tubes in JFE steel. *JFE Tech. Rep.* **2006**, *7*, 1–6.

13.	Shikanai, N.; Mitao, S.; Endo, S. Recent development in microstructural control technologies through the thermo-mechanical control process (TMCP) with JFE steel's high-performance plates. *JFE Tech. Rep.* **2008**, *11*, 1–6.

14.	Tian, Y.; Wang, B.X.; Tang, S.; Wang, Z.D.; Wang, G.D. Development and industrial application of ultra-fast cooling technology. *Sci. China Technol. Sci.* **2012**, *55*, 2748–2751. [CrossRef]

15.	Rosado, D.B.; Waele, W.D.; Hertelé, S.; Vanderschueren, D. Latest developments in mechanical properties and metallurgical features of high strength line pipe steels. *Int. J. Sustain. Constr. Des.* **2013**, *4*. [CrossRef]

16.	Okatsu, M.; Shikanai, N.; Kondo, J. Development of a high-deformability linepipe with resistance to strain-aged hardening by HOP® (heat-treatment on-line process). *JFE Tech. Rep.* **2008**, *12*, 8–14.

17.	Nishioka, K.; Ichikawa, K. Progress in thermomechanical control of steel plates and their commercialization. *Sci. Technol. Adv. Mater.* **2012**, *13*, 023001. [CrossRef] [PubMed]

18.	Standards, B. Petroleum and Natural Gas Industries-Steel Pipes for Use as Casing or Tubing for Wells. Available online: https://www.iso.org/standard/64101.html (accessed on 17 August 2016).

19.	Chen, Z.J.; Han, H.Q.; Ren, W.; Huang, G.J. Heat transfer modeling of an annular on-line spray water cooling process for electric-resistance-welded steel pipe. *PLoS ONE* **2014**, *10*, e0131574. [CrossRef] [PubMed]

20.	Han, H.Q.; Hu, J.P.; Wang, Q. Effect of water flux on heat transfer coefficient for steel pipe cooling. *Iron Steel* **2014**, *3*, 55–58.

21.	Han, Y. Properties research of hot-rolled steel strip of steel grade J55 applied in oil welded casing. *Weld. Pipe Tube* **2003**, *26*, 19–22.

22.	Salganik, V.M.; Sychev, O.N. Modeling and development of an efficient technology for the controlled rolling of steels with a prescribed set of mechanical characteristics. *Metallurgist* **2009**, *53*, 283–289. [CrossRef]

23.	Šarković, Ž.; Arsić, M.; Vulićević, L.L. Mechanical properties of welded pipes produced by high frequency welding of the steel API J55. *Zavar. I Zavarene Konstr.* **2010**, *55*, 137–142.

24.	Shukla, R.; Ghosh, S.K.; Chakrabarti, D.; Chatterjee, S. Microstructure, texture, property relationship in thermo-mechanically processed ultra-low ca rbon microalloyed steel for pipeline application. *Mater. Sci. Eng. A Struct.* **2013**, *587*, 201–208. [CrossRef]

25.	Liu, S.C. Study on Continuous Cooling Transformation and Hot Deformation Behaviors of J55 Steel for Oil Casing Pipe. Master's Thesis, Yanshan University, Qinhuangdao, China, 2011.

26.	Offerman, S.E.; Dijk, N.H.V.; Rekveldt, M.T.; Sietsma, J.; Zwaag, S.V.D. Ferrite/pearlite band formation in hot rolled medium carbon steel. *Mater. Sci. Technol.* **2002**, *18*, 297–303. [CrossRef]

27.	Krauss, G.; Thompson, S.W. Ferritic microstructures in continuously cooled low-carbon and ultralow-carbon steels. *ISIJ Int.* **1995**, *35*, 937–945. [CrossRef]

28. Beladi, H.; Timokhina, I.B.; Xiong, X.Y.; Hodgson, P.D. A novel thermomechanical approach to produce a fine ferrite and low-temperature bainitic composite microstructure. *Acta Mater.* **2013**, *61*, 7240–7250. [CrossRef]

29. Zhao, M.C.; Yang, K.; Xiao, F.R.; Shan, Y.Y. Continuous cooling transformation of undeformed and deformed low carbon pipeline steels. *Mater. Sci. Eng. A Struct.* **2003**, *355*, 126–136. [CrossRef]

30. Xiao, F.R.; Liao, B.; Shan, Y.Y.; Qiao, G.Y.; Zhong, Y.; Zhang, C.L.; Yang, K. Challenge of mechanical properties of an acicular ferrite pipeline steel. *Mater. Sci. Eng. A Struct.* **2006**, *431*, 41–52. [CrossRef]

31. Khlestov, V.M.; Konopleva, E.V.; McQueen, H.J. Kinetics of austenite transformation during thermomechanical processes. *Can. Metall. Q.* **1998**, *37*, 75–89. [CrossRef]

32. Liu, S.X.; Chen, Y.; Liu, G.Q.; Zhang, Y.G.; Huang, J.K. Effect of intermediate cooling on precipitation behavior and austenite decomposition of V-Ti-N steel for non-quenched and tempered oil-well tubes. *Mater. Sci. Eng. A Struct.* **2008**, *485*, 492–499. [CrossRef]

33. Xiao, F.R.; Liao, B.; Ren, D.L.; Shan, Y.Y.; Yang, K. Acicular ferritic microstructure of a low-carbon Mn-Mo-Nb microalloyed pipeline steel. *Mater. Charact.* **2005**, *54*, 305–314. [CrossRef]

34. Huang, Y.D.; Froyen, L. Important factors to obtain homogeneous and ultrafine ferrite-pearlite microstructure in low carbon steel. *J. Mater. Process. Technol.* **2002**, *124*, 216–226. [CrossRef]

metals

MDPI

Article

In Situ Observation of Crystal Rain and Its Effect on Columnar to Equiaxed Transition

Honggang Zhong [1,2,*], Yunhu Zhang [1], Xiangru Chen [1,*], Congsen Wu [1], Zhiqiang Wei [1] and Qijie Zhai [1,2]

[1] State Key Laboratory of Advanced Special Steel, Shanghai Key Laboratory of Advanced Ferrometallurgy, School of Materials Science and Engineering, Shanghai University, Shanghai 200072, China; yunhuzhang@shu.edu.cn (Y.Z.); qiuxiai@shu.edu.cn (C.W.); ZQW212133@163.com (Z.W.); qjzhai@shu.edu.cn (Q.Z.)
[2] Shanghai Institute of Materials Genome, Shanghai 200444, China
*   Correspondence: hgzhong@shu.edu.cn (H.Z.); cxr16@shu.edu.cn (X.C.); Tel.: +86-21-5633-4042 (H.Z.); +86-21-6613-5030 (X.C.)

Academic Editor: Soran Birosca
Received: 30 August 2016; Accepted: 3 November 2016; Published: 8 November 2016

Abstract: The investigation of a columnar to equiaxed transition (CET) and grain refinement is of high commercial importance for the improvement of the solidification structure of metal castings. The crystal rain from the free surface is frequently generated to produce grain refinement and promote a CET in alloys under the application of electromagnetic fields. However, the mechanism underlying the CET influenced by the generated crystal rain is not clear because the employed metallic alloys are opaque. In the present paper, the crystal rain in a transparent NH_4Cl–H_2O solution is produced by blowing a cooled nitrogen gas on the free surface to observe in situ its impact on the occurrence of a CET. The results show that the crystal rain can significantly promote a CET even in a high temperature gradient and that a CET only can occur when the temperature gradient is almost close to zero in the reference experiment. Finally, the most likely mechanism is discussed and clarified.

Keywords: NH_4Cl–H_2O; crystal rain; CET; refinement; physical simulation

1. Introduction

Solidified structures and segregation of castings are of great importance for its mechanical, chemical, and processing properties. Generally, the solidified structure in ingots and billets consists of columnar and equiaxed grains. Columnar grains show anisotropic properties and can induce macrosegregation, whereas fine equiaxed grains can improve the homogeneity of the castings. Hence, the most effective way to improve the homogeneity of ingots and billets is to generate large amounts of equiaxed grains and promote a columnar to equiaxed transition (CET).

According to previous research, the origin of equiaxed grains in ingots is mainly attributed to the free grains detached from the mold walls [1] and the melt surface [2], as well as the nucleation under the constitutional undercooling [3]. Technically, more equiaxed grains can be obtained by adding nucleating agents [4]. However, this chemical method has limitations in many cases. The development of effective nucleation agents in steels has not been very effective yet. Hence, many other methods have been employed to cause grain refinement in solidifying metal alloys, such as electromagnetic stirring (EMS) [5], ultrasonic vibration [6], electric current pulse (ECP) [7,8], pulsed electromagnetic fields [9], and process control methods [10–12]. Recently, Zhai [13] developed a pulse magneto-oscillation (PMO) technique, which could significantly increase the number of free grains and promote a CET in ingots and billets. In particular, the surface pulse magneto-oscillation (SPMO) method [14], putting the electromagnetic coil above the free surface of ingots, can stimulate grain nucleation and detachment from the surface of liquid metal to cause crystal rain. The crystal rain can block the growth of columnar dendrites and refine the solidification structure significantly. However, the CET mechanism underlying this phenomenon is not clear.

It is unfortunate that the solidification process in metal materials can hardly be observed in situ because they are opaque. Recently, in situ observation of solidifying metals by a synchrotron X-ray technique [15,16] was successfully performed to study the solidification behavior of alloys. However, crystal rain is difficult to generate in a solidification cell because the gap between quartz plates is only 200–300 μm, which can significantly limit the crystal sedimentation and convection in bulk melt. Jackson [17] found several transparent compounds that show similar solidification morphology to metals. Thus, transparent model materials have become ideal simulation materials for the study of the solidification of metals. Peppin [18] observed the steady-state solidification of aqueous ammonium chloride and discussed the effect of the freezing rate and initial concentration of the solution on morphological transitions. Beckermann et al. [19–21] investigated the equiaxed dendritic solidification of a $NH_4Cl–H_2O$ solution with convection. Nayak [22,23] observed and presented numerically the solidification behavior of aqueous ammonium chloride on an inclined cooling plate. Liu [24] observed the detachment of dendrite side arms induced by deceleration in $NH_4Cl–H_2O$ and $[CH_2CN]_2–H_2O$ solutions. The effects of vibrational frequency and roughness of the substrate on nucleation were studied using a $NH_4Cl–H_2O$ solution by Wang [25]. Zhang et al. studied the effect of substrate surface structure on the nucleation with NH_4Cl and 70 wt. % H_2O [26,27]. Nonetheless, the effect of crystal rain on the CET has been paid little attention.

In the present paper, the sedimentation of free crystals from the melt surface (crystal rain) was generated and observed in situ in a $NH_4Cl–H_2O$ solution. The influence of the generated crystal rain on the occurrence of a CET was considered, and the related mechanism is herein discussed.

2. Experimental Procedure

A solution of NH_4Cl and 65 wt. % H_2O was selected as the experimental material. Figure 1 shows the setup for the NH_4Cl aqueous solution crystallization. Two quartz tubes (Beijing Zhong Cheng Quartz Glass Co. Ltd., Beijing, China) (open at both ends) with inner diameter of 26 mm and length of 260 mm were used as mold. A cylindrical copper block was inserted into the tube bottom and cooled by an ice water mixture (see Figure 1a). Three K-type thermocouples (Anhui Tian Kang Co. Ltd., Chuzhou, China) were fixed in each mold to measure the temperature evolution, as shown in Figure 1. In order to generate crystal rain in the bulk solution, nitrogen gas was cooled in a spiral copper tube, which was immersed in the ice water mixture, and then directly blown onto the free surface of the solution. As shown in Figure 1c, the cooled nitrogen gas generated a thin solid layer at the free surface and caused certain amounts of free crystals to detach from the solid layer. The gas flux was controlled by a rotameter in a range from 4 L/min to 8 L/min. Since the gas was blown onto the surface of solution, global forced convection was observed inside the bulk solution. A reference experiment was also performed at the same time to show the solidification of a solution of NH_4Cl and 65 wt. % H_2O without the influence of nitrogen gas or crystal rain.

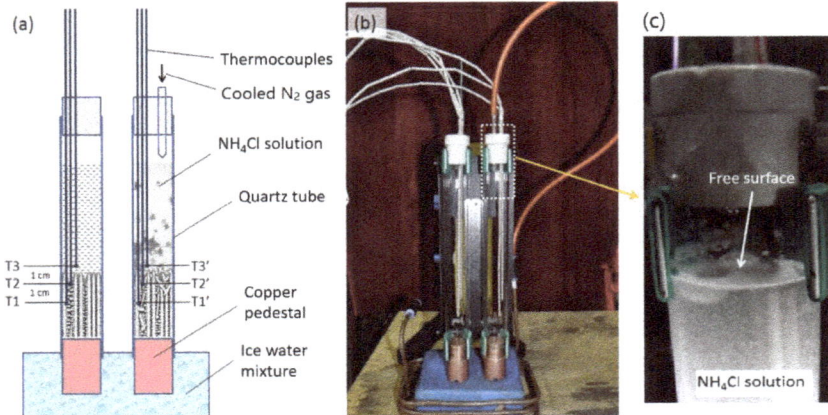

Figure 1. Setup of NH₄Cl aqueous solution crystallization. (**a**) Schematic view; (**b**) device picture; (**c**) picture of free liquid surface and crystals generated by cooled N_2 gas.

The experimental procedure is as follows. The solution of NH_4Cl and 65 wt. % H_2O was heated to 85 °C (or 95 °C, to investigate the effect of superheat on the CET), and it was then poured into the molds to directionally solidify. When the top of the columnar dendrites grew over the bottom thermocouple (T1 and T1′), the cooled nitrogen gas was blown onto the free surface of solution in one of the quartz tubes to cause crystal rain and forced convection inside the bulk solution. The cooling curves were measured via thermocouples. A temperature gradient that divides the temperature difference by the distance difference between the employed thermocouples T2 and T1 (or T2′ and T1′) was defined. The dendritic growth and crystal rain were photographed by a digital single lens reflex camera (Nikon D90) (Nikon Corporation, Tokyo, Japan).

3. Results and Discussion

3.1. Crystal Growth and the CET in the Reference Sample

Figure 2 shows the columnar grains directionally solidified from the bottom of the reference sample, 600 s after the solution was poured into the tube. The coarse columnar dendrite morphology can be observed at the solid–liquid (S–L) interface. It took 750 s for the interface to reach thermocouple T1. In the solidified structure (about 1300 s after pouring), shown in Figure 3, a few small equiaxed crystals (about 0.1 mm in diameter) were clearly observed ahead of the S–L interface as the columnar crystals gradually grew. However, these small equiaxed crystals were not able to stop the growth of columnar crystals and fell into the columnar crystal branches (see in Figure 4, in black dash line circles). When the S–L interface moved over thermocouple T2 (about 1650 s after pouring), numerous larger equiaxed grains formed at the front of the S–L interface to block the growth of columnar crystals. However, no sharp columnar-equiaxed interface was observed, and the CET occurred in a zone (between T2 and T3) where columnar and equiaxed crystals co-exist, which is in agreement with the studies by Ares and Gueijman [28,29]. Generally, there are examples of both kinds of CETs in the literature [30].

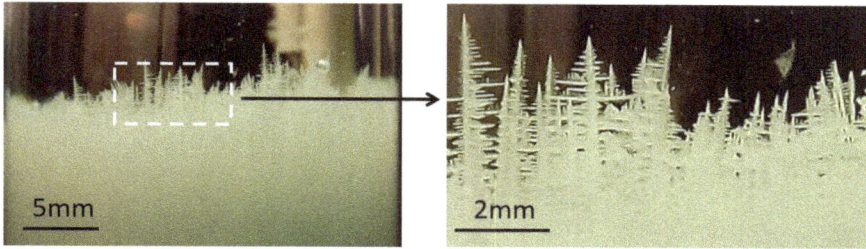

Figure 2. The columnar dendrites of ammonium chloride without the impact of nitrogen gas (600 s after pouring).

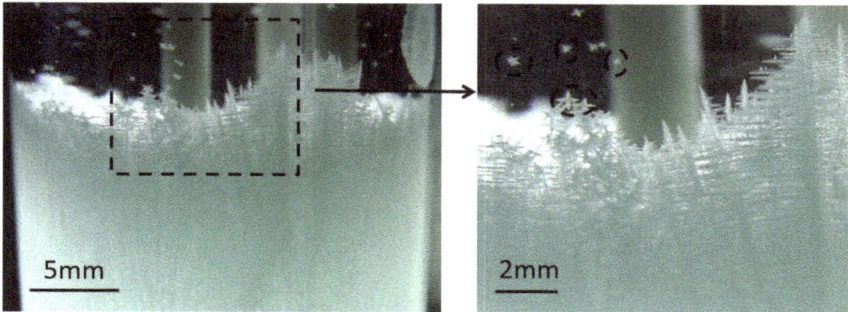

Figure 3. A few equiaxed crystals nucleated in front of the S–L interface and fell in the columnar crystal branches in the reference sample.

Figure 4. The image of equiaxed crystals falling into the columnar dendritic branches in the reference sample. The equiaxed crystals nucleating in front of the S–L interface are too small in both quantity and size to block the columnar crystals.

3.2. Crystal Rain and Its Effect on the CET

Coarse columnar dendrites with the same growth rate as shown in the reference sample were also achieved in the sample before the nitrogen gas was blown onto the liquid surface. When the S–L interface arrived at thermocouple T1, the cooled nitrogen gas was triggered to cause crystal rain (as shown in Figure 1c). Figure 5 shows the generated crystal rain above the S–L interface in the samples for different nitrogen gas fluxes. A large amount of free equiaxed crystals and the agglomeration phenomenon of free crystals (see Figure 5b) can be observed in the samples with different nitrogen gas fluxes rather than in the reference sample. Moreover, it can be found that the amount of free crystals remarkably increased, and their size reduced significantly when the higher gas flux was employed. In addition, the sedimentation rate of free crystals accelerated with increasing gas flux. When the flow rate was 6–8 L/min (see Figure 5c,d), plenty of fine equiaxed crystals fell onto the S–L interface like a snowstorm. The corresponding sedimentation rates of equiaxed crystals adjacent to the zone above the S–L interface were measured, showing a linear correlation to the intensity of nitrogen gas flux, as shown in Figure 6.

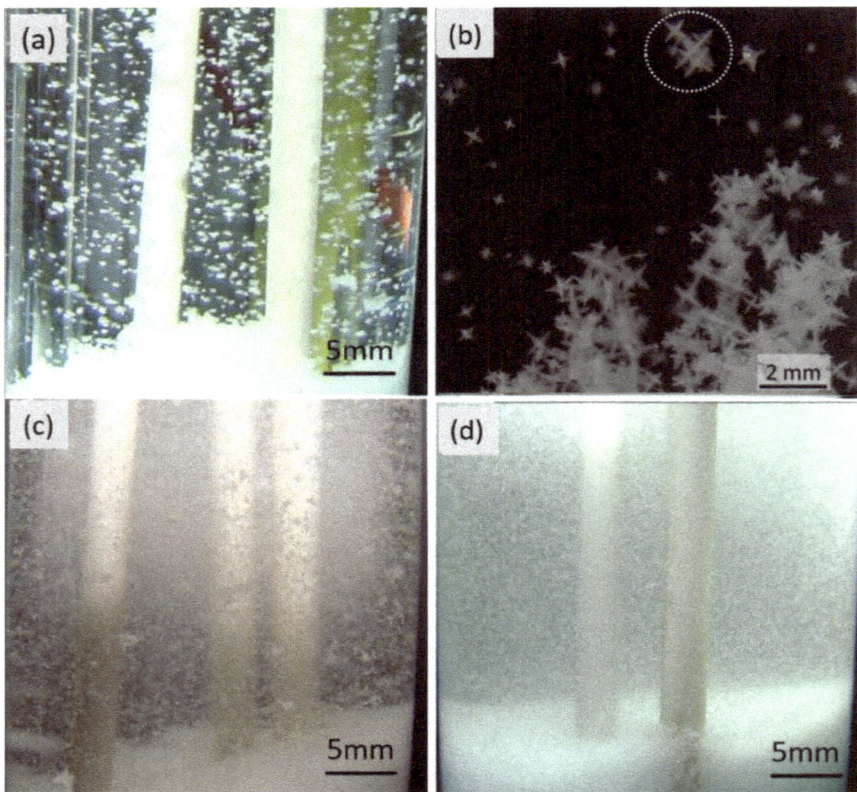

Figure 5. Crystal rain in aqueous solution of ammonium chloride. (**a**) Crystal rain under nitrogen gas flux of 4 L/min; (**b**) some equiaxed crystals agglomerating to each other (nitrogen gas flux = 4 L/min); (**c**) crystal rain under nitrogen gas flux of 6 L/min; (**d**) crystal rain under nitrogen gas flux of 8 L/min.

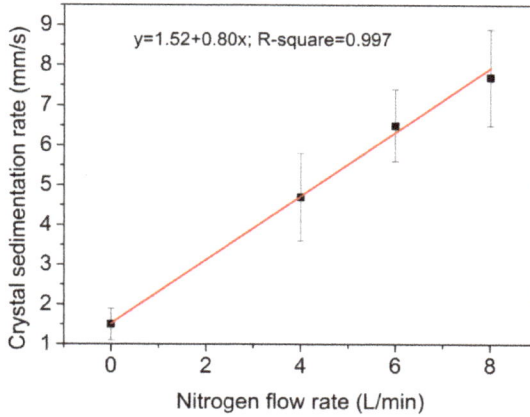

Figure 6. Influence of nitrogen gas flux on the sedimentation rate of equiaxed crystal grains.

Figure 7 shows the measured cooling curves and the corresponding mean temperature gradients accompanying the occurrence of the CET in the samples with and without the influence of the cooled nitrogen gas. The CET in the reference sample occurs in a period after the S–L interface grew over the thermocouple T2. The value of the temperature gradient during the CET period was almost zero—sometimes even negative. This means that the solution just above the S–L interface is undercooled to promote the heterogeneous nucleation of equiaxed grains, which locks the growth of columnar crystals. In comparison with the reference sample, the cooled nitrogen gas causing crystal rain significantly promotes the occurrence of CET. The clear CET interface is formed when the S–L interface grows over the lowest thermocouple T1′. Moreover, it should be noted here that the temperature gradient ahead of the S–L interface is as high as 6 °C/cm when the CET occurs. This indicates that the crystal rain can trigger the CET at one elevated temperature gradient ahead of the S–L interface. In order to confirm this phenomenon, 11 solidification experiments were performed under the application of the cooled nitrogen gas (see Figure 8). The figure shows that the generated crystal rain can still significantly promote a CET, even when the temperature gradient is in the 5–6.5 °C/cm range. Furthermore, it was found that the pouring temperature and the intensity of nitrogen gas flux has no significant influence on the temperature gradient when the CET occurs.

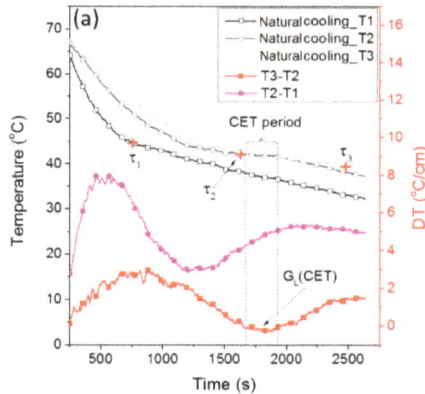

Figure 7. *Cont.*

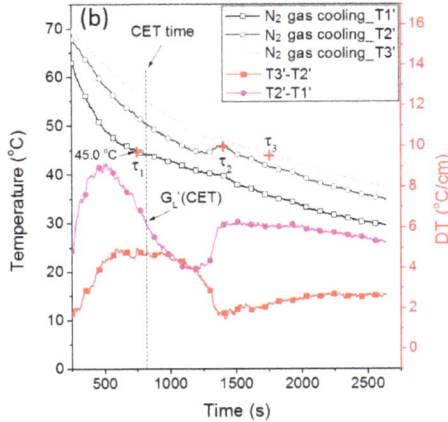

Figure 7. The cooling curves and temperature gradients in the CET zone (pouring temperature: 85 °C, the pouring time is defined as $t = 0$), (**a**) reference sample; (**b**) sample with crystal rain.

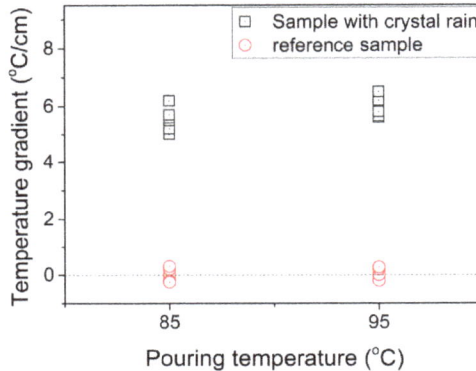

Figure 8. The temperature gradient ahead of the S–L interface during the CET.

3.3. The Mechanism of the CET with Crystal Rain

According to previous investigations, the occurrence of a CET depends on the amount of equiaxed grains, the solute distribution, and the temperature gradient in front of the S–L interface. Nguyen-Thi [31] observed in situ the solute enrichment area surrounding the dendritic microstructure of the Al–Ni alloy. It was found that equiaxed grains appeared in the solute enrichment layer after the pulling rate was increased, and the columnar as well as equiaxed dendrites generally cease to grow long before effective contact, which confirmed that the blocking mechanism in the CET was likely to be solutal [32]. The CET in the reference sample is most likely due to the significant heterogeneous nucleation caused by the constitutional undercooling in the low temperature gradient region.

In case of the sample under the influence of crystal rain, the temperature gradient ahead of the S–L interface was far higher than that in the reference sample when the CET occurs (as shown in Figure 7). It has been shown in the reference sample that few equiaxed grains formed at front of the S–L interface due to the constitutional undercooling when the temperature is about 6 °C/cm. These few nuclei just fell into the columnar dendritic branches and cannot block the growth of columnar grains. In addition, the convection forced by the nitrogen gas flow could reduce the thickness of the solute-enriched layer and the temperature gradient at the front of the S–L interface [33,34]. Hence, the most likely mechanism of the CET is that the generated crystal rain supplies a sufficient amount of equiaxed crystals to block the growth of columnar crystals.

4. Conclusions

The influence of crystal rain on CET was investigated in situ in a solution of NH_4Cl and 65 wt. % H_2O. The crystal rain can be generated by blowing the cooled nitrogen gas on the free surface of the solution. It was found that nitrogen gas with a higher flux can produce a significantly larger amount of free equiaxed crystals in the bulk solution, and can increase the sedimentation velocity of free crystals at the S–L interface. As a consequence, the CET is immediately provoked by the generated crystal rain, even when the temperature gradient is as high as 6 °C/cm. However, in comparison with the reference experiment, the CET only occurs when the temperature gradient decreases to almost zero. It was deduced that the most likely mechanism underlying the CET promoted by crystal rain is the growth of columnar dendrites mechanically blocked by the generated free equiaxed crystals.

Acknowledgments: The authors acknowledge the financial supports from the NSFC (No. 51227803, 51404150 & 51504148), Science and Technology Commission of Shanghai Municipality (No. 14DZ2261200).

Author Contributions: Honggang Zhong and Qijie Zhai conceived and designed the experiments; Consen Wu performed the experiments and calculation; Yunhu Zhang, Xiangru Chen, Honggang Zhong, and Qijie Zhai analyzed the results and fostered the interpretation; all authors contributed equally to the discussion and writing the paper.

Conflicts of Interest: The authors declare no conflict of interest.

References

1. Ohno, A.; Motegi, T.; Soda, H. Origin of the equiaxed crystals in castings. *Tran. Iron Steel Inst. Jpn.* **1971**, *11*, 18–23.
2. Southin, R.T. Nucleation of the equiaxed zone in cast metals. *Trans. Metall. Soc. AIME* **1967**, *239*, 220–225.
3. Winegard, W.C.; Chalmers, B. Supercooling and dendritic freezing in alloys. *Trans. Am. Soc. Met.* **1954**, *46*, 1214–1224.
4. Sigworth, G.K. The grain refining of aluminum and phase relationships in the Al–Ti–B system. *Metall. Trans. A* **1984**, *15*, 277–282. [CrossRef]
5. Tzavaras, A.A.; Brody, H.D. Electromagnetic stirring and continuous casting—Achievements, problems, and goals. *JOM* **1984**, *36*, 31–37. [CrossRef]
6. Ma, Q.; Ramirez, A.; Das, A. Ultrasonic refinement of magnesium by cavitation: Clarifying the role of wall crystals. *J. Cryst. Growth* **2009**, *311*, 3708–3715.
7. Liao, X.; Zhai, Q.; Luo, J.; Chen, W.; Gong, Y. Refining mechanism of the electric current pulse on the solidification structure of pure aluminum. *Acta Mater.* **2007**, *55*, 3103–3109. [CrossRef]
8. Nakada, M.; Shiohara, Y.; Flemings, M.C. Modification of solidification structures by pulse electric discharging. *Tran. Iron Steel Inst. Jpn.* **1990**, *30*, 27–33. [CrossRef]
9. Kolesnichenko, A.F.; Podoltsev, A.D.; Kucheryavaya, I.N. Action of pulse magnetic field on molten metal. *Tran. Iron Steel Inst. Jpn.* **1994**, *34*, 715–721. [CrossRef]
10. Campbell, J. Grain refinement of solidifying metals by vibration: A review. In Proceedings of the Solidification Technology in the Foundry and Cast House, Proceedings of an International Conference of the Applied Metallurgy & Metals Technology Group, Coventry, UK, 15–17 September 1983; Metals Society: London, UK, 1983; pp. 61–64.
11. Easton, M.A.; Kaufmann, H.; Fragner, W. The effect of chemical grain refinement and low superheat pouring on the structure of NRC castings of aluminium alloy Al–7Si–0.4Mg. *Mater. Sci. Eng. A* **2006**, *420*, 135–143. [CrossRef]
12. Wang, X.; Zhang, H.; Zuo, Y.; Zhao, Z.; Zhu, Q.; Cui, J. Experimental investigation of heat transport and solidification during low frequency electromagnetic hot-top casting of 6063 aluminum alloy. *Mater. Sci. Eng. A* **2008**, *497*, 416–420. [CrossRef]

13. Gong, Y.Y.; Zhai, Q.J.; Li, B.; Li, R.X.; Yin, Z.X. A Method of Refining Solidificaiton Microstructure with Surface Pulsed Magneto-Oscillation. CN Patent 201,010,167,538.3, 5 June 2010.

14. Zhao, J.; Yu, J.; Li, Q.; Zhong, H.; Song, C.; Zhai, Q. Structure of slowly solidified 30Cr2Ni4MoV casting with surface pulsed magneto-oscillation. *Mater. Sci. Technol.* **2015**, *31*, 1589–1594. [CrossRef]

15. Nguyen-Thi, H.; Salvo, L.; Mathiesen, R.H.; Arnberg, L.; Billia, B.; Suery, M.; Reinhart, G. On the interest of synchrotron X-ray imaging for the study of solidification in metallic alloys. *C. R. Phys.* **2012**, *13*, 237–245. [CrossRef]

16. Mathiesen, R.H.; Arnberg, L.; Nguyen-Thi, H.; Billia, B. In Situ X-ray video microscopy as a tool in solidification science. *JOM* **2012**, *64*, 76–82. [CrossRef]

17. Jackson, K.A.; Hunt, J.D. Transparent compounds that freeze like metals. *Acta Metall.* **1965**, *13*, 1212–1215. [CrossRef]

18. Peppin, S.; Huppert, H.E.; Worster, M.G. Steady-state solidification of aqueous ammonium chloride. *J. Fluid Mech.* **2008**, *599*, 465–476. [CrossRef]

19. Beckermann, C.; Wang, C.Y. Equiaxed dendritic solidification with convection. 3. Comparisons with NH_4Cl–H_2O experiments. *Metall. Mater. Trans. A* **1996**, *27*, 2784–2795. [CrossRef]

20. Ramani, A.; Beckermann, C. Dendrite tip growth velocities of settling NH_4Cl equiaxed crystals. *Scr. Mater.* **1997**, *36*, 633–638. [CrossRef]

21. Badillo, A.; Ceynar, D.; Beckermann, C. Growth of equiaxed dendritic crystals settling in an undercooled melt, Part 1: Tip kinetics. *J. Cryst. Growth* **2007**, *309*, 197–215. [CrossRef]

22. Nayak, A.K.; Barman, N.; Chattopadhyay, H. Solidification of a binary solution (NH_4Cl + H_2O) on an inclined cooling plate: A parametric study. *Procedia Mater. Sci.* **2014**, *5*, 454–463. [CrossRef]

23. Mohanty, D.; Nayak, A.K.; Barman, N. Studies on transport phenomena during solidification of a binary solution (NH_4Cl + H_2O) on an inclined cooling plate. *Trans. Indian Inst. Met.* **2012**, *65*, 801–807. [CrossRef]

24. Liu, S.; Lu, S.; Hellawell, A. Dendritic array growth in the systems NH_4Cl–H_2O and $[CH_2CN]_2$–H_2O: The detachment of dendrite side arms induced by deceleration. *J. Cryst. Growth* **2002**, *234*, 740–750. [CrossRef]

25. Wang, W.; Wang, K.; Lin, X. Crystal nucleation and detachment from a chilling metal surface with vibration. *Mater. Chem. Phys.* **2009**, *117*, 199–203. [CrossRef]

26. Zhang, Y.; Wang, M.; Lin, X.; Huang, W. Effect of substrate surface microstructure on heterogeneous nucleation behavior. *J. Mater. Sci. Technol.* **2012**, *28*, 67–72. [CrossRef]

27. Zhang, Y.; Wang, M.; Lin, X.; Huang, W. Effect of substrate wettability and surface structure on nucleation of crystal. *J. Mater. Sci. Technol.* **2012**, *28*, 859–864. [CrossRef]

28. Ares, A.E.; Gassa, L.M.; Gueijman, S.F.; Schvezov, C.E. Correlation between thermal parameters, structures, dendritic spacing and corrosion behavior of Zn–Al alloys with columnar to equiaxed transition. *J. Cryst. Growth* **2008**, *310*, 1355–1361. [CrossRef]

29. Ares, A.E.; Gueijman, S.F.; Schvezov, C.E. An experimental investigation of the columnar-to-equiaxed grain transition in aluminum-copper hypoeutectic and eutectic alloys. *J. Cryst. Growth* **2010**, *312*, 2154–2170. [CrossRef]

30. Liu, D.R.; Mangelinck-Noël, N.; Gandin, C.A.; Zimmermann, G.; Sturz, L.; Nguyen-Thi, H.; Billia, B. Structures in directionally solidified Al–7 wt.% Si alloys: Benchmark experiments under microgravity. *Acta Mater.* **2014**, *64*, 253–265. [CrossRef]

31. Nguyen-Thi, H.; Reinhart, G.; Mangelinck-NoëL, N.; Jung, H.; Billia, B.; Schenk, T.; Gastaldi, J.; Härtwig, J.; Baruchel, J. In-situ and real-time investigation of columnar-to-equiaxed transition in metallic alloy. *Metall. Mater. Trans. A* **2007**, *38*, 1458–1464. [CrossRef]

32. Martorano, M.; Beckermann, C.; Gandin, C. A solutal interaction mechanism for the columnar-to-equiaxed transition in alloy solidification. *Metall. Mater. Trans. A* **2003**, *34*, 1657–1674. [CrossRef]

33. Willers, B.; Eckert, S.; Michel, U.; Haase, I.; Zouhar, G. The columnar-to-equiaxed transition in Pb–Sn alloys affected by electromagnetically driven convection. *Mater. Sci. Eng. A* **2005**, *402*, 55–65. [CrossRef]

34. Banaszek, J.; Browne, D.J. Modelling columnar dendritic growth into an undercooled metallic melt in the presence of convection. *Mater. Trans.* **2005**, *46*, 1378–1387. [CrossRef]

metals

MDPI

Article

Microstructure and Mechanical Properties of an Al-Li-Mg-Sc-Zr Alloy Subjected to ECAP

Anna Mogucheva * and Rustam Kaibyshev

Laboratory of Mechanical Properties of Nanostructured Materials and Superalloys, Belgorod State University, 85 Pobedy, Belgorod 308015, Russia; rustam_kaibyshev@bsu.edu.ru
* Correspondence: mogucheva@bsu.edu.ru; Tel.: +7-4722-58-54-17

Academic Editor: Soran Birosca
Received: 31 August 2016; Accepted: 17 October 2016; Published: 25 October 2016

Abstract: The effect of post-deformation solution treatment followed by water quenching and artificial aging on microstructure and mechanical properties of an Al-Li-Mg-Sc-Zr alloy subjected to equal-channel angular pressing (ECAP) was examined. It was shown that the deformed microstructure produced by ECAP remains essentially unchanged under solution treatment. However, extensive grain refinement owing to ECAP processing significantly affects the precipitation sequence during aging. In the aluminum-lithium alloy with ultrafine-grained (UFG) microstructure, the coarse particles of the S_1-phase (Al_2LiMg) precipitate on high-angle boundaries; no formation of nanoscale coherent dispersoids of the δ'-phase (Al_3Li) occurs within grain interiors. Increasing the number of high-angle boundaries leads to an increasing portion of the S_1-phase. As a result, no significant increase in strength occurs despite extensive grain refinement by ECAP.

Keywords: aluminum alloys; equal channel angular pressing; thermomechanical processing; microstructure; mechanical properties; precipitation

1. Introduction

The formation of a ultra-fine grained (UFG) microstructure through severe plastic deformation (SPD) is an advanced approach to produce semi-finished products from aluminum alloys with a high strength and endurance limit, good ductility and sufficient fracture toughness [1]. It was postulated [1] that the enhanced strength of aluminum alloys is attributed to structural strengthening due to the formation of a UFG microstructure that was defined as a homogeneous and reasonably equiaxed microstructure with average grain sizes of less than ~1 μm; the fraction of high angle boundaries (HAGBs) is higher than 56%. From a practical point of view, a UFG structure can be easily produced in aluminum alloys by equal channel angular pressing (ECAP) [1]. This technique can be scaled up fairly easily for the fabrication of high-volume billets with commercially viable dimensions [1–4].

Most high-performance aluminum alloys are age-hardenable. Precipitation strengthening is a dominant strengthening mechanism for these alloys. It is obvious that only the combination of grain size strengthening with precipitation hardening can significantly affect the strength of these aluminum alloys. The formation of a UFG microstructure significantly affects the size of second-phase precipitations and their distribution within the aluminum matrix [5–10]. Therefore, mechanical properties of age-hardenable aluminum alloys subjected to ECAP depend critically on the aging response including precipitation kinetics and sequence, origin of dispersoids and the precipitate morphology.

Unfortunately, despite numerous studies [11–15] dealing with the examination of extensive grain refinement through ECAP on mechanical properties of these precipitation-hardened aluminum alloys, the effect of a UFG microstructure on precipitation sequence under post-deformation aging is relatively poorly known [5,9,16–18]. It was shown that the formation of a UFG microstructure changes the precipitation sequence: stable phases precipitate on high-angle boundaries [5] or even within the aluminum matrix [6] that suppresses the appearance of a metastable phase. It is well-known that the strength of an age-hardenable aluminum alloy is a function of grain size strengthening, dislocation strengthening and precipitation hardening. The last strengthening mechanism is most important for this type of aluminum alloy; decrease in the precipitation strengthening due to lack of uniform distribution of metastable phases with coherent or semi-coherent interface boundaries could not be compensated by the two other strengthening mechanisms. As a result, the achievement of increased yield stress (YS) in age-hardenable aluminum alloys is not a trivial task. A positive effect of ECAP on YS is reported in a limited number of studies [5,7–9,17,19,20]. It is worth noting that in all these works the temperature of final aging was lower in comparison with aging temperature used to achieve optimal YS in their coarse grained (CG) counterparts. The low aging temperature was used to prevent over-aging due to precipitations of stable phase coarse particles; application of conventional aging temperature to ECAP processed aluminum alloys leads to rapid softening [9,21].

ECAP Processing of Age Hardenable Aluminum Alloys

Currently, there exist four different routes of thermomechanical processing (TMP) through ECAP for achieving high strength in age hardenable bulk aluminum alloys (Figure 1). First, aluminum alloys usually belonging to 6XXX series are subjected to ECAP at $T \leq 100\,^{\circ}\text{C}$ in fully annealed [22] or quenched [23,24] conditions without any subsequent artificial aging [22–24] (Figure 1a); post-ECAP natural aging usually occurs at room temperature. An increase in YS varies from ~25% [13] to ~40% [23] in comparison with values specified for these commercial alloys [17]. It seems that an increase in YS is attributed to a combination of dislocation strengthening attributed to low temperature ECAP with solid solution strengthening or precipitation strengthening attributed to natural aging. However, a remarkable reduction in ductility [22–24] is the main disadvantage of this TMP route.

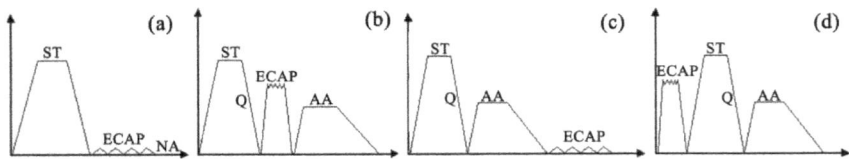

Figure 1. Scheme of thermomechanical treatment. ST—solution treatment, ECAP–equal channel angular pressing, NA—natural aging, Q—Quenching, AA—artificial aging. (**a**) ECAP between ST and NA; (**b**) ECAP between ST and AA; (**c**) ECAP after ST and AA; (**d**) ECAP before ST and AA.

Second, an aluminum alloy in a condition of supersaturated solid solution produced by initial water or oil quenching is processed by ECAP at a temperature below, equal to or slightly higher than the aging temperature (Figure 1b). Finally, the processed alloy is subjected to artificial aging [6–9,17,19,21,25–27]. This is the most popular TMP route. It was found that this route results in the 18%–40% increase in YS. High YS originated from a combination of dislocation strengthening and dispersion strengthening. However, the total elongation decreases by almost a factor of two [6–9,17,21,26,27]. Decreased ductility is attributed to low work-hardening rate [6]. There exists no potential for further work-hardening after this TMP. As a result, uniform elongation does not exceed 3%–5% [7,17]. It was shown that it is possible to increase strength by ECAP with ductility remaining essentially unchanged [19]. However, even in this case, the uniform elongations in an Sc modified AA7055 aluminum alloy subjected to T6 heat treatment were much larger (~15% vs. 2%–3%) in comparison with the material subjected to ECAP processing. It seems that low value of uniform elongation is the main disadvantage of age-hardenable aluminum alloys processed through this route.

Third, aluminum alloys can be subjected to ECAP in aged condition (Figure 1c) [5]. Under ECAP, the mobile lattice dislocation cuts off dispersoids of a metastable phase with semi-coherent or/and coherent boundaries that leads to their gradual fragmentation into nanoscale particles. The last are thermodynamically unstable due to high surface energy [5]. As a result, a complete dissolution of this particle occurs. Next, the Genie-Preston Zones (GPZ) form providing an increase in hardness [5]. This TMP route is very exotic.

It is worth noting that these three routes of TMP are not viable to produce true grains entirely delimited by HAGBs. Therefore, the deformed structure resulting from these TMP routes could not be interpreted in terms of UFG structure [1]. Careful inspection of structural characterization data [6,8,9,17,19,23,26] showed that, for the most part, crystallites evolved under ECAP are delimited by low-angle boundaries (LAGBs); under SPD, the lattice dislocations arrange into cells and LAGBs or form the dislocation network within initial grains [17]. Therefore, it is not obvious that the grain size strengthening is important for significant increase in YS of aluminum alloys subjected to ECAP through the three aforementioned TMP routes.

Only the fourth route (Figure 1d) of TMP, which was initially developed to process Al-Li alloys [16,28], provides the formation of UFGs entirely delimited by HAGBs. In this route, the post-ECAP standard heat treatment consisting of solution treatment, quenching and final aging can be applied to the aluminum alloy with UFG structure. A critical condition for viability of this method is the stability of UFG structure under solution treatment. Materials such as AA5091 [28], in which abnormal grain growth in UFG structure occurs under solution treatment, are not suitable for this processing method. The fourth route of TMP can be applied to materials such as Weldalite (AA2X95) alloy [16], in which normal grain growth took place; increase in grain size under solution treatment is insignificant. Therefore, it is reasonable to apply the fourth route of TMP only to Sc-bearing aluminum alloys, in which coherent Al_3Sc dispersoids effectively suppress the migration of HAGBs under conditions of solution treatment that retains a UFG structure evolved under prior ECAP [2,29,30]. It was found that a 9% increase in YS of Al-Li alloys processed through this TMP route could be achieved [16]. Concurrently, a 50% increase in ductility and uniform elongation takes place due to the formation of UFG structure and elimination of precipitation-free zones [16].

Thus, the positive effect of UFG structure on mechanical properties of precipitation-hardened aluminum alloys can be achieved only in certain cases:

(I) a UFG structure evolved during ECAP processing must be stable under solution treatment;
(II) aging of a material with UFG structure must be capable to provide significant precipitation hardening.

Accordingly, the present work was conducted to evaluate the feasibility of applying the fourth route of TMP to an Al-Li-Mg-Sc-Zr alloy (designated in Russia as 1421Al). It was shown previously [31] that ECAP at 300 °C of this alloy provides the formation of the well-defined subgrain structure (ε~2), partially recrystallized structure (ε~4) and fully recrystallized structure (ε~8). The last structure is uniform with an average grain size of ~1 µm [2,29–31]; most of the crystallites are true grains entirely outlined by HAGBs. This structure is essentially stable under annealing at a temperature of 450 °C [2,30,32]. Therefore, this alloy can be subjected to solution treatment followed by quenching with a UFG structure. In contrast [29], the partially recrystallized structure in the 1421Al could be unstable under conditions of solution treatment. The 1421Al was developed for application as a structural material for airframes [33]. As a result, the feasibility to use this material is dependent on a combination of strength, toughness and fatigue resistance. From this point of view, it is quite reasonable to examine mechanical properties of the 1421Al in all of the aforementioned structural states; we initially expected that the partially recrystallized structure provides the highest toughness, and the 1421Al with a fully recrystallized structure would exhibit the highest strength. Thus, the aim of this study is to examine the effect of ECAP followed by conventional hardening heat treatment on strength and ductility of this material.

In context of this goal, the effect of post-deformation heat treatment on structure will be examined. In addition, the aging response of the 1421Al with different deformation structure will be analyzed in relationships with its mechanical properties, which are highly dependent on dispersion of the secondary phase precipitated under aging [33].

The precipitation sequence is indicated in the 1421Al by [33]:

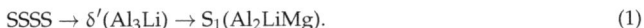

$$SSSS \rightarrow \delta'(Al_3Li) \rightarrow S_1(Al_2LiMg). \tag{1}$$

High strength is provided by precipitation of coherent the δ'-phase dispersoids with an average size of 6 nm [33,34]; high volume fraction of the δ'-phase is a prerequisite condition for achieving the highest strength in the 1421Al. Accordingly, it is very important to dissolve whole Li into solid solution under solution treatment. Next, Li must precipitate in the form of the metastable δ'-phase under aging. Precipitations of the incoherent equilibrium S_1-phase lead to material softening because the amount of Li available for the formation of the δ'-phase in solid solution decreases. The S_1-phase tends to precipitate on HAGBs [35]. Under certain conditions, the precipitations of the S_1-phase can almost suppress the formation of the δ'-phase [33,35]. This leads to a minor strengthening increment by precipitation hardening during aging. It is worth noting that the aging behavior of Al-Li-Mg alloys with a grain size higher than 6 μm is well-known [35]. However, aging behavior of an Al-Li-Mg alloy with UFG microstructure has never been examined.

2. Materials and Methods

The details of the chemical composition of the 1421Al, fabrication procedure and experimental techniques were reported elsewhere [2,32]. It is necessary to additionally report the following. ECAP of samples cut from hot extruded bar [32] was carried out in isothermal conditions at 325 °C using a die with a rectangular shape of channel cross-section [2,32]. The samples strained up to $\varepsilon\sim2$ and $\varepsilon\sim4$ were rotated by 90° around the *z*-axis in the same direction and by 180° around the *x*-axis between each pass, i.e., the modified route B_{CZ} [1] was used. The samples strained up to $\varepsilon\sim8$ were rotated by 180° around the *x*-axis. The pressing speed was approximately 3 mm/s. Next, the samples were subjected to solution treatment at 460 °C followed by oil quenching. Part of these samples was additionally aged at 120 °C for 6 h. Structural characterizations were only performed on samples subjected to post-ECAP or post-hot extrusion heat treatment.

Misorientations were determined using a FEI Quanta 200 3D SEM (FEI company, Hillsboro, OR, USA) fitted with an automated indexing of electron back scattering diffraction (EBSD) pattern collection system provided by EDAX (Mahwah, NJ, USA). Notably, thick and thin lines on the EBSD maps indicate the HAGBs ($\geq15°$) and LAGBs ($2°-15°$), respectively. Terms "grain" and "subgrain" are used for definition of crystallites, which are entirely delimited by HAGBs and LAGBs, respectively [31]. Term "(sub)grain" is used for definition of crystallites which are bounded partly by LAGBs and partly by HAGBs [31]. Thin foils were examined using a Jeol-2100 (Tokyo, Japan) transmission electron microscope (TEM) with a double-tilt stage and equipped with energy dispersive spectrum analysis (EDS) produced by Oxford Instruments, Ltd. (Oxfordshire, UK) at an accelerating potential of 200 kV. The dislocation density was estimated by counting the individual dislocations within (sub)grain interiors using a method, which is described in previous work [22] in detail, on at least six arbitrary selected typical TEM images for each sample.

Tensile specimens of 6 mm gauge length and 1.4×3 mm^2 cross-section were machined from the plates or initial extruded bar [32] with tension axis lying parallel to the last *x*-axis (Figure 2). Both types of samples were subjected to final T6 heat treatment or without heat treatment. These samples were tensioned to failure at ambient temperature at an initial strain rate of 5.6×10^{-3} s^{-1} using an Instron 5882 testing machine (Instron Ltd., Norwood, UK).

Precipitation sequence was examined using a SDT Q600 TA differential scanning calorimeter (DSC) (New Castle, PA, USA) where the specimens after the oil quenching were heated at rates of 10 K/min from 20 °C to 490 °C.

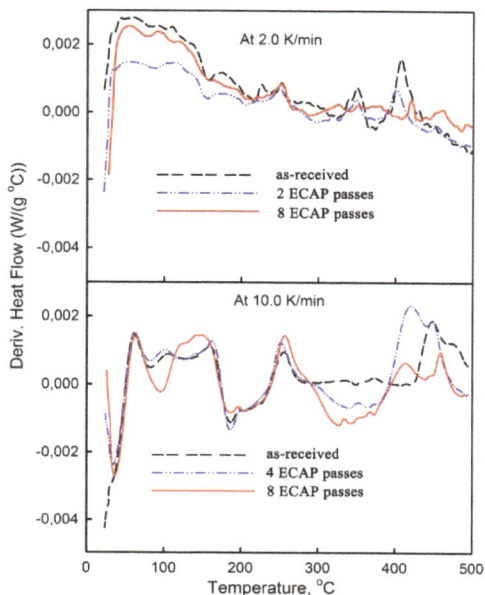

Figure 2. Differential scanning calorimeter thermograms of the solution-treated 1421 alloy at different heating rates.

3. Results

3.1. DSC Analysis

The DSC results are shown in Figure 2. For initial extruded 1421Al and this alloy subjected to ECAP up to total strains of ~2 and ~4, four exothermic peaks at ~60 °C, 90 °C–140 °C, 140 °C–170 °C and 230 °C–280 °C appear during the DSC scan. Therefore, the aging sequence Equation (1) is proposed.

The second and third exothermic peaks are usually attributed to precipitations of the δ'-phase and S_1-phase [33], respectively. It is worth noting that precipitation reactions in the 1421Al subjected to ECAP are shifted to lower temperatures. In contrast, the 1421Al subjected to ECAP up to a total strain of ~8 shows one exothermic peak at 120 °C–180 °C. This peak can be interpreted in terms of precipitation of S_1-phase. However, it is necessary to keep in mind that these DSC experiments are merely indications, and they cannot prove or disprove the formation of different phases.

3.2. Microstructure after Extrusion and Heat Treatment

Hot extrusion resulted in partially recrystallized microstructure in the 1421Al alloy (Figure 3a,b). Coarse grains with dimensions of ~171 and ~21 μm in longitudinal and transverse directions, respectively, are elongated toward the extrusion axis (Figure 3a) after final heat treatment. Chains of recrystallized grains having equiaxed shape and an average size of ~5 μm are located along separate initial HAGBs (Figure 3a). The portion of LAGBs is approximately 54%. It is worth noting the relatively high portion of HAGBs of 46% associated with mainly longitudinal initial boundaries. A well-defined subgrain structure was observed within highly elongated grains (Figure 3c); lattice dislocation density (ρ~4.5 × 10^{13} m^{-2}) is not so high (Figure 3c). Three different types of secondary phases were found after the T6 heat treatment. Coherent dispersoids of Al$_3$(Sc,Zr)-phase with an average size of ~20 nm and equiaxed shape are uniformly distributed within grain interiors [31,32]. Particles of the S_1-phase with an average size ranging from ~0.5 to ~0.7 μm and equiaxed shape are predominantly located on HAGBs or LAGBs (Figure 3c). In addition, coarse particles of the S_1-phase with an average size ranging from ~2 to ~4 μm comprise chains along extrusion axis [32]. Thus, bimodal size distribution of the S_1-phase particles takes place in the 1421Al in hot extruded conditions. Volume fraction of the S_1-phase is ~2 pct. Evidence for a dispersion of very fine coherent δ'-particles with a size of ~5 nm was found due to the TEM study at high magnification (Figure 3d). These particles exhibit specific contrast because the shear modulus of the δ-phase is less than that of the aluminum matrix, and, as a result, coherent stress induces lattice distortion within the soft dispersoids rather than in the strong matrix [33,34]. Thus, this structure is characterized by relatively low dislocation density, large grain size and high volume fraction of the δ'-phase.

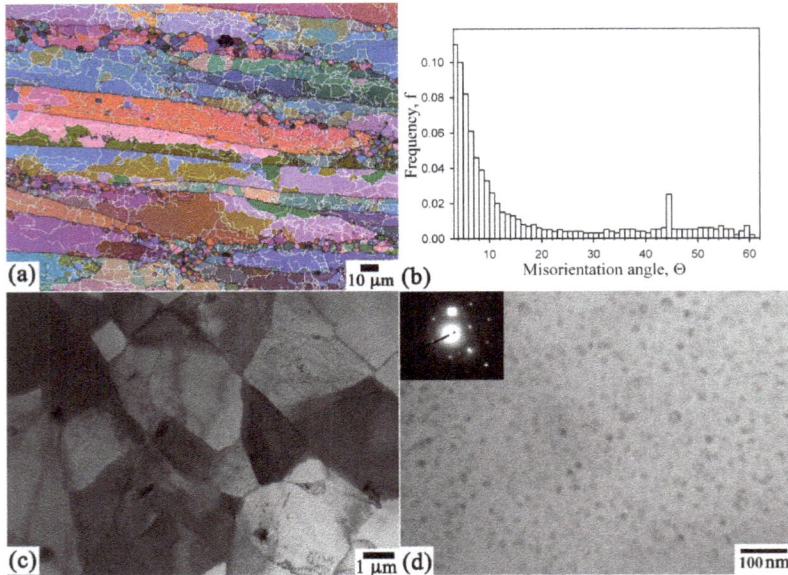

Figure 3. Typical microstructures of the 1421Al as-received and after standard heat treatment: (**a**) misorientation map; (**b**) distribution of misorientations; (**c**) transmission electron microscope (TEM) image of structure; and (**d**) δ'-phase particles (TEM).

3.3. Microstructure after ECAP and Heat Treatment

After $\varepsilon \sim 2$, a partially recrystallized microstructure can be observed (Figure 4a,b). The formation of this structure after a relatively low total strain compared with work [31] is aided by the use of initially extruded billet, a plate shape of samples and an optimal deformation temperature. Areas of grains with an average size of ~1 μm and (sub)grains alternate with areas of subgrains (Figure 4a). The second structure component is dominant. Volume fraction of true grains is relatively low (~20%). The portion of HAGBs is ~55%, and average misorientation is ~18° (Figure 4b). Grains and (sub)grains exhibit essentially equiaxed shape (the aspect ratio (*AR*) is ~1.25). In contrast, subgrains are elongated along the last extrusion direction; dimensions of subgrains in longitudinal and transverse directions are ~3.7 and ~2 μm, respectively (*AR* is ~1.9). The volume fraction of the S_1-phase is approximately 5%, and their size is the same as in the initial hot extruded state after aging (Figure 4e). Coarse particles of the S_1-phase with an average size ranging from ~2 to ~4 μm comprise chains along extrusion axis (Figure 4) as in the hot extruded condition. Particles of the S_1-phase located at triple junctions and on HAGBs are dominant (Figure 4e). However, these particles could be observed in separate grains; most HAGBs are free from the particles of the S_1-phase. A high density of dispersoids of the δ'-phase with an average size of ~6 nm was found within grain interiors (Figure 4f). No δ'-phase was found near coarse particles of the S_1-phase. Relatively high density of lattice dislocations (ρ~6 × 10^{14} m^{-2}) was observed. The number of dislocations within subgrain structure is higher than that within grains by a factor of ~4. It is worth noting that evidence for grain boundary dislocation was found. Therefore, high temperature annealing did not lead to their adsorption by grain boundaries [35]. Thus, this structure is characterized by high dislocation density, moderate grain size and moderate density of the δ'-phase.

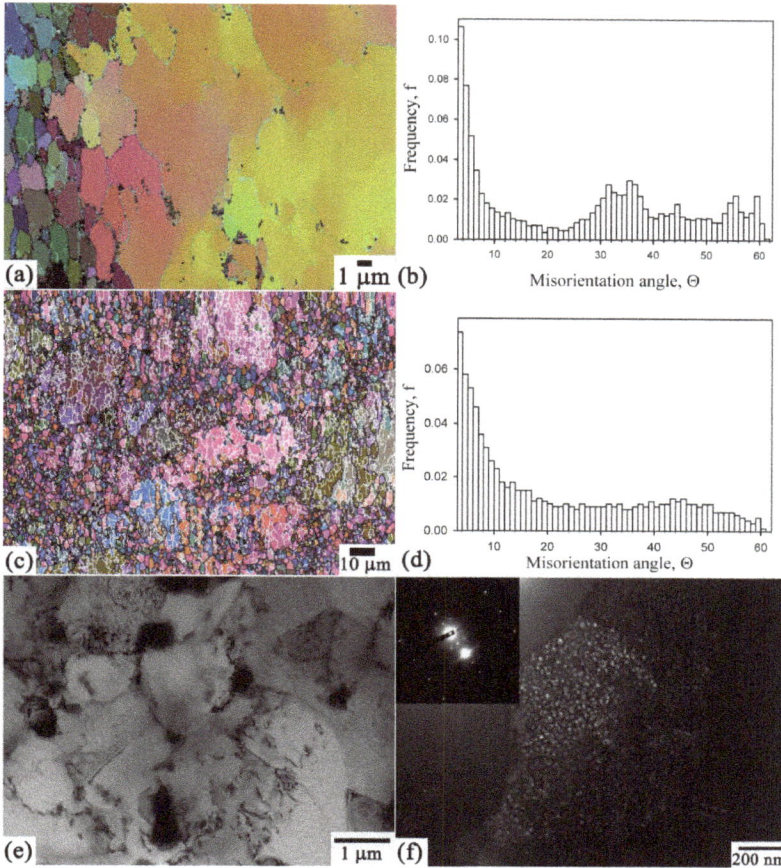

Figure 4. (**a,c**) Electron back scattering diffraction maps showing the high and low angle boundaries (dark and light lines, respectively) and boundary misorientation histograms (**b,d**) for the 1421 Al alloy deformed by equal-channel angular pressing to strains of (**a,b**) 2; (**c,d**) two and standard heat treatment; and (**e,f**) transmission electron microscope image of structure ε~2 and standard heat treatment.

Pressing to a strain of ε~4 results in a recrystallized structure, the portion of unrecrystallized grains, which are highly elongated along the last extrusion direction, is minor (Figure 5a,b). The fraction of recrystallized grains with an average size of ~1 μm is very high, ~90%. It seems that the effect of annealing under solution treatment on this UFG is insignificant. It seems that the structure evolved during ECAP remains nearly unchanged. The population of HAGBs is ~82%; the average misorientation is 31° (Figure 5a,b). There exists a large difference in dislocation density within crystallites. True grains contain a low dislocation density (ρ~4 × 10^{13} m^{-2}), whereas high dislocation density of ($\rho\sim1\times10^{15}$ m^{-2}) is observed within (sub)grains and unrecrystallized grains. The formation of recrystallized structure strongly affects the origin and distribution of precipitations. S$_1$-phase particles with a size ranging from 0.2 to ~0.4 µm are uniformly distributed within areas of fine grains. S$_1$-phase particles are located on HAGBs and even within grain interiors (Figure 5e). It is worth noting that well-defined shells of unknown origin can be observed around S$_1$-phase particles (Figure 5e), In addition, coarse S$_1$-phase particles are also uniformly distributed within the aluminum matrix (Figure 5). Their overall volume fraction attains ~16%. There is weak evidence for the appearance of dispersions of the δ′-phase within micron scale grains (Figure 5f); their volume fraction is negligible. Thus, this structure is characterized by moderate dislocation density, UFG structure and low density of the δ′-phase.

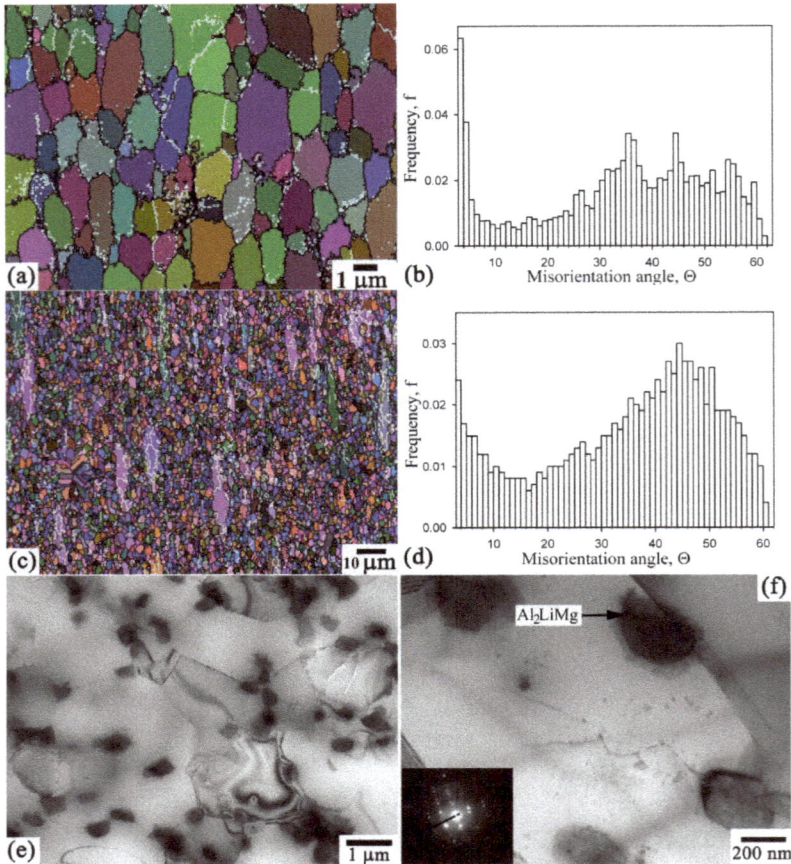

Figure 5. (**a,c**) Electron back scattering diffraction maps showing the high and low angle boundaries (dark and light lines, respectively) and boundary misorientation histograms (**b,d**) for the 1421Al alloy deformed by equal-channel angular pressing to strains of (**a,b**) 4; (**c,d**) 4 and standard heat treatment; and (**e,f**) transmission electron microscope image of structure ε~4 and standard heat treatment.

After ε~8, the volume fraction of grains is approximately 85%, the population of HAGBs is ~72%, the average misorientation is ~29°, and the volume fraction of the S$_1$-phase is ~18% (Figure 6). Most of the crystallites are grains containing low density of lattice dislocations ($\rho\sim2\times10^{13}$ m^{-2}) (Figure 6c).

Dimensions of grains and (sub)grains are essentially the same (1.4 μm, $AR \sim 1.2$). The structures after $\varepsilon \sim 4$ and $\varepsilon \sim 8$ are distinctly distinguished by distribution and size of the S_1-phase (Figure 6c). Particles of this phase have an average size of ~0.8 μm. They are uniformly distributed within the aluminum matrix located both on HAGBs and within grain interiors as well (Figure 6c). As a result, this structure can be considered to be a duplex one. Dispersion of the δ'-phase was not found within grains (Figure 6c). Thus, this structure is characterized by low dislocation density, fine grains and lack of δ'-phase dispersoids within crystallites.

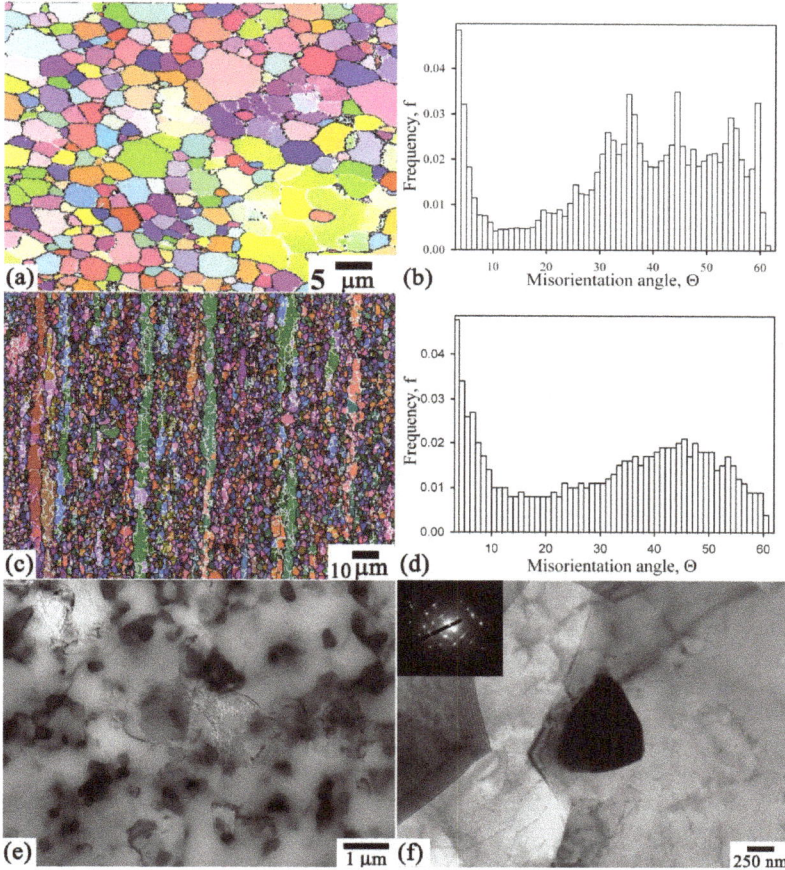

Figure 6. (**a,c**) Electron back scattering diffraction maps showing the high and low angle boundaries (dark and light lines, respectively) and boundary misorientation histograms (**b,d**) for the 1421 alloy deformed by equal-channel angular pressing to strains of (**a,b**) 8; (**c,d**) 8 and standard heat treatment; and (**e,f**) transmission electron microscope image of structure $\varepsilon \sim 8$ and standard heat treatment.

3.4. Mechanical Properties

Figure 7a shows the engineering stress-strain curves of the 1421Al in the as-received state and after ECAP subjected to final T6 heat treatment. Values of YS, ultimate tensile strength (UTS), total elongation and uniform elongation assessed from the strain-stress curves in Figure 7a are summarized in Table 1. It is clear that the shapes of the σ-ε curves for all states of the 1421Al are similar. Extensive strain hardening takes place after reaching a yield stress. As a result, uniform elongation occurs up to failure providing total elongation higher than 10% for all samples subjected to ECAP with different strains. It is worth noting that this type of σ-ε curves is attributed to materials with a grain size close to 1 μm having low dislocation density ($\rho < 10^{14}$ m^{-2}) and no residual stress. The values of ductility obtained can be considered sufficient [31] for Al-Li-Mg alloys. It is possible to conclude that only the T6 heat treatment following ECAP (Figure 1c) provides moderate ductility associated with significant strain hardening [24]; values of uniform elongation and elongation-to-failure are almost the same.

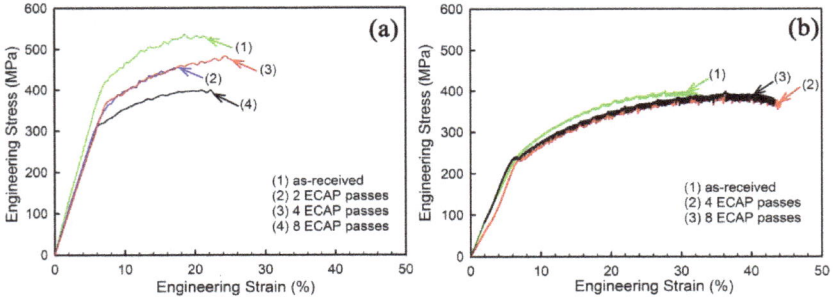

Figure 7. The engineering stress-strain curves of the 1421Al alloy at room temperature: (**a**) without and (**b**) with standard heat treatment.

Table 1. Yield stress (YS, MPa), ultimate tensile strength (UTS, MPa), ductility (δ, %) and uniform elongation (e_u, %) of the 1421Al alloy subjected to T6 heat treatment.

Conditions	YS	UTS	YS/UTS	δ	e_u
As-received	400	522	0.76	14.3	13.5
2 ECAP passes	329	480	0.68	11.3	11.3
4 ECAP passes	378	490	0.77	13.1	12
8 ECAP passes	322	451	0.71	15	15

It is observed that ECAP results in decreased yield stress and ultimate strength in the heat treated state in comparison with conventional extrusion. The 1421Al in the as-received condition exhibits the highest YS and UTS (Table 1). ECAP with a total strain of ~4 provides 6% decreases in YS and UTS (Table 1). The YS decrease is 21% and the UTS decrease is 8% in the material subjected to ECAP with a total strain of ~2 (Table 1). The effect of ECAP with a total strain of ~8 on YS is almost the same; the UTS decrease is 14% (Table 1). Thus ECAP processing of the 1421Al through the fourth route of TMP leads to minor degradation in strength; ductility remains almost unchanged. It is worth noting that the value of YS/UTS for all states of the 1421Al is close to 0.6 (Table 2) that is typical for natural aging of age-hardenable aluminum alloys if the formation of GP zones takes place.

Table 2. Yield stress (YS, MPa), ultimate strength (UTS, MPa), ductility (δ, %) and uniform elongation (e_u, %) of the 1421Al alloy without heat treatment.

Conditions	YS	UTS	YS/UTS	δ	e_u
As-received	241	398	0.6	19	17
4 ECAP passes	228	388	0.59	35	34
8 ECAP passes	226	397	0.57	29	28

Microstructural observations of the samples strained up to failure showed that there exist two types of crack initiation. In the samples subjected to ECAP with ε~2, the crack nucleation occurs in the vicinity of coarse S$_1$-phase particles comprising chains (Figure 8a). As a result, an interlinkage of cracks easily occurs resulting in premature fracture. In the samples subjected to ECAP with $\varepsilon \sim 4$, most of the coarse S_1-phase particles dissolved and separate particles of this phase are distributed uniformly in the aluminum matrix. Crack nucleation occurs in the vicinity of these particles (Figure 8b) or primary precipitations of $Al_3(Sc,Zr)$-phase (Figure 8c). However, no crack interlinkage takes place (Figure 8b,c); this material exhibits increased ductility.

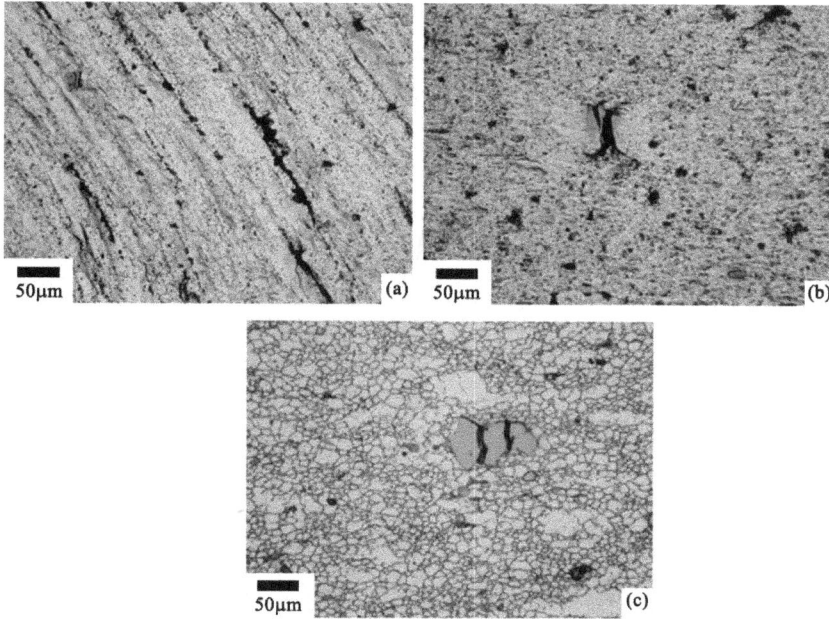

Figure 8. Typical microstructures of the samples strained up to failure: (**a**) after equal-channel angular pressing (ECAP) with $\varepsilon \sim 2$; and (**b**,**c**) after ECAP with $\varepsilon \sim 4$.

Figure 7b shows the engineering stress-strain curves of the 1421Al without heat treatment, and values of YS, UTS, total elongation and uniform elongation are presented in Table 2. All of the curves essentially have a similar shape; extensive strain hardening takes place up to failure. The serrated flow is manifested as the repeating oscillations on the stress-strain curves (Figure 7b). This phenomenon is generally associated with the Portevin-Le Chatelier (PLC) effect attributed to dynamic strain aging [29]. Characteristics of strength are almost the same for all states of the 1421Al (Table 2). However, ECAP provides 100% increases in ductility. Thus, the difference in YS and UTS between different states of the 1421Al associates mainly with differences in precipitation behavior. Notably, the value of YS/UTS for all states of the 1421Al is close to 0.75 (Table 1), which is typical for artificial aging of age-hardenable aluminum alloys if the formation of nanoscale dispersoids with coherent interface boundaries takes place.

4. Discussion

It is observed that the deformed structure developed in the 1421Al containing coherent Al3Sc dispersoids under ECAP is retained under solution treatment. The Al_3Sc nanoscale particles are very effective in pinning HAGBs and LAGBs, suppressing recrystallization and recovery processes at 460 °C. Therefore, the fourth route of TMP is suitable for application to this alloy. However, ECAP provided no positive influence on mechanical properties. This fact is caused by a strong effect of UFG on aging response in Al-Li-Mg alloys.

It is apparent that the 1421Al with different types of deformation structures without heat treatment shows essentially similar strength due to dynamic strain aging. It is known [36] that jerky flow can be associated with the formation of clouds of solutes around the dislocations (normal PLC) or precipitations of coherent dispersoids on dislocation (pseudo-PLC). The origin of PLS effect was not examined in this study. However, it is assumed that the appearance of jerky flow is attributed to precipitation of the δ'-phase on dislocations [16]. As a result, dislocations become immobile, and LAGBs with misorientation higher than 6° start to play a role of effective barrier for gliding dislocation. In this case, the subgrains with the LAGBs can be considered as grains in the Hall-Petch equation, which states that the YS, σ_y, is given by

$$\sigma_y = \sigma_0 + k_y d^{-1/2} \tag{2}$$

where σ_0 is termed the friction stress and k_y is a constant of yielding [1]. However, subgrain structure as well as UFG structure provide nearly the same grain size strengthening in accordance with Equation (2). Perhaps this is why the YS values for 1421Al with well-defined subgrain structure, partially recrystallized structure and fully recrystallized structure are almost the same in initial condition without heat treatment. The difference in the YS arises due to heat treatment.

After heat treatment, no jerky flow was observed in all states of the 1421Al because all Li precipitated in the form of different phases indicated in the first scheme (Figure 1a). In an extruded bar of the 1421Al, the precipitation sequence is described by Equation (1). As a result, at 120 °C, the precipitation of nanoscale particles of coherent δ'-phase provides extensive dispersion strengthening. δ'-phase free zones are evolved only in the vicinity of coarse S_1-phase particles. The last phase is a stable phase in comparison with the δ'-phase. As a result, under any aging conditions, the δ'-phase dispersoids will be dissolved around HAGBs where stable S_1-phase nucleates. The rate of this process is dependent on diffusion path, and, therefore, aging temperature and distribution of S_1-phase particles within aluminum matrix will strongly affect this process.

The S_1-phase tends to precipitate in a strong heterogeneous manner on HAGBs. As a result, in the 1421Al subjected to ECAP with $\varepsilon\sim4$, the S_1-phase precipitates uniformly on boundaries of micron scale grains. Diffusion path is very short and almost all dispersoids of the δ'-phase dissolve under aging even at low temperature. In this state of the 1421Al, the dispersion strengthening associated with the S_1-phase plays an unimportant role in the strength of this alloy; particles of this phase are relatively coarse, and, in addition, most of them are located on HAGBs. As a result, these particles are not able to impede dislocation motion. The heat treatment provides a 62% increase in YS; however, the origin of this increase is unknown. We can assume only that S_1-phase particles may hinder operation of grain boundary dislocation sources, and sparse dispersoids of the δ'-phase effectively impede dislocation glide within micron scale grains. It is worth noting that the value of YS/UTS~0.76 (Table 1) for this state of the 1421Al and for the hot extruded state is almost the same despite significant differences in origin and distribution of secondary phases.

The formation of coarser particles of the S_1-phase in the 1421Al subjected to ECAP with $\varepsilon\sim8$ provides a 42% increase in YS. Extensive localization of plastic deformation resulting in cracking along the chains of coarse S_1-phase particles takes place in the 1421Al subjected to ECAP with $\varepsilon\sim2$. This localization occurs within the δ'-phase free zones around the S_1-phase particles. As a result, the heat treatment provides a 44% increase in YS that is significantly less than that (66%) for the hot extruded state of the 1421Al.

Thus, to achieve a positive effect from extensive grain refinement by ECAP on the strength of age-hardenable aluminum alloys, it is necessary to modify chemical composition and aging regimes to prevent the formation of stable phases on HAGBs undergoing final heat treatment.

5. Conclusions

It was shown that the post-ECAP standard heat treatment consisting of solution treatment, quenching and final aging can be applied to the 1421Al with different types of deformation structures including an ultra-fine grained structure produced by ECAP. It was found that deformation structures evolved during ECAP processing are stable under solution treatment. However, aging of the 1421Al with these structures is not capable of providing significant precipitation hardening. The formation of coarse particles of the stable S1-phase on boundaries of micron scale grains and even within interiors of these grains prevents the precipitation of nanoscale dispersoids of the δ'-phase under aging conditions that leads to a decrease in strength of the alloy.

Acknowledgments: The financial support received from the Ministry of Education and Science, Moscow, Russia, (Belgorod State University project No. 14.587.21.0018 (RFMEFI58715X0018)) is acknowledged. The main results were obtained by using equipment from the Joint Research Center, Belgorod State University.

Author Contributions: A.M. and R.K. conceived and designed the experiments; A.M. performed the experiments. A.M. and R.K. discussed and analyzed the obtained results.

Conflicts of Interest: The authors declare no conflict of interest.

References

1. Valiev, R.Z.; Langdon, T.G. Principles of equal-channel angular pressing as a processing a processing tool for grain refinement. *Prog. Mater. Sci.* **2006**, *51*, 881–981. [CrossRef]
2. Kaibyshev, R.; Tagirov, D.; Mogucheva, A. Cost-Affordable Technique Involving Equal Channel Angular Pressing for the Manufacturing of Ultrafine Grained Sheets of an Al-Li-Mg-Sc Alloy. *Adv. Eng. Mater.* **2010**, *12*, 735–739. [CrossRef]
3. Ferrasse, S.; Segal, V.M.; Alford, F.; Kardokus, J.; Strothers, S. Scale up and application of equal-channel angular extrusion for the electronics and aerospace industries. *Mater. Sci. Eng. A* **2008**, *493*, 130–140. [CrossRef]
4. Yuzbekova, D.; Mogucheva, A.; Kaibyshev, R. Superplasticity of ultrafine-grained Al-Mg-Sc-Zr alloy. *Mater. Sci. Eng. A.* **2016**, *675*, 228–242. [CrossRef]
5. Murayama, M.; Horita, Z.; Hono, K. Microstructure of two-phase Al-1.7 at. % Cu alloy deformed by equal-channel angular pressing. *Acta Mater.* **2001**, *49*, 21–29. [CrossRef]
6. Radetic, T.; Popovic, M.; Romhanji, E.; Verlinden, B. The effect of ECAP and Cu addition on the aging response and grain substructure evolution in an Al-4.4 wt. % Mg alloy. *Mater. Sci. Eng. A* **2010**, *527*, 634–644. [CrossRef]
7. Kim, J.K.; Kim, H.K.; Park, J.W.; Kim, W.J. Large enhancement in mechanical properties of the 6061 Al alloys after a single pressing by ECAP. *Scr. Mater.* **2005**, *53*, 1207–1211. [CrossRef]
8. Kim, W.J.; Wang, J.Y. Microstructure of the post-ECAP aging processed 6061 Al alloys. *Mater. Sci. Eng. A* **2007**, *464*, 23–27. [CrossRef]
9. Kim, W.J.; Chung, C.S.; Ma, D.S.; Hong, S.I.; Kim, H.K. Optimization of strength and ductility of 2024 Al by equal channel angular pressing (ECAP) and post-ECAP aging. *Scr. Mater.* **2003**, *49*, 333–338. [CrossRef]
10. Shaterani, P.; Zarei-Hanzaki, A.; Fatemi-Varzaneh, S.M.; Hassas-Irani, S.B. The second phase particles and mechanical properties of 2124 aluminum alloy processed by accumulative back extrusion. *Mater. Des.* **2014**, *58*, 535–542. [CrossRef]
11. Cabibbo, M. Partial dissolution of strengthening particles induced by equal channel angular pressing in an Al-Li-Cu alloy. *Mater. Charact.* **2012**, *68*, 7–13. [CrossRef]
12. Namdar, M.; Jahromi, S.A.J. Influence of ECAP on the fatigue behavior of age-hardenable 2xxx aluminum alloy. *Int. J. Miner. Metall. Mater.* **2015**, *22*, 285–291. [CrossRef]
13. Kotan, G.; Tan, E.; Kalay, Y.E.; Gür, C.H. Homogenization of ECAPed Al 2024 alloy through age-hardening. *Mater. Sci. Eng. A* **2013**, *559*, 601–606. [CrossRef]
14. Roshan, M.R.; Jahromi, S.A.J.; Ebrahimi, R. Predicting the critical pre-aging time in ECAP processing of age-hardenable aluminum alloys. *J. Alloy. Compd.* **2011**, *509*, 7833–7839. [CrossRef]
15. Shaeri, M.H.; Shaeri, M.; Ebrahimi, M.; Salehi, M.T.; Seyyedein, S.H. Effect of ECAP temperature on microstructure and mechanical properties of Al-Zn-Mg-Cu alloy. *Prog. Nat. Sci. Mater. Int.* **2016**, *26*, 182–191. [CrossRef]

16. Salem, H.G.; Goforth, R.E.; Hartwig, K.T. Influence of intense plastic straining on grain refinement, precipitation, and mechanical properties of Al-Cu-Li-based alloys. *Metall. Mater. Trans.* **2003**, *34A*, 1153–1161. [CrossRef]

17. Zhao, Y.H.; Liao, X.Z.; Jin, Z.; Valiev, R.Z.; Zhu, Y.T. Microstructures and mechanical properties of ultrafine grained 7075 Al alloy processed by ECAP and their evolutions during annealing. *Acta Mater.* **2004**, *52*, 4589–4599. [CrossRef]

18. Mohamed, I.F.; Yonenaga, Y.; Lee, S.; Edalati, K.; Horita, Z. Age hardening and thermal stability of Al-Cu alloy processed by high-pressure torsion. *Mater. Sci. Eng. A* **2015**, *627*, 111–118. [CrossRef]

19. Kim, W.J.; Kim, J.K.; Kim, H.K.; Park, J.W.; Jeong, Y.H. Effect of post equal-channel-angular-pressing aging on the modified 7075 Al alloy containing Sc. *J. Alloy. Compd.* **2008**, *450*, 222–228. [CrossRef]

20. Hirosawa, S.; Hamaoka, T.; Horita, Z.; Lee, S.; Matsuda, K.; Terada, D. Methods for Designing Concurrently Strengthened Severely Deformed Age-Hardenable Aluminum Alloys by Ultrafine-Grained and Precipitation Hardenings. *Metall. Mater. Trans. A* **2013**, *44A*, 3921–3933. [CrossRef]

21. Roven, H.J.; Nesboe, H.; Werenskiold, J.C.; Seibert, T. Mechanical properties of aluminium alloys processed by SPD: Comparison of different alloy systems and possible product areas. *Mater. Sci. Eng. A* **2005**, *410–411*, 426–429. [CrossRef]

22. Mallikarjuna, C.; Shashidhara, S.M.; Mallik, U.S. Evaluation of grain refinement and variation in mechanical properties of equal-channel angular pressed 2014 aluminum alloy. *Mater. Des.* **2009**, *30*, 1638–1642. [CrossRef]

23. Sabirov, I.; Estrin, Y.; Barnett, M.R.; Timokhina, I.; Hodgson, P.D. Enhanced tensile ductility of an ultra-fine-grained aluminum alloy. *Scr. Mater.* **2008**, *58*, 163–166. [CrossRef]

24. Valiev, R.Z.; Murashkin, M.Y.; Bobruk, E.V.; Raab, G.I. Grain refinement and mechanical behavior of the Al Alloy, subjected to the new SPD technique. *Mater. Trans.* **2009**, *50*, 87–91. [CrossRef]

25. Chinh, N.Q.; Gubicza, J.; Czeppe, T.; Lendvai, J.; Xu, C.; Valiev, R.Z.; Langdon, T.G. Developing a strategy for the processing of age-hardenable alloys by ECAP at room temperature. *Mater. Sci. Eng. A* **2009**, *516*, 248–252. [CrossRef]

26. Kim, J.K.; Jeong, H.G.; Hong, S.I.; Kim, Y.S.; Kim, W.J. Effect of aging treatment on heavily deformed microstructure of a 6061 aluminum alloy after equal channel angular pressing. *Scr. Mater.* **2001**, *45*, 901–907. [CrossRef]

27. Kim, W.J.; Kim, J.K.; Park, T.Y.; Hong, S.I.; Kim, D.I.; Kim, Y.S.; Lee, J.D. Enhancement of strength and superplasticity in a 6061 Al alloy processed by equal-channel-angular-pressing. *Metall. Mater. Trans. A* **2002**, *33*, 3155–3164. [CrossRef]

28. Wang, Z.C.; Prangnell, P.B. Microstructure refinement and mechanical properties of severely deformed Al-Mg-Li alloys. *Mater. Sci. Eng A* **2002**, *328*, 87–97. [CrossRef]

29. Kaibyshev, R.; Shipilova, K.; Musin, F.; Motohashi, Y. Continuous dynamic recrystallization in an Al-Li-Mg-Sc alloy during equal-channel angular extrusion. *Mater. Sci. Eng.* **2005**, *21*, 408–418. [CrossRef]

30. Musin, F.; Kaibyshev, R.; Motohashi, Y.; Sakuma, T.; Itoh, G. High strain rate superplasticity in an Al-Li-Mg alloy subjected to equal-channel angular extrusion. *Mater. Trans.* **2002**, *43*, 2370–2377. [CrossRef]

31. Kaibyshev, R.; Shipilova, K.; Musin, F.; Motohashi, Y. Continuous dynamic recrystallization in an Al-Li-Mg-Sc alloy during equal-channel angular extrusion. *Mater. Sci. Eng. A* **2005**, *396*, 341–351. [CrossRef]

32. Mogucheva, A.A.; Kaibyshev, R.O. Structure and properties of aluminum alloy 1421 after equal-channel angular pressing and isothermal rolling. *Phys. Met. Metall.* **2008**, *106*, 424–433. [CrossRef]

33. Fridlyander, I.N.; Chuistova, K.V.; Berezina, A.L.; Kolobnev, N.I. *Aluminum–Lithium Alloys. Structure and Properties*; Naukova Dumka: Kiev, Ukraine, 1992; p. 192.

34. Lee, S.; Berbon, P.B.; Furukawa, M.; Horita, Z.; Nemoto, M.; Tsenev, N.K.; Valiev, R.Z.; Langdon, T.G. Developing superplastic properties in an aluminum alloy through severe plastic deformation. *Mater. Sci. Eng. A* **1999**, *272*, 63–72. [CrossRef]

35. Kaibyshev, O.A. *Superplasticity of Alloys, Intermetallics, and Ceramics*; Springer: Berlin, Germany, 1992; p. 149.

36. Brechet, Y.; Estrin, Y. On the influence of precipitation on the Portevin-Le Chatelier effect. *Acta Metall. Mater.* **1995**, *43*, 955–963. [CrossRef]

metals

MDPI

Article

Deformation Characteristic and Constitutive Modeling of 2707 Hyper Duplex Stainless Steel under Hot Compression

Huabing Li [1,*], Weichao Jiao [1], Hao Feng [1,*], Xinxu Li [1], Zhouhua Jiang [1], Guoping Li [2], Lixin Wang [2], Guangwei Fan [2] and Peide Han [3]

[1] School of Metallurgy, Northeastern University, Shenyang 110819, China; jiaowcneu@163.com (W.J.);
    lxx20110180@163.com (X.L.); jiangzh@smm.neu.edu.cn (Z.J.)
[2] Technology Center of Taiyuan Iron and Steel Croup Co. Ltd., Taiyuan 030001, China;
    ligp@tisco.com.cn (G.L.); wanglx@tisco.com.cn (L.W.); fangw@tisco.com.cn (G.F.)
[3] College of Materials Science and Engineering, Taiyuan University of Technology, Taiyuan 030024, China;
    hanpeide@126.com
*   Correspondence: lihb@smm.neu.edu.cn (H.L.); fenghao241@163.com (H.F.);
    Tel.: +86-24-8368-9580 (H.L. & H.F.); Fax: +86-24-2389-0559 (H.L. & H.F.)

Academic Editor: Soran Birosca
Received: 18 August 2016; Accepted: 5 September 2016; Published: 12 September 2016

Abstract: Hot deformation behavior and microstructure evolution of 2707 hyper duplex stainless steel (HDSS) were investigated through hot compression tests in the temperature range of 900–1250 °C and strain rate range of 0.01–10 s^{-1}. The results showed that the flow behavior strongly depended on strain rate and temperature, and flow stress increased with increasing strain rate and decreasing temperature. At lower temperatures, many precipitates appeared in ferrite and distributed along the deformation direction, which could restrain processing of discontinuous dynamic recrystallization (DRX) because of pinning grain boundaries. When the temperature increased to 1150 °C, the leading softening behaviors were dynamic recovery (DRV) in ferrite and discontinuous DRX in austenite. When the temperature reached 1250 °C, softening behavior was mainly DRV in ferrite. The increase of strain rate was conducive to the occurrence of discontinuous DRX in austenite. A constitutive equation at peak strain was established and the results indicated that 2707 HDSS had a higher Q value (569.279 kJ·mol^{-1}) than other traditional duplex stainless steels due to higher content of Cr, Mo, Ni, and N. Constitutive modeling considering strain was developed to model the hot deformation behavior of 2707 HDSS more accurately, and the correlation coefficient and average absolute relative error were 0.992 and 5.22%, respectively.

Keywords: 2707 hyper duplex stainless steel; hot deformation behavior; hot compression; microstructure; constitutive modeling

1. Introduction

Duplex stainless steel (DSS) with approximately equal volume fractions of ferrite (α) and austenite (γ) phases exhibits a good combination of mechanical properties and corrosion resistance. For this reason, DSSs are considered an excellent choice in various industrial applications, particularly in the oil and gas industry, petrochemical, and chemical processing [1–3]. With the development of industry, the requirement for the performance of DSSs becomes higher. 2707 hyper duplex stainless steel (HDSS) is a special class of DSS that contains high levels of Cr, Mo, Ni, and N, contributing to its excellent mechanical properties and a high value (45–50) of pitting resistance equivalent number (PREN) [4]. Therefore, 2707 HDSS can extend the application range of DSS into even more aggressive conditions [5].

However, the coexistence of austenite and ferrite with different deformation responses in these alloys makes the hot processing complicated [6].

Hot working is an important step in producing DSSs. It is known that ferrite is characterized by high stacking fault energy (SFE) and, therefore, undergoes dynamic recovery (DRV). On the contrary, austenite, having low SFE, undergoes only limited DRV, and dynamic recrystallization (DRX) comes into operation when the dislocation density reaches a critical value [7–10]. In this respect, the incoherence containing thermal expansion coefficient and flow response mechanism between the two constituents easily reduces hot workability of DSSs and then leads to the formation of defects on the surface of products [11]. Moreover, higher alloy content than traditional DSSs and higher sensitivity of secondary phase precipitation in the hot deformation process of 2707 HDSS further increase its processing difficulty. Therefore, it is necessary to investigate the hot deformation behavior of 2707 HDSS.

High temperature flow behavior, as the macro-reflection of the mechanisms of microscopic deformation and microstructure evolution, is often represented by making use of constitutive modeling [12]. In recent years, various analytical, phenomenological, and empirical models have been developed to predict the constitutive behavior of a wide range of metals and alloys [13–16]. Among these models, the hyperbolic sine law in the Arrhenius equation has been widely applied for engineering applications. However, it lacks suitability for materials having obvious dynamic softening during hot deformation, since the influence of strain is not included. Accordingly, the researchers have put their efforts on the strain compensated Arrhenius model [17,18], and the prediction accuracy is greatly enhanced.

In the present study, hot deformation behavior and microstructure evolution of 2707 HDSS were investigated by conducting hot compression tests with varying deformation temperatures and strain rates. The effects of hot working parameters (strain rate and temperature) on the mechanical properties and microstructures were analyzed. Additionally, the experimentally measured true stress-strain curves were employed to establish a constitutive model using the strain compensated Arrhenius model. Moreover, the prediction accuracy of this model was evaluated by comparing the calculated and experimental flow stresses.

2. Experimental Procedures

The experimental 2707 HDSS was melted using a 25 kg vacuum induction melting furnace (Jinzhou Yuanteng Electric Furnace Technology Co., Ltd., Shenyang, China). The chemical composition (wt %) of the 2707 HDSS is as follows: 0.0044 C, 0.42 Si, 1.11 Mn, 26.83 Cr, 7.14 Ni, 4.88 Mo, 0.39 N, 0.97 Cu, 0.97 Co, 0.005 P, 0.003 S and Fe in balance. The ingot was forged into 110 mm × 30 mm billet in the temperature range of 1100–1200 °C. The cylindrical specimens with 8 mm diameter and 12 mm height were prepared from the hot forged billet with the longitudinal axis parallel to the deformation direction. A chromel-alumel thermocouple (Shenyang Dongda Sensor Technology Co., Ltd., Shenyang, China) was embedded in the middle of the specimen to monitor the temperature during hot compression. In order to reduce friction between the sample and thermal simulation testing machine squeeze head, graphite flakes with 0.25 mm thickness and 12 mm diameter were clamped between them. And tantalum sheets (Changsha Nanfang Tantalum Niobium Co., Ltd., Changsha, China) were used to prevent adhesion between the sample and the squeeze head [19].

Before the compression, all specimens were reheated to 1250 °C and held for 4 min to achieve homogenized microstructure. The microstructure of 2707 HDSS after reheating is exhibited in Figure 1, in which the volume fraction of ferrite and austenite is about 54% and 45%, respectively. Subsequently, the specimens were cooled to compression temperature at the rate of 5 °C/s and held for 30 s to eliminate temperature gradient. To investigate the effects of deformation temperature and strain rate, the uniaxial compression tests were performed on a Thermecmaster-Z tester (Fuji Electronic Industrial Co., Ltd., Tsurugashima, Japan) in the temperature range of 900–1250 °C with intervals of 50 °C and strain rate range of 0.01–10 s^{-1} with intervals of an order of magnitude, respectively. The specimens were compressed up to a total strain of 0.8 followed by immediate quenching with forced gas to retain the deformed microstructure. Hot deformed specimens were cut along the longitudinal axis, mechanically ground, and polished. Following these steps, electrolytic etching in 10 wt % KOH solution was adopted to reveal microstructures. The microstructures of deformed specimens were characterized using an optical microscope (OM, Olympus Corporation, Tokyo, Japan) and scanning electron microscope (SEM, Japan Electron Optics Laboratory, Tokyo, Japan), and the composition of the constituents was analyzed using an energy dispersive spectroscope (EDS, Bruker, Berlin, Germany). In addition, a transmission electron microscope (TEM, FEI, Hillsboro, OR, USA) was used to confirm the precipitates and substructure.

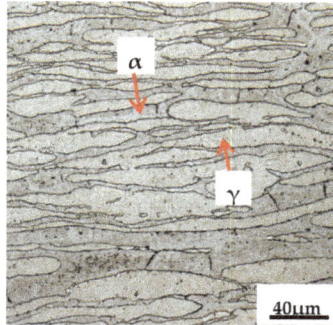

Figure 1. Optical microstructure of 2707 hyper duplex stainless steel (HDSS) after reheating.

3. Results and Discussions

3.1. Flow Stress-Strain Cures

The flow stress-strain curves of 2707 HDSS in the temperature range of 900–1250 °C and strain rate range of 0.01–10 s^{-1} are shown in Figure 2. Obviously, the flow stress was strongly dependent on both the deformation temperature and strain rate. Meanwhile, the peak stress increased with increasing strain rate and decreasing temperature. The results revealed that all the curves showed similar characteristics of work hardening at the early stage of deformation. The initial increment of stress as a result of work hardening is attributed to an increase of dislocation density [20].

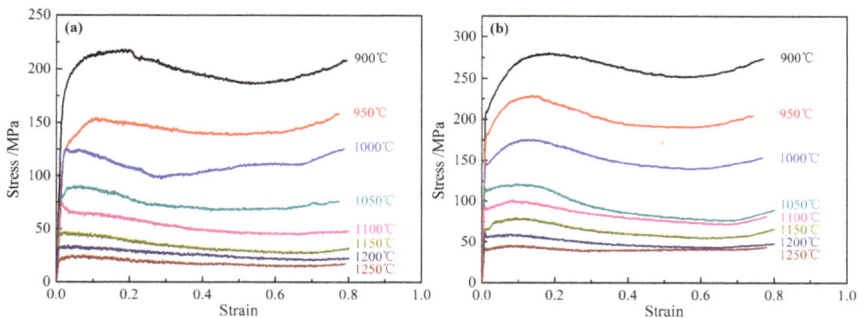

Figure 2. *Cont.*

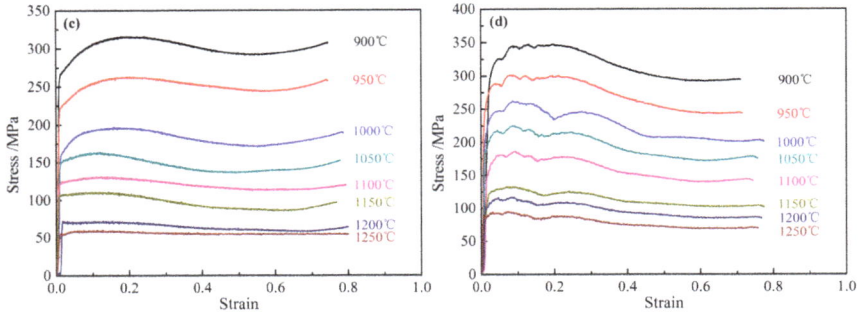

Figure 2. Flow stress vs. strain curves of 2707 HDSS under different strain rates: (**a**) 0.01 s^{-1}; (**b**) 0.1 s^{-1}; (**c**) 1 s^{-1}; (**d**) 10 s^{-1}.

After work hardening at the early stage of deformation, the curves presented different features in the different deformation conditions. The different features of flow curves can be interpreted considering the coexistence of different constituents, i.e., ferrite and austenite with different softening mechanisms and interaction of strain partitioning [21]. When the temperature was higher than 1100 °C and strain rate was less than 1 s^{-1}, the true stress-strain curves exhibited DRV characteristics without peak stress, as shown in Figure 2a–c. It could be explained that strain was mostly accommodated in ferrite because ferrite had higher SFE so that arrangement and annihilation of dislocations would occur more easily through DRV [22]. Flow curves with peak stress could be observed at lower temperatures, which may be attributed to the DRX or flow instability [23]. When strain rate was 10 s^{-1}, the flow curves exhibited significant waviness at strain lower than 0.2 as shown in Figure 2d, which was attributed to severely uneven distribution of strain in the ferrite and austenite phases at very high strain rates [24]. At the last stage of deformation, flow stress rose again because the effect of softening behavior was weaker than work hardening.

3.2. Analysis of Microstructure Evolution

The microstructures of 2707 HDSS under different deformation temperatures with strain of 0.8 and strain rate of 0.01 s^{-1} are shown in Figure 3. Figure 3a,b show that a large number of precipitates with the streamline shape appeared in ferrite and distributed along the deformation direction at 950 and 1050 °C. In order to identify the precipitates, SEM and TEM observation were performed as shown in Figures 4 and 5. SEM image shows that the ferrite was decomposed into one phase with coral-like structure and another phase with irregular island structure. And EDS analysis shows that the former phase was σ phase containing high content of Cr and Mo and the latter phase was secondary austenite with high Ni content, as reported by Paulraj et al. [25]. TEM image and selected area electron diffraction (SAED) pattern indicate σ phase with a tetragonal crystal structure presenting round type. The results of SEM and TEM analysis proved the occurrence of $\alpha \rightarrow \gamma_2 + \sigma$. It is reported that precipitates could restrain processing of discontinuous DRX because of precipitates pinning grain boundaries [26–29]. Therefore, it is difficult for discontinuous DRX to occur under these conditions. TEM image of the dislocation cells in ferrite is shown in Figure 6. The dislocation cells formed by tangled dislocations in ferrite indicate the occurrence of DRV in ferrite [30].

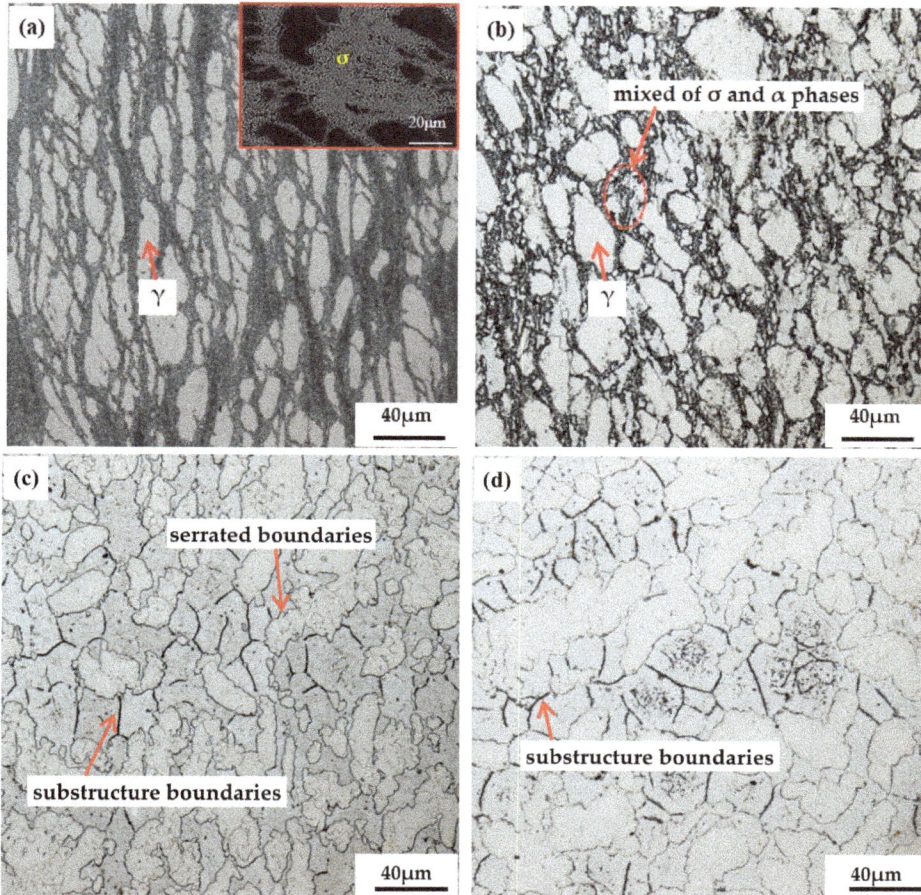

Figure 3. Microstructures of 2707 HDSS deformed at 0.01 s^{-1} and different temperatures: (a) 950 °C; (b) 1050 °C; (c) 1150 °C; (d) 1250 °C.

When the deformation temperature increased to 1150 °C, precipitates disappeared and a large amount of equiaxed ferrite appeared (Figure 3c). As indicated by arrows, sharper substructure in ferrite and some serrated boundaries in austenite were observed. It is considered that the leading softening mechanisms were DRV in ferrite and discontinuous DRX in austenite at the temperature of 1150 °C and strain rate of 0.01 s^{-1} [10,31]. When the deformation temperature reached 1250 °C, sharper substructure in ferrite indicated a well-developed DRV, as shown in Figure 3d. In addition, austenite had no obvious serrated boundaries and austenite islands became larger, indicating that strain was accumulated in the ferrite and did not spread to the austenite [32], and grain growth had occurred. These phenomena are consistent with the true stress-strain curves.

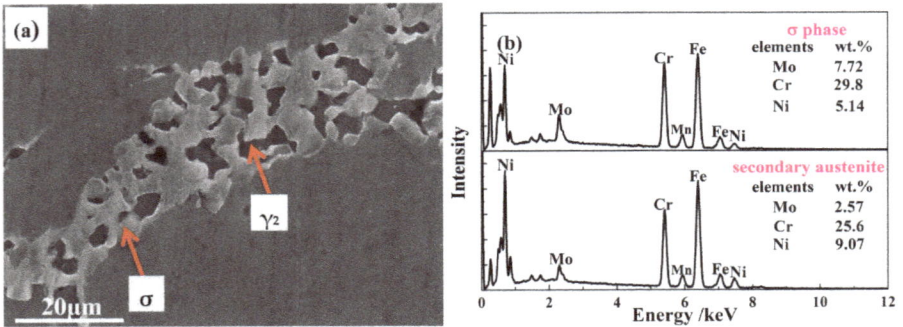

Figure 4. (**a**) SEM micrographs; and (**b**) energy dispersive spectroscope (EDS) analysis of precipitates and secondary austenite in 2707 HDSS deformed at 900 °C and 0.01 s^{-1}.

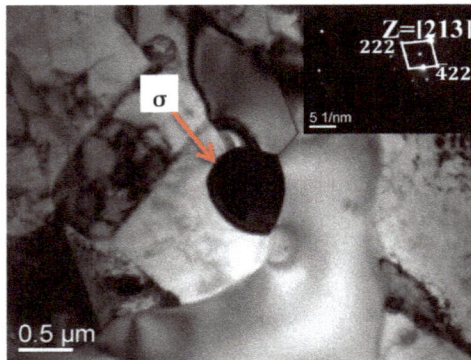

Figure 5. TEM micrograph and selected area electron diffraction (SAED) of σ in 2707 HDSS deformed at 900 °C and 0.01 s^{-1}.

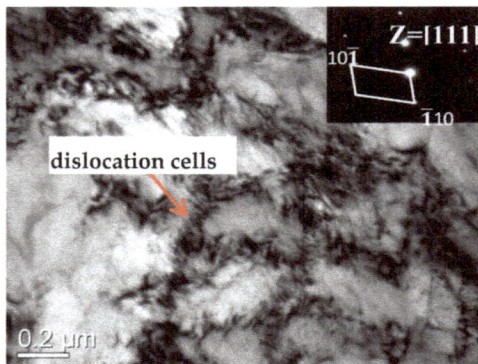

Figure 6. The dislocation cells in ferrite of 2707 HDSS deformed at 1050 °C and 0.01 s^{-1}.

Figure 7 shows the microstructures of 2707 HDSS deformed at 1250 °C and different strain rates. As shown in Figure 7a,b, the sharper substructure in ferrite can be observed. There were no obvious serrated boundaries in elongated austenite islands implying the restriction of discontinuous DRX.

Therefore, DRV in ferrite was the main softening mechanism. Consequently, strain was accumulated in the ferrite and did not spread to the austenite [32]. When the compression was conducted at 1 s^{-1}, a weak substructure in ferrite and serrated boundaries in austenite could be observed in Figure 7c. It indicted that the load was transformed from ferrite to austenite, leading to stain accumulation in austenite in latter stages triggering discontinuous DRX [33]. When the strain rate increased to 10 s^{-1}, the same phenomenon of discontinuous DRX could also be observed and the size of austenite islands was smaller because grains did not have enough time to grow, as shown in Figure 7d. A new DRX grain of austenite in 2707 HDSS, observed by TEM, is shown in Figure 8. The austenite developed into equiaxed grain and dislocation in the austenite interiors disappeared. Based on the above analysis, it could be concluded that the increase of strain rate was conducive to the occurrence of discontinuous DRX of austenite in 2707 HDSS.

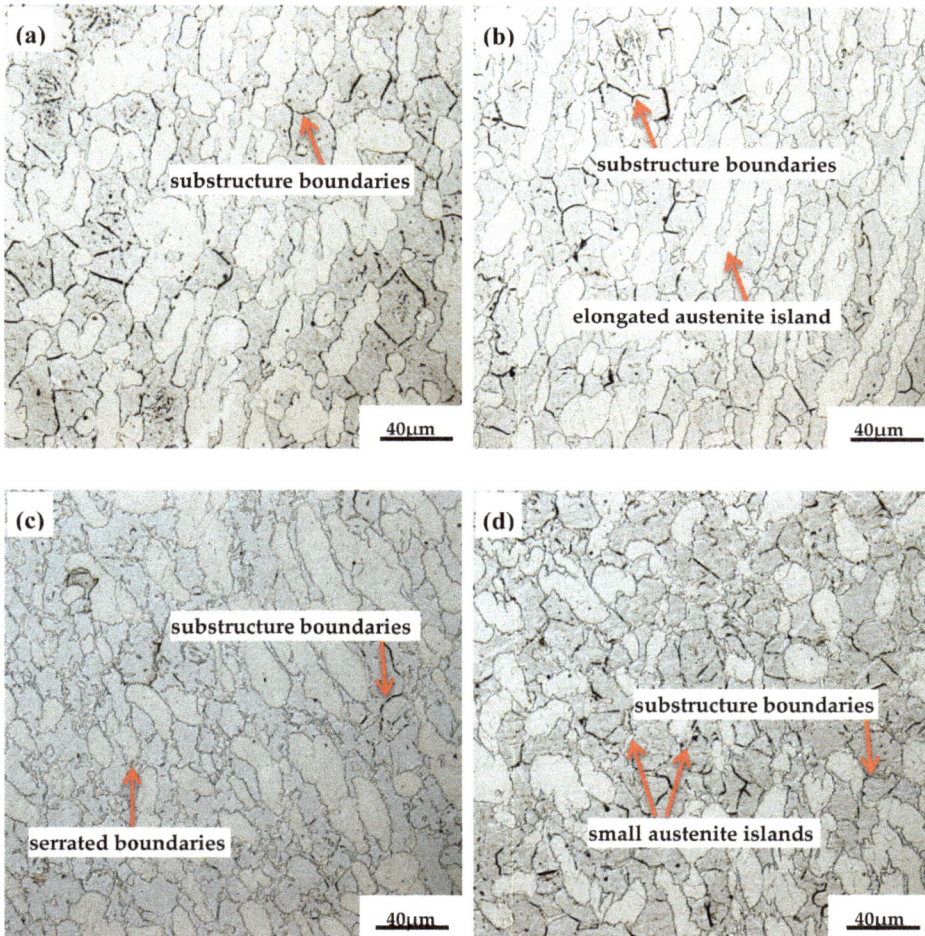

Figure 7. Microstructures of 2707 HDSS deformed at 1250 °C and different strain rates: (**a**) 0.01 s^{-1}; (**b**) 0.1 s^{-1}; (**c**) 1 s^{-1}; (**d**) 10 s^{-1}.

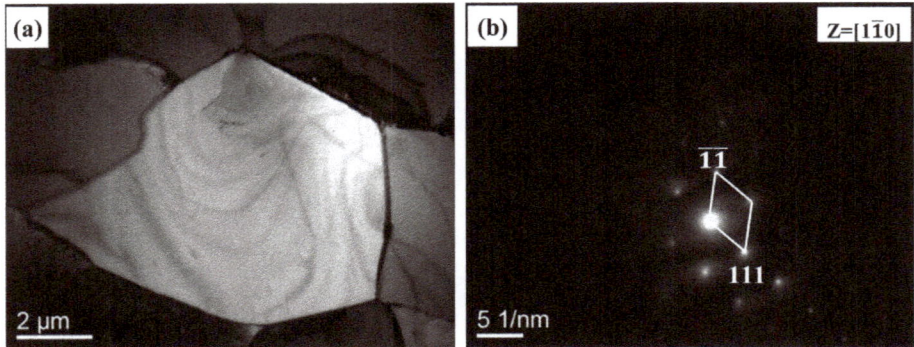

Figure 8. TEM images of specimens deformed at 1250 °C and 10 s^{-1}: (**a**) dynamic recrystallization (DRX) grain of austenite; (**b**) SAED result.

Besides, the existence of DRX grains in 2707 HDSS deformed at 1050 °C and 10 s^{-1} observed by TEM is shown in Figure 9. The new austenite grains were surrounded by dislocations, and the high dislocation density meant more significant plastic deformation characteristic [34]. However, discontinuous DRX phenomenon could not be observed, as shown in Figure 3b. From these, it could be concluded that the driving force of discontinuous DRX was the reduction of storage energy [35], and high strain rate could promote discontinuous DRX of austenite in 2707 HDSS.

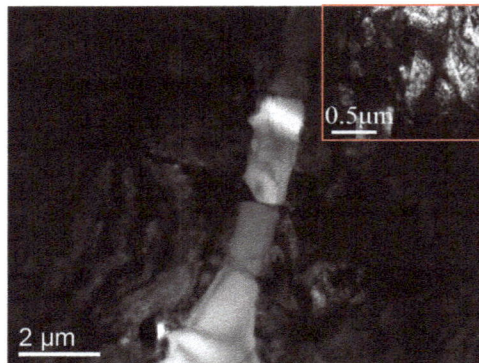

Figure 9. Dislocation structure around new DRX grains in 2707 HDSS deformed at 1050 °C and 10 s^{-1}.

In summary, all the curves showed similar characteristics of work hardening at the early stage of deformation. With the strain increasing, the curves presented different features in the different deformation conditions. At low temperatures, a large number of precipitates appeared in ferrite and distributed along the deformation direction, which could restrain processing of DRX and lead to flow instability. When the temperature increased, precipitates disappeared and the extent of DRV in ferrite enhanced, so the flow curves exhibited DRV characteristics without peak stress. Additionally, strain rate had important effects on the restoration process. At low strain rate, strain was accumulated in the ferrite and did not spread to the austenite, and DRV in ferrite was the main softening mechanism. When the strain rate increased to 1 s^{-1}, the load was transformed from ferrite to austenite, leading to strain accumulation in austenite in latter stages until the triggering of DRX. Consequently, the flow curves exhibited DRX characteristics with peak stress.

3.3. Establishment of Constitutive Relationship at Peak Strain

The constitutive equation is an important mathematical model to predict and analyze the relationship among temperature, strain rate, and flow stress in a wide hot deformation temperature range [36,37]. The dependence of flow stress to hot deformation variables can be analyzed by the well-known hyperbolic sine function [38] incorporated with the definition of the Zener-Hollomon parameter as follows:

$$Z = \dot{\varepsilon}\exp(Q/RT) \tag{1}$$

where Q, T, and R donate the apparent activation energy (J·mol^{-1}), absolute temperature (K), and universal gas constant (8.314 J·mol^{-1}·K^{-1}), respectively.

Based on the analysis of a large number of metals and alloys, many researchers have proposed the constitutive equations describing the thermal deformation of different metals and alloys [7,12,39,40]:

$$\dot{\varepsilon} = A_1\sigma^{n_1}\exp\left[-Q/(RT)\right] \tag{2}$$

$$\dot{\varepsilon} = A_2\exp(\beta\sigma)\exp\left[-Q/(RT)\right] \tag{3}$$

$$\dot{\varepsilon} = A\left[\sinh(\alpha\sigma)\right]^n\exp\left[-Q/(RT)\right] \tag{4}$$

where $\dot{\varepsilon}$ is the strain rate, σ is the flow stress, and A, A_1, A_2, n_1, n, β, and α (β/n_1) are material constants, respectively.

Taking the natural logarithm of Equations (2) and (3) yields:

$$\ln\dot{\varepsilon} = \ln A_1 + n_1\ln\sigma - Q/(RT) \tag{5}$$

$$\ln\dot{\varepsilon} = \ln A_2 + \beta\sigma - Q/(RT) \tag{6}$$

By plotting $\ln\dot{\varepsilon} - \sigma$ and $\ln\dot{\varepsilon} - \ln\sigma$, β and n_1 could be obtained from the slope of the linear regression lines, as shown in Figure 10a,b, respectively. The average value of n_1 and β were calculated to be 8.1803 and 0.0665, respectively, and then $\alpha = \beta/n_1 = 0.00813$ MPa^{-1}.

Taking the natural logarithm of Equation (4) yields:

$$\ln\dot{\varepsilon} = \ln A + n\ln\left[\sinh(\alpha\sigma)\right] - Q/(RT) \tag{7}$$

By plotting $\ln\dot{\varepsilon}$ against $\ln[\sinh(\alpha\sigma)]$ at constant temperatures, the material constant n can be calculated from the slope of the linear regression lines, as shown in Figure 10c. The average value of n was 5.514 based on the slope of four linear regression lines at different temperatures. Therefore, in the same way, the apparent activation energy (569.279 kJ·mol^{-1}) can be derived from the average slope of the $\ln[\sinh(\alpha\sigma)] - 1000/T$, as shown in Figure 10d. The value of A was 1.34×10^{21} in the present study based on the intercept of the $\ln Z - \ln[\sinh(\alpha\sigma)]$, as shown in Figure 10e. Filling the obtained material constants (α, n, Q, and A) into Equations (1) and (4), the hyperbolic sine equation embracing the Z parameter for the 2707 HDSS was developed as follows:

$$Z = \dot{\varepsilon}\exp(\frac{569279}{RT}) = 1.34 \times 10^{21}\left[\sinh(0.00813 \times \sigma)\right]^{5.514} \tag{8}$$

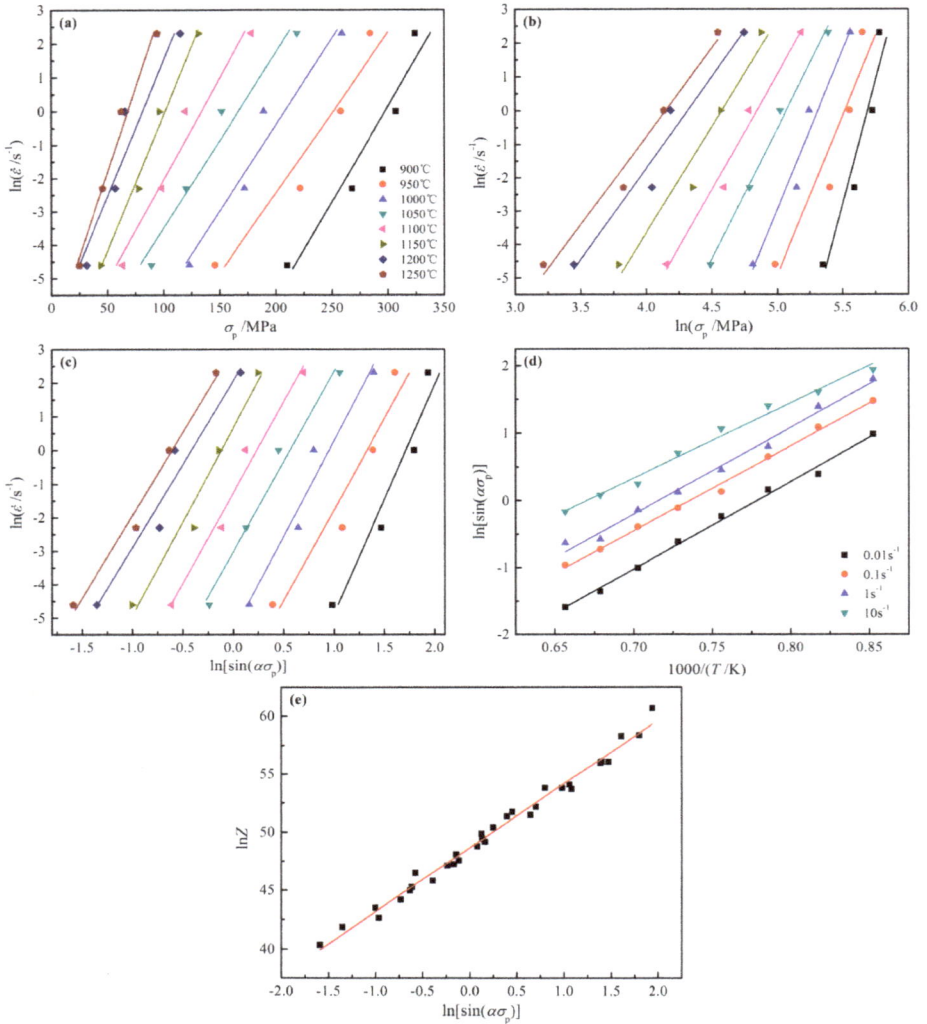

Figure 10. Relationships among flow strain rate and temperature for 2707 HDSS under hot compression: (a) $\ln\dot{\varepsilon} - \sigma$; (b) $\ln\dot{\varepsilon} - \ln\sigma$; (c) $\ln\dot{\varepsilon} - \ln[\sinh(\alpha\sigma)]$; (d) $\ln[\sinh(\alpha\sigma)] - 1000/T$; (e) $\ln Z - \ln[\sinh(\alpha\sigma)]$.

Many researchers [11,21,41] studied hot deformation activation energy of various typical DSSs, such as 2101, 2205 and 2507. In order to facilitate comparison, the content of main alloying elements and hot deformation activation energy of those steels reported in the literature are listed in Table 1.

Table 1. Chemical composition and apparent activation energy of duplex stainless steel (DSS)/wt %.

Steel Grades	C	Si	Mn	Cr	Ni	Mo	N	$Q/\text{kJ}\cdot\text{mol}^{-1}$
as-21Cr EDSS [11]	0.02	0.58	4.65	21.03	1.26	0.03	0.24	401.6
2205 DSS [21]	0.023	0.001	1.35	22.07	4.83	2.37	0.19	460.9
2507 [41]	0.03	0.37	1.08	25.21	7.08	4.23	0.26	493.0
2707 HDSS	0.0044	0.42	1.11	26.83	7.14	4.88	0.37	569.279

It can be seen from Table 1 that hot deformation activation energy of 2707 HDSS is the highest among these DSSs. The Q value describes the activation barrier that atoms need to overcome during the deformation procedure [42]. It is generally believed that the higher alloy steels have higher strain activation energy. This is because of the solid solution strengthening effect of the alloying elements, the drag and drop effect of these solute atoms hindering slipping of dislocation, and the movement of the grain boundary [36]. Among the elements, Cr and Mo provide more contribution to thermal deformation activation energy [39]. Ni and N could also promote the value of Q [43]. In the present research, the content of Cr, Mo, Ni, and N were higher in 2707 HDSS than traditional DSSs, which increased hot deformation activation energy significantly.

3.4. Constitutive Modeling Considering Effect of Strain

It is considered that the effect of strain on flow stress at elevated temperature is insignificant and thus would usually be ignored in Equation (1) [44]. However, it could be found that the strain also has an important effect on the flow stress besides strain rate and deformation temperature as shown in Figure 2. Meanwhile, the effect of the strain on the material constants (i.e., α, β, n, Q, and $\ln A$) is significant in the entire strain range as shown in Figure 11. Therefore, compensation of strain should be taken into account in order to derive constitutive modeling to predict the flow stress more precisely. The effect of strain was represented by presuming that different material constants were polynomial functions of strain [45–47]. In the present study, the values of material constants of the constitutive relationships were calculated under various true strains within the range of 0.05–0.75 and the intervals of 0.05. The order of the fitted polynomial was varied from two to eight for selecting the best fitted model. As shown in Equation (9), a sixth order polynomial was found to represent the influence of strain on material constants with a good correlation and generalization, and the fitted curves are shown in Figure 11. A higher order (i.e. >6) polynomial would over-fit, thus losing the ability of true representation and generalization. The polynomial fit results of α, β, n, Q, and $\ln A$ of 2707 HDSS are provided in Table 2.

$$\alpha = \alpha_0 + \alpha_1\varepsilon + \alpha_2\varepsilon^2 + \alpha_3\varepsilon^3 + \alpha_4\varepsilon^4 + \alpha_5\varepsilon^5 + \alpha_6\varepsilon^6$$
$$\beta = \beta_0 + \beta_1\varepsilon + \beta_2\varepsilon^2 + \beta_3\varepsilon^3 + \beta_4\varepsilon^4 + \beta_5\varepsilon^5 + \beta_6\varepsilon^6$$
$$n = n_0 + n_1\varepsilon + n_2\varepsilon^2 + n_3\varepsilon^3 + n_4\varepsilon^4 + n_5\varepsilon^5 + n_6\varepsilon^6 \tag{9}$$
$$Q = Q_0 + Q_1\varepsilon + Q_2\varepsilon^2 + Q_3\varepsilon^3 + Q_4\varepsilon^4 + Q_5\varepsilon^5 + Q_6\varepsilon^6$$
$$\ln A = A_0 + A_1\varepsilon + A_2\varepsilon^2 + A_3\varepsilon^3 + A_4\varepsilon^4 + A_5\varepsilon^5 + A_6\varepsilon^6$$

Table 2. Coefficients of the polynomial for α, β, n, Q, and $\ln A$ for 2707 HDSS.

α	β	n	Q	$\ln A$
$\alpha_0 = 0.00778$	$\beta_0 = 0.05921$	$n_0 = 5.38072$	$Q_0 = 433.7134$	$A_0 = 37.6406$
$\alpha_1 = 0.01391$	$\beta_1 = 0.24588$	$n_1 = 4.23215$	$Q_1 = 3340.689$	$A_1 = 267.325$
$\alpha_2 = -0.13113$	$\beta_2 = -2.35617$	$n_2 = -58.3197$	$Q_2 = -26,985.42$	$A_2 = -2155.24$
$\alpha_3 = 0.68476$	$\beta_3 = 9.23339$	$n_3 = 145.989$	$Q_3 = 96,937.91$	$A_3 = 7660.88$
$\alpha_4 = -1.61171$	$\beta_4 = -16.0984$	$n_4 = 5.81170$	$Q_4 = -171,312.5$	$A_4 = -13,301.7$
$\alpha_5 = 1.76105$	$\beta_5 = 12.4872$	$n_5 = -350.183$	$Q_5 = 146,514.1$	$A_5 = 11,097.16$
$\alpha_6 = -0.73869$	$\beta_6 = -3.37538$	$n_6 = 277.087$	$Q_6 = -48,226.32$	$A_6 = -3528.39$

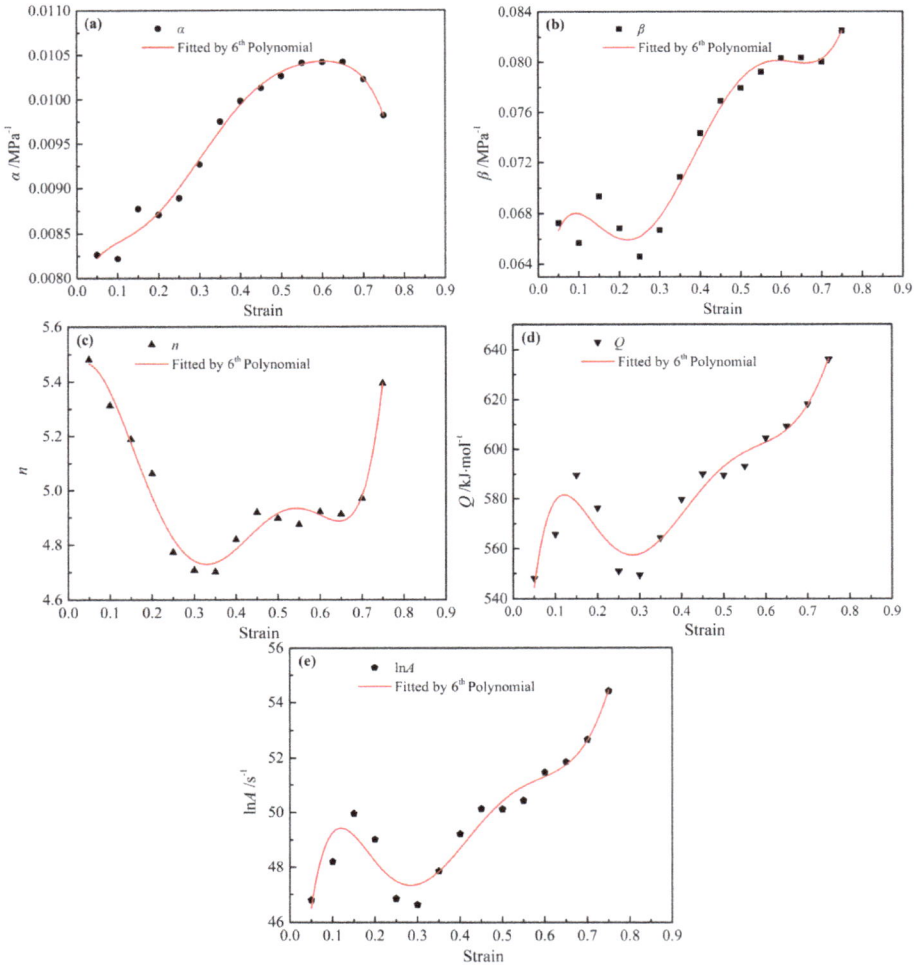

Figure 11. Relationship between material constants and strains: (a) α; (b) β; (c) n; (d) Q; (e) $\ln A$.

Once the polynomial coefficients of all material constants were evaluated, the flow stress at a particular true strain can be estimated. Using the expression of hyperbolic sine function, the constitutive equation that relates to flow stress and the Zener-Holloman parameter can be written in the following form [48]:

$$\sigma = \frac{1}{\alpha} \ln \left\{ \left(\frac{Z}{A} \right)^{\frac{1}{n}} + \left[\left(\frac{Z}{A} \right)^{\frac{2}{n}} + 1 \right]^{\frac{1}{2}} \right\} \tag{10}$$

3.5. Verification of the Developed Constitutive Modeling

The experimental and calculated values were compared to evaluate the accuracy of the developed constitutive modeling in predicting the hot deformation behavior of 2707 HDSS, as illustrated in Figure 12. The calculated stress increased with increasing strain rate and decreasing deformation temperature. It can be observed that an agreement between the experimental and calculated values is satisfactory for most of the experimental conditions. Only in some deformation conditions (i.e., at 900 °C in 0.1, 1, and 10 s^{-1}) was a notable deviation between experimental and calculated values observed. The fitting of material constants might result in the deviation between experimental and calculated values of flow stress. For example, as shown in Figure 10a, experimental data shows some aberration, particularly at the lower temperatures. Additionally, Equation (2) is suitable for lower stress as mentioned in a previous paragraph. Therefore, some errors might be introduced in calculating the values of n_1 using Equation (2) for higher stress (i.e., at lower temperature and higher strain rate) and finally affect the accuracy of constitutive modeling [44]. Similar errors were reported by Mandal et al. [48] in Ti-modified austenitic stainless steel and Zou et al. [12] in as-cast 21Cr economical duplex stainless steel.

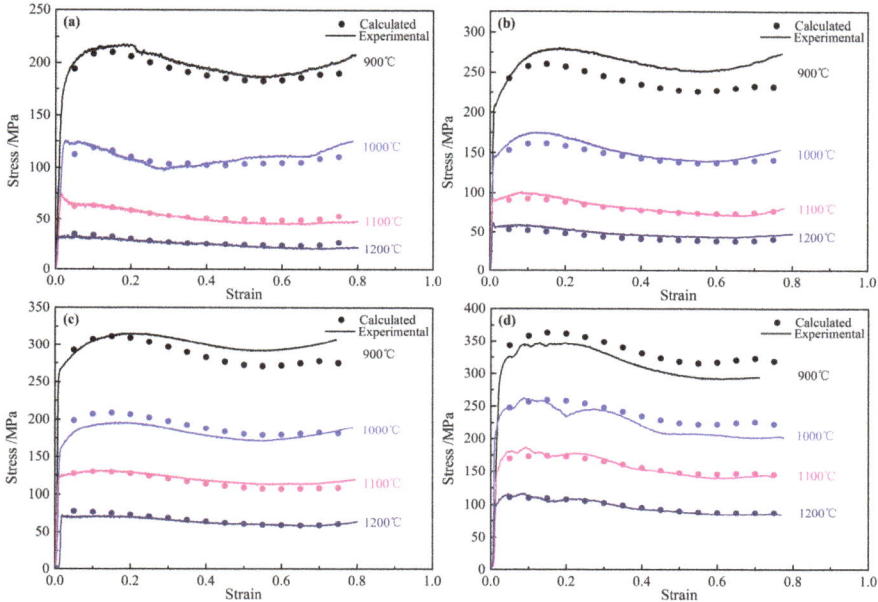

Figure 12. Comparison between calculated and experimental flow stress curves at strain rates: (a) 0.01 s^{-1}; (b) 0.1 s^{-1}; (c) 1 s^{-1}; (d) 10 s^{-1}.

With the aim to evaluate the accuracy of the developed constitutive modeling, standard statistical parameters such as correlation coefficient (R) and average absolute relative error (AARE) are adopted [7,44]. The parameters are expressed by the following equations:

$$R = \frac{\sum_{i=1}^{N} (E_i - \overline{E})(P_i - \overline{P})}{\sqrt{\sum_{i=1}^{N} (E_i - \overline{E})^2 (P_i - \overline{P})^2}} \tag{11}$$

$$AARE = \frac{1}{N} \sum_{i=1}^{N} \left| \frac{E_i - P_i}{E_i} \right| \times 100\% \tag{12}$$

where E_i is the experimental flow stress; P_i is the calculated flow stress; $\overline{E}$ and $\overline{P}$ are the mean values of E_i and P_i, respectively; N is the number of data points used in this investigation. The correlation coefficient R is used to reflect the strength of the linear relationship between the calculated and experimental values. Sometimes, higher values of R may not always indicate a better predictability and reliability because the tendency of the constitutive modeling could be biased towards higher or lower values [49]. However, the AARE value is computed in terms of relative error and therefore is an impartial value for evaluating the capability of the modeling in the prediction of flow stress [47,50]. Figure 13 shows the plots of the calculated values of flow stress against the experimental values at various processing conditions. It is obvious that most of the data points lie very close to the line, and the correlation coefficient for the developed constitutive modeling of 2707 HDSS is 0.992. Meanwhile, the AARE value for the developed constitutive equation of 2707 HDSS alloy is 5.22%. These results show that the proposed constitutive modeling considering strain compensation provided a precise estimate of the flow stress of 2707 HDSS.

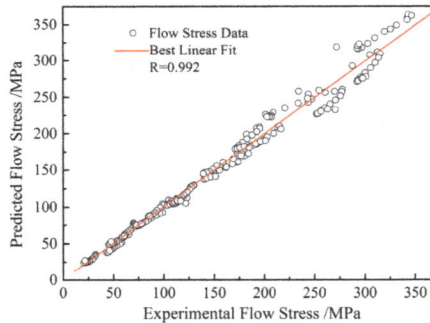

Figure 13. Correlation between the experimental and calculated flow stress data.

4. Conclusions

The hot deformation behavior, microstructure evolution, and constitutive analysis of 2707 HDSS were investigated by hot compression tests in the temperature range of 900–1250 °C and strain rate range of 0.01–10 s^{-1}. The main results are listed below:

(1) The flow stress was especially sensitive to deformation temperature and strain rate as it increased with increasing strain rate and decreasing deformation temperature. When the temperature was higher than 1100 °C and strain rate was less than 1 s^{-1}, the true stress-strain curves exhibited DRV characteristics without peak stress. Flow curves with peak stress could be observed at lower temperatures, which may be attributed to DRX or flow instability.

(2) At lower temperatures, a large number of precipitates appeared in ferrite and distributed along the deformation direction, which could restrain processing of discontinuous DRX because of pinning grain boundaries. When the temperature increased to 1150 °C, precipitates disappeared and sharper substructure in ferrite and some serrated boundaries in austenite were observed, indicating the leading softening behaviors were DRV in ferrite and discontinuous DRX in austenite. When the temperature reached 1250 °C, sharper substructure in ferrite and no obvious serrated boundaries in austenite showed that softening behavior was mainly DRV in ferrite. With the increasing of strain rate, the load was transformed from ferrite to austenite, leading to strain accumulation in austenite in latter stages until the triggering of discontinuous DRX.

(3) The peak stress of 2707 HDSS was well fitted by the constitutive modeling of the hyperbolic sine function. Due to the higher content of Cr, Mo, Ni, and N, 2707 HDSS processed higher apparent activation energy (569.279 kJ· mol^{-1}) than traditional DSSs.

(4) The hyperbolic sine constitutive model considering the compensation of strain was developed. A sixth order polynomial was found to be most suitable for describing the relationship between the material constants and the strains. The developed constitutive relationships of flow stress with strain, strain rate, and temperature were successfully used to predict the flow stresses under various deformation conditions, and the correlation coefficient and average absolute relative error were 0.992 and 5.22%, respectively.

Acknowledgments: The present research was financially supported by the High Technology Research and Development Program of China (No. 2015AA034301) and the Fundamental Research Funds for the Central Universities (Grant No. N150204007).

Author Contributions: Huabing Li, Zhouhua Jiang, Lixin Wang, and Guangwei Fan conceived and designed the experiments; Weichao Jiao, Hao Feng, Xinxu Li, and Guoping Li performed the experiments; Weichao Jiao, Hao Feng, Xinxu Li, and Huabing Li analyzed the data; Huabing Li, Weichao Jiao, Xinxu Li, Hao Feng, and Peide Han contributed to the writing and editing of the manuscript.

Conflicts of Interest: The authors declare no conflict of interest.

References

1. Örnek, C.; Idris, S.A.M.; Reccagni, P.; Engelberg, D.L. Atmospheric-induced stress corrosion cracking of grade 2205 duplex stainless steel-effects of 475 °C embrittlement and process orientation. *Metals* **2016**, *6*, 167. [CrossRef]
2. Kim, S.K.; Kang, K.Y.; Kim, M.S.; Lee, J.M. Low-temperature mechanical behavior of super duplex stainless steel with sigma precipitation. *Metals* **2016**, *5*, 1732–1745. [CrossRef]
3. Zhao, H.; Zhang, Z.Y.; Zhang, H.Z.; Hu, J.; Li, J. Effect of aging time on intergranular corrosion behavior of a newly developed LDX 2404 lean duplex stainless steel. *J. Alloys Compd.* **2016**, *672*, 147–154. [CrossRef]
4. Li, H.B.; Zhou, E.Z.; Zhang, D.W.; Xu, D.K.; Xia, J.; Yang, C.; Feng, H.; Jiang, Z.H.; Li, X.G.; Gu, T.Y.; et al. Microbiologically influenced corrosion of 2707 hyper-duplex stainless steel by marine *pseudomonas aeruginosa* biofilm. *Sci. Rep.-UK* **2016**, *6*, 20190. [CrossRef] [PubMed]
5. Stenvall, P.; Holmquist, M. Weld properties of Sandvik SAF 2707. *Metall. Ital.* **2008**, *10*, 11–18.
6. Momeni, A.; Kazemi, S.; Bahrani, A. Hot deformation behavior of microstructural constituents in a duplex stainless steel during high-temperature straining. *Int. J. Miner. Metall. Mater.* **2013**, *20*, 953–960. [CrossRef]
7. Yang, L.C.; Pan, Y.T.; Chen, I.G.; Lin, D.Y. Constitutive relationship modeling and characterization of flow behavior under hot working for Fe–Cr–Ni–W–Cu–Co super-austenitic stainless steel. *Metals* **2015**, *5*, 1717–1731. [CrossRef]
8. Farnoush, H.; Momeni, A.; Dehghani, K.; Mohandesi, J.A.; Keshmiri, H. Hot deformation characteristics of 2205 duplex stainless steel based on the behavior of constituent phases. *Mater. Des.* **2010**, *31*, 220–226. [CrossRef]
9. Liu, Y.Y.; Yan, H.T.; Wang, X.H.; Yan, M. Effect of hot deformation mode on the microstructure evolution of lean duplex stainless steel 2101. *Mater. Sci. Eng. A* **2013**, *575*, 41–47. [CrossRef]
10. Momeni, A.; Dehghan, K. Hot working behavior of 2205 austenite–ferrite duplex stainless steel characterized by constitutive equations and processing maps. *Mater. Sci. Eng. A* **2011**, *528*, 1448–1454. [CrossRef]
11. Zou, D.N.; Wu, K.; Han, Y.; Zhang, W.; Cheng, B.; Qiao, G.J. Deformation characteristic and prediction of flow stress for as-cast 21Cr economical duplex stainless steel under hot compression. *Mater. Des.* **2013**, *51*, 975–982. [CrossRef]
12. Ashtiania, H.R.R.; Parsab, M.H.; Bisadia, H. Constitutive equations for elevated temperature flow behavior of commercial purity aluminum. *Mater. Sci. Eng. A* **2012**, *545*, 61–67. [CrossRef]
13. Abbasi-Bani, A.; Zarei-Hanzaki, A.; Pishbin, M.H.; Haghdadi, N. A comparative study on the capability of Johnson–Cook and Arrhenius-type constitutive equations to describe the flow behavior of Mg–6Al–1Zn alloy. *Mech. Mater.* **2014**, *71*, 52–61. [CrossRef]
14. Lin, Y.C.; Chen, X.M. A critical review of experimental results and constitutive descriptions for metals and alloys in hot working. *Mater. Des.* **2011**, *32*, 1733–1759. [CrossRef]
15. Haghdadi, N.; Zarei-Hanzaki, A.; Khalesian, A.R.; Abedi, H.R. Artificial neural network modeling to predict the hot deformation behavior of an A356 aluminum alloy. *Mater. Des.* **2013**, *49*, 389–361. [CrossRef]
16. Haghdadi, N.; Zarei-Hanzaki, A.; Abedi, H.R. The flow behavior modeling of cast A356 aluminum alloy at elevated temperatures considering the effect of strain. *Mat. Sci. Eng. A* **2012**, *535*, 252–257. [CrossRef]
17. Li, H.P.; He, L.F.; Zhao, G.Q.; Zhang, L. Constitutive relationships of hot stamping boron steel B1500HS based on the modified Arrhenius and Johnson–Cook model. *Mater. Sci. Eng. A* **2013**, *580*, 330–348. [CrossRef]
18. Lin, Y.C.; Xia, Y.C.; Chen, X.M.; Chen, M.S. Constitutive descriptions for hot compressed 2124-T851 aluminum alloy over a wide range of temperature and strain rate. *Comput. Mater. Sci.* **2010**, *50*, 227–233. [CrossRef]
19. Sun, C.Y.; Zuo, X.; Xiang, Y.; Yang, J. Investigation on hot deformation behavior and hot processing map of BSTMUF601 super-alloy. *Metals* **2016**, *70*, 1–8. [CrossRef]

20. Pu, E.X.; Feng, H.; Liu, M.; Zheng, W.J.; Dong, H.; Song, Z.G. Constitutive modeling for flow behaviors of super austenitic stainless steels S32654 during hot deformation. *J. Iron Steel Res. Int.* **2016**, *23*, 178–184. [CrossRef]

21. Yang, Y.H.; Yan, B. The microstructure and flow behavior of 2205 duplex stainless steels during high temperature compression deformation. *Mater. Sci. Eng. A* **2013**, *579*, 194–201. [CrossRef]

22. Haghdadi, N.; Martin, D.; Hodgson, P. Physically-based constitutive modelling of hot deformation behavior in a LDX 2101 duplex stainless steel. *Mater. Des.* **2016**, *106*, 420–427. [CrossRef]

23. Si, J.Y.; Liao, X.H.; Xie, L.Q.; Lin, K.H. Flow behavior and constitutive modeling of delta-processed Inconel 718 alloy. *J. Iron Steel Res. Int.* **2015**, *22*, 837–845. [CrossRef]

24. Han, Y.; Zou, D.N.; Chen, Z.Y.; Fan, G.W.; Zhang, W. Investigation on hot deformation behavior of 00Cr23Ni4N duplex stainless steel under medium–high strain rates. *Mater. Charact.* **2011**, *62*, 198–203. [CrossRef]

25. Paulraj, P.; Garg, R. Effect of intermetallic phases on corrosion behavior and mechanical properties of duplex stainless steel and super duplex stainless steel. *Adv. Sci. Technol. Res. J.* **2015**, *9*, 87–105. [CrossRef]

26. Doherty, R.D.; Hughes, D.A.; Humphreys, F.J.; Jonas, J.J.; Juul Jensen, D.; Kassner, M.E.; King, W.E.; McNelley, T.R.; McQueen, H.J.; Rollett, A.D. Current issues in recrystallization: A review. *Mater. Sci. Eng. A* **1997**, *238*, 219–274. [CrossRef]

27. Smith, C.S. Grains, phases, and interfaces—An interpretation of microstructure. *Trans. Metall. Soc. AIME* **1948**, *175*, 15–51.

28. Zurob, H.S.; Brechet, Y.; Purdy, G. A model for the competition of precipitation and recrystallization in deformed austenite. *Acta Mater.* **2001**, *49*, 4183–4190. [CrossRef]

29. Vervynckt, S.; Verbeken, K.; Thibauxc, P.; Houbaert, Y. Recrystallization–precipitation interaction during austenite hot deformation of a Nb microalloyed steel. *Mater. Sci. Eng. A* **2011**, *528*, 5519–5528. [CrossRef]

30. Timokhina, I.; Pereloma, E.V.; Hodgson, P. The formation of complex microstructures after different deformation modes in advanced high-strength steels. *Metall. Mater. Trans. A* **2014**, *45*, 4247–4256. [CrossRef]

31. Ma, X.C.; An, Z.J.; Chen, L.; Mao, T.Q.; Wang, J.F.; Long, H.J.; Xue, H.Y. The effect of rare earth alloying on the hot workability of duplex stainless steel—A study using processing map. *Mater. Des.* **2015**, *86*, 848–854. [CrossRef]

32. Balancin, O.; Hoffmann, W.A.M.; Jonas, J.J. Influence of microstructure on the flow behavior of duplex stainless steels at high temperatures. *Metall. Mater. Trans. A* **2000**, *31*, 1353–1364. [CrossRef]

33. Tehovnik, F.; Arzenšek, B.; Arh, B.; Skobir, D.; Pirnar, B.; Žužek, B. Microstructure evolution in SAF 2507 super duplex stainless steel. *Mater. Technol.* **2011**, *45*, 339–345.

34. Ma, M.; Ding, H.; Tang, Z.Y.; Zhao, J.T.; Jiang, Z.H.; Li, G. Effect of strain rate and temperature on hot workability and flow behaviour of duplex stainless steel. *Ironmak. Steelmak.* **2016**, *43*, 88–96. [CrossRef]

35. Wang, S.L.; Zhang, M.X.; Wu, H.C.; Yang, B. Study on the dynamic recrystallization model and mechanism of nuclear grade 316LN austenitic stainless steel. *Mater. Charact.* **2016**, *118*, 92–101. [CrossRef]

36. Xi, T.; Yang, C.G.; Shahzad, M.B.; Yang, K. Study of the processing map and hot deformation behavior of a Cu-bearing 317LN austenitic stainless steel. *Mater. Des.* **2015**, *87*, 303–312. [CrossRef]

37. Sun, C.Y.; Xiang, Y.; Zhou, Q.J.; Politis, D.J.; Sun, Z.H.; Wang, M.Q. Dynamic recrystallization and hot workability of 316LN stainless steel. *Metals* **2016**, *6*, 152. [CrossRef]

38. McQueen, H.J.; Ryan, N.D. Constitutive analysis in hot working. *Mater. Sci. Eng. A* **2002**, *322*, 43–63. [CrossRef]

39. McQueen, H.J.; Yue, A.S.; Ryan, N.D.; Fry, E. Hot working characteristics of steels in austenitic state. *J. Mater. Process. Technol.* **1995**, *53*, 293–310. [CrossRef]

40. Zener, C.; Hollomon, J.H. Effect of strain rate upon plastic flow of steel. *J. Appl. Phys.* **1944**, *22*, 22–32. [CrossRef]

41. Ma, M.; Ding, H.; Tang, Z.Y.; Zhao, J.W.; Jiang, Z.H.; Fan, G.W. Effects of temperature and strain rate on flow behavior and microstructural evolution of super duplex stainless steel under hot deformation. *J. Iron Steel Res. Int.* **2016**, *23*, 244–252. [CrossRef]

42. Farabi, E.; Zarei-Hanzaki, A.; Pishbin, M.H.; Moallemi, M. Rationalization of duplex brass hot deformation behavior: The role of microstructural components. *Mater. Sci. Eng. A* **2015**, *641*, 360–368. [CrossRef]

43. Pu, E.X.; Zheng, W.J.; Xiang, J.Z.; Song, Z.G.; Li, J. Hot deformation characteristic and processing map of superaustenitic stainless steel S32654. *Mater. Sci. Eng. A* **2014**, *598*, 174–182. [CrossRef]

44. Cai, J.; Li, F.G.; Liu, T.Y.; Chen, B.; He, M. Constitutive equations for elevated temperature flow stress of Ti–6Al–4V alloy considering the effect of strain. *Mater. Des.* **2011**, *32*, 1144–1151. [CrossRef]
45. Samantaray, D.; Mandal, S.; Bhaduri, A.K. Constitutive analysis to predict high-temperature flow stress in modified 9Cr–1Mo (P91) steel. *Mater. Des.* **2010**, *31*, 981–984. [CrossRef]
46. Lin, Y.C.; Chen, M.S.; Zhong, J. Constitutive modeling for elevated temperature flow behavior of 42CrMo steel. *Comput. Mater. Sci.* **2008**, *42*, 470–477. [CrossRef]
47. Mandal, S.; Rakesh, V.; Sivaprasad, P.V.; Venugopal, S.; Kasiviswanathan, K.V. Constitutive equations to predict high temperature flow stress in a Ti-modified austenitic stainless steel. *Mater. Sci. Eng. A* **2009**, *500*, 114–121. [CrossRef]
48. Li, H.Y.; Wei, D.D.; Hu, J.D.; Li, Y.H.; Chen, S.L. Constitutive modeling for hot deformation behavior of T24 ferritic steel. *Comput. Mater. Sci.* **2012**, *53*, 425–430. [CrossRef]
49. Phaniraj, M.P.; Lahiri, A.K. The applicability of neural network model to predict flow stress for carbon steels. *J. Mater. Process. Technol.* **2003**, *141*, 219–227. [CrossRef]
50. Xiao, M.L.; Li, F.G.; Zhao, W.; Yang, G.L. Constitutive equation for elevated temperature flow behavior of TiNiNb alloy based on orthogonal analysis. *Mater. Des.* **2012**, *35*, 184–193. [CrossRef]

metals

MDPI

Article

Influence of Grain Growth Inhibitors and Powder Size on the Properties of Ultrafine and Nanostructured Cemented Carbides Sintered in Hydrogen

Tamara Aleksandrov Fabijanić [1],*, Suzana Jakovljević [1], Mladen Franz [1] and Ivan Jeren [2]

[1]   Faculty of Mechanical Engineering and Naval Architecture, University of Zagreb, Ivana Lučića 5,
      Zagreb 10000, Croatia; suzana.jakovljevic@fsb.hr (S.J.); mladen.franz@fsb.hr (M.F.)
[2]   Alfa tim d.o.o., Čulinečka cesta 25, Zagreb 10040, Croatia; info@alfatim.hr
*   Correspondence: tamara.aleksandrov@fsb.hr; Tel.: +385-1-6168-389

Academic Editor: Soran Birosca
Received: 8 July 2016; Accepted: 19 August 2016; Published: 24 August 2016

Abstract: The influence of grain growth inhibitors and powder size on the microstructure and mechanical properties of ultrafine and nanostructured cemented carbides was researched. Three different WC powders, with an addition of different type and content of grain growth inhibitors GGIs, VC and Cr_3C_2 and with d_{BET} grain sizes in the range from 95 to 150 nm were selected as starting powders. Four different mixtures with 6 and 9 wt. % Co were prepared. The consolidated samples are characterized by different microstructural and mechanical properties with respect to the characteristics of starting powders. Increased sintering temperatures led to microstructural irregularities in the form of a discontinuous WC growth, carbide agglomerates and abnormal grain growth as a consequence of coalescence via grain boundary elimination. The addition of 0.45% Cr_3C_2 contributed to microstructure homogeneity, reduced discontinuous and continuous grain growth, and increased Vickers hardness by approximately 70 HV and fracture toughness by approximately $0.15 \, MN/m^{3/2}$. The reduction of the starting powder to a real nanosize of 95 nm resulted in lower densities, and significant hardness increase, with a simultaneously small increase in fracture toughness. The consolidation of real nanopowders ($d_{BET} < 100$ nm) solely by conventional sintering in hydrogen without isostatic pressing is not preferred.

Keywords: ultrafine and nanostructured cemented carbides; grain growth inhibitors; powder size; sintering temperature; microstructural characteristics; mechanical properties

1. Introduction

Ultrafine cemented carbides with a d_{WC} grain size in the range from 0.2 to 0.500 μm and nanostructured cemented carbides with a d_{WC} grain size < 0.200 μm are consolidated from ultrafine and nanosized WC starting powders, which are characterized by a big surface area and a very high sintering activity. One of the biggest challenges of sintering ultrafine and nanoscaled powders is the retention of a small WC grain size in the sintered product [1,2]. Numerous attempts to achieve nanostructured cemented carbides have failed due to high sintering activity of WC nanopowders [2,3]. For that reason, the nano- and near nanopowders with very low sintering activity with respect to the re-crystallization during liquid phase sintering were developed [2,4]. Furthermore, powders with small amounts of the so-called grain growth inhibitors, GGIs, homogeneously distributed in powder mixtures, were used. GGIs are dissolved in the Co matrix, segregated at WC-Co interfaces, thereby forming a solid solution with a lower melting temperature in the range from 1200 to 1250 °C, saturated with GGIs [5,6]. The most common GGIs are vanadium carbide, VC, chromium carbide, Cr_3C_2, tantalum carbide, TaC, titanium carbide, TiC and niobium carbide NbC, and due to their effectiveness, they can be ranked in the following order: VC > Cr_3C > NbC > TaC [6]. Their primary task is to preserve the particle size of starting powders in the sintered product, while at the same time influencing the properties of consolidated samples: decreasing density, increasing the value of hardness at room temperature, and also affecting toughness, hardness and creep resistance at elevated temperatures. The combination of VC-Cr_3C_2 (TaC) proved the most effective due to its high solubility and mobility in the Co binder, which results in an optimal combination of hardness and toughness [7,8]. The process of grain growth in cemented carbides can be characterized as both continuous and discontinuous. Continuous growth is characterized by uniform growth of WC grains, while discontinuous growth characterizes growth of isolated WC grains or groups of WC grains, which grow faster and larger compared to the surrounding grains [8–10]. Both growth processes must be avoided in order to obtain satisfactory properties in the sintered products since mechanical properties of ultrafine and nanostructured cemented carbides are directly dependent on the developed microstructure: WC grain size, mean free path of Co and contiguity of WC grains [2,10]. Homogeneous microstructure without grain growth and homogeneous distribution of cobalt between carbide grains lead to optimal strength and improvements of hardness and toughness, while microstructural irregularities, like porosity and especially abnormal grain growth, result in poor transverse rupture strength [11]. Abnormal grain growth is characteristic of a system with faceted grains, which possess a singular interface and anisotropic surface energy.

Microstructure characteristics and mechanical properties of ultrafine and nanostructured cemented carbides were developed by conventional liquid phase sintering in hydrogen with respect to characteristics of the starting powder, and GGIs and the grain size of starting powders were researched. The influence of sintering temperature on microstructure and mechanical properties of sintered samples was analyzed. The research was performed to examine the possibility of consolidation by sintering in hydrogen atmosphere.

2. Materials and Methods

Near nano- and nanopowders were used as starting materials. According to the producers specification (HC Starck, Goslar, Germany) the powders have an average d_{BET} grain size in the range from 95 to 150 nm and a specific surface area (BET) in the range from 2.5 to 4.0 m^2/g. The selected WC powders had an addition of a different type and content of GGIs; VC and Cr_3C_2. The Co powder used was HMP Co, produced by Umicore (Bruxelles, Belgium). It has a d_{BET} of 210 nm and a specific surface area (BET) of 2.96 m^2/g. The characteristics of starting powders are presented in Table 1. Powders with different characteristics were selected in order to investigate the influence of grain size and GGIs on the behavior during sintering and the microstructure and mechanical properties of sintered samples.

Table 1. Characteristics of starting powders and mixtures.

Mixture	Powder	Grain Size, d_{BET}, nm	Specific Surface, m^2/g	GGI, wt. %	Co, %
WC-9Co/1	WC DN 2-5/1	150	2.57	0.26% VC, 0.45% Cr_3C_2	9
WC-9Co/2	WC DN 2-5/2	150	2.59	0.27% VC	9
WC-6Co/1	WC DN 2-5/1	150	2.57	0.26% VC, 0.45% Cr_3C_2	6
WC-6Co/2	WC DN 4-0	95	3.92	0.41%VC, 0.80% Cr_3C_2	6

WC powders were mixed with 6 and 9 wt. % Co and paraffin wax dissolved in hexane. WC-9Co mixtures had the same grain size of the starting WC powder, 150 nm, but different GGIs in order to examine their influence on the properties of sintered samples. WC-6Co mixtures had the same type of grain growth inhibitors, VC + Cr_3C_2, but different grain sizes of starting powders so as to analyze the influence of grain size on the properties of sintered samples. Mixing was carried out in a horizontal ball mill for the purpose of WC and Co homogenization. After subsequent drying of the slurry and granulation by sieving, the samples were compacted by uniaxial die pressing at room temperature on a hydraulic press. Different pressures were used for the compaction of green parts. The compaction pressure of 180 MPa was applied for the mixtures with 150 nm starting powders while the compaction pressure of 250 MPa was applied for the mixture with 95 nm starting powder. Dewaxing of green compacts was carried out in a separate cycle at 800 °C. Before entering the sintering furnace, green compacts were placed on graphite trays and covered with aluminum oxide, Al_2O_3, which creates an atmosphere around the green compacts and separates each green compact. The samples were sintered by conventional liquid phase sintering in hydrogen. Sintering was performed at temperatures ranging from 1360 to 1420 °C in order to research the influence of sintering temperature on the microstructure and mechanical properties. Twelve samples overall were consolidated. The parameters of consolidation are presented in Table 2.

Table 2. Parameters of consolidation.

Mixture	Sample	Milling Time, h	Compaction Pressure, MPa	$T_{sint.}$, °C	Holding Time, min	Heating Rate, °C/min
WC-9Co/1	1-1	72	180	1360	80	5
	1-2	72	180	1400	80	5
	1-3	72	180	1420	80	5
WC-9Co/2	2-1	72	180	1360	80	5
	2-2	72	180	1400	80	5
	2-3	72	180	1420	80	5
WC-6Co/1	3-1	48	180	1360	80	5
	3-2	48	180	1400	80	5
	3-3	48	180	1420	80	5
WC-6Co/2	4-1	48	250	1360	80	5
	4-2	48	250	1400	80	5
	4-3	48	250	1420	80	5

The density of the consolidated samples was determined using the Archimedes method, weighing the samples in air and in liquid according to EN ISO 3369:2006. The degree of porosity and uncombined carbon was determined by comparing the polished surface with photo micrographs from the standard (ISO 4505:1978). The analysis was performed using an optical microscope, a scanning electron microscope, or SEM (Tescan, Brno, Czech Republic), and a field emission electron microscope, or FESEM (Tescan, Brno, Czech Republic). The measurement of WC grain size was carried out by a conventional linear intercept method according to EN ISO 4499-2:2011 at very high magnification in order to isolate each individual grain at FESEM ($50000\times$, $60000\times$). During the microstructure analysis on the field emission scanning electron microscope, it was found that for the analysis of such small grain sizes it is not recommended to etch the surface. Due to very small grain size it was necessary to evaporate the sample surface with carbon. Vickers hardness HV30 and Palmqvist toughness were measured simultaneously according to ISO 28079:2009 on the basis of five indentations performed on each sample.

3. Results and Discussion

The characteristics of sintered samples are presented in Table 3.

Table 3. Characteristics of sintered samples.

Mixture	Sample	Porosity A	Porosity B	Porosity C	ρ, %	d_{WC}, μm	HV 30	W_k, MN/m$^{3/2}$
WC-9Co/1	1-1	A < 02, A02	B00, B02	C00	99.2	0.213	1880.1	9.52
	1-2	A < 02, A02	B00, B02	C00	99.6	0.249	1834.6	9.58
	1-3	A < 02, A02	B00, B02	C00	99.5	0.261	1801.3	9.54
WC-9Co/2	2-1	A < 02, A02	B00, B02	C00	99.2	0.251	1808.3	9.41
	2-2	A < 02, A02	B00, B02	C00	99.4	0.265	1763.1	9.33
	2-3	A < 02, A02	B00, B02	C00	99.5	0.316	1729.3	9.42
WC-6Co/1	3-1	A04	B00, B02	C00	99.2	0.209	2041.1	9.05
	3-2	A02	B00, B02	C00	99.0	0.235	2034.4	9.01
	3-3	A02	B00, B02	C00	99.1	0.251	1988.2	9.10
WC-6Co/2	4-1	A06	B02, B04	C00	98.6	0.166	2196.1	9.14
	4-2	A04	B02, B04	C00	98.6	0.175	2164.9	9.12
	4-3	A04	B02, B04	C00	98.9	0.195	2103.9	9.14

3.1. Density and Microstructural Characteristics

Lower degree of porosity was achieved for samples with higher Co content. The degree of porosity of WC-9Co samples is predominantly A02, partially A < 02, predominantly B00, partially B02, without uncombined carbon or η-phase. The degree of porosity of WC-6Co samples ranges from A02 to A06, predominantly B02, without uncombined carbon or η-phase. Additionally, small cracks occurred on several samples resulting from one of the following technological operations: milling, waxing or granulation. Achieved densities of all samples are lower than the theoretical density. Densities measured for WC-9Co samples range from 99.2% to 99.6% of the theoretical density and the maximal density is obtained for sample 1-2 sintered at 1400°C. Densities measured for WC-6Co samples are lower compared to the WC-9Co mixture and range from 98.6% to 99.2%. Even with higher compaction pressures used during uniaxial die pressing, the high degree of porosity with pore sizes larger than 2 μm was obtained for the WC-6Co/2 mixture (Figure 1).

SEM MAG: 2.13 kx   DET: BSE Detector
HV: 30.0 kV   DATE: 12/16/11   20 um
Name: 377-1-1   Vega ©Tescan
Digital Microscopy Imaging
Faculty of Mechanical Engineering, Zagreb
377-1

Figure 1. Porosity of sample 4-2.

The influence of sintering temperature on the density of consolidated samples is presented graphically in Figure 2.

WC-9Co mixtures had the same trend, and they recorded an increase in density with increasing sintering temperature. The highest density of WC-9Co mixtures was obtained at 1400 °C. The difference between the measured density with respect to sintering temperature is small and amounts to only 0.3% to 0.4%, which may be due to compaction. The maximal density achieved for WC-9Co samples is 99.6% of the theoretical density, which is a highly relevant result for the process of sintering in hydrogen atmosphere. Sun published that the density of WC-Co samples with a grain sizes of approximately 0.250 μm is significantly influenced by Cr_3C_2 content. The increase of 0.4 wt. % Cr_3C_2 content considerably reduces the density; when the Cr_3C_2 content is small the change is slight [8]. The addition of 0.45% Cr_3C_2 in WC-9Co/1 did not lead to lower density compared to WC-9Co/2, which has only 0.27% VC. WC-6Co mixtures recorded lower values of measured density compared to WC-9Co mixtures. The density of WC-6Co/1 ranges from 99.0% to 99.2% of the theoretical density. The maximal value is achieved at the lowest sintering temperature, meaning that full densification was achieved. The density decreases with increasing temperature, which can be explained by the fact that the samples were compacted by uniaxial die pressing, the lack whereof is density non-uniformity. The density of WC-6Co/2 is increasing with increasing sintering temperature. WC-6Co/2 has a GGIs content two times higher compared to WC-6Co/1, which may indicate that GGIs influenced the achieved densities. In general, it can be concluded that the mixtures with lower Co content achieved lower densities. The achieved densities are quite low compared to densities which could be achieved by sinter-HIP process.

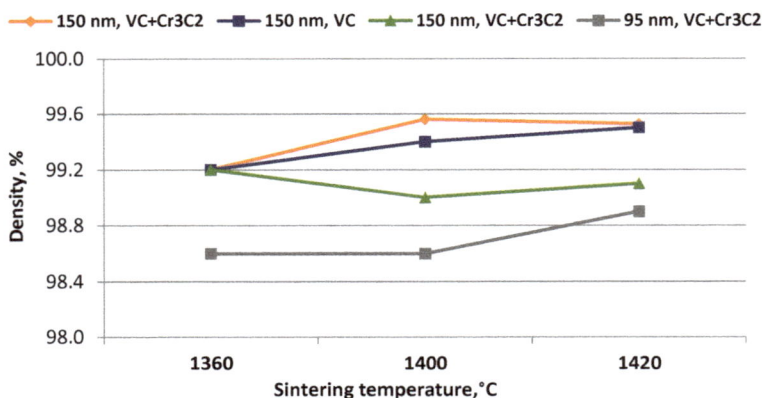

Figure 2. Correlation of densities and sintering temperature.

The size of the starting powders is almost retained at the lowest sintering temperatures. Accordingly, some of the consolidated cemented carbides can be classified as nanostructured WC-Co ($d_{WC} < 0.200$ μm) where d_{WC} represents the mean WC grain size determined by linear intercept method. The microstructure of samples obtained by SEM at the lowest sintering temperature, 1360 °C, is presented in Figure 3.

The microstructures are fine, with uniform grain size distribution, homogeneous and without abnormal grain growth at lower sintering temperatures. Nevertheless, a difference in microstructure between samples was noted. The smallest grain size of $d_{WC} = 0.166$ nm was measured for sample 4-1 (Figure 3d). The most homogenous microstructure was noted for sample 3-1 (Figure 3c). The biggest grain size of $d_{WC} = 0.251$ was measured for sample 2-1 (Figure 3b), where a discontinuous growth of single carbides in the microstructure, even at the lowest sintering temperature, was noted. Microstructural defects in the form of discontinuous growth of individual WC grains and abnormal grain growth of carbide agglomerates occurred, especially for mixture 2, at higher sintering temperatures. Discontinuous growth of individual WC grains, which grows more dramatically, faster and larger compared to surrounding grains, and the grouping of carbides are presented in Figure 4.

In general, such behavior represents a classical theory called Ostwald ripening (solution/re-precipitation process), where large WC grains grow at the expense of small WC grains that dissolve in Co binder spread around the grains. This process is called coarsening of WC grains. Still, scientists have found that Ostwald ripening/coarsening is not sufficient to explain grain growth in nanostructured cemented carbides with lower Co content where WC grains are not surrounded by Co binder but with one another. A new mechanism suggested by numerous scientists [6,11,12] is coalescence. Coalescence is the process by which two or more faceted WC grains merge during contact and consequently form a larger grain by diffusion and grain rotation thought rapid mass transport [11]. This mechanism can take place only if the grains are already bonded to each other, with a grain boundary and preferred orientations, as presented in Figure 5. Further holding at the sintering temperature would result in grain boundaries movement, the consequence of which would be further grain growth. Another reason for discontinuous grain growth can be chemical inhomogenities (i.e., S, Ca, C, etc.) and areas without GGIs.

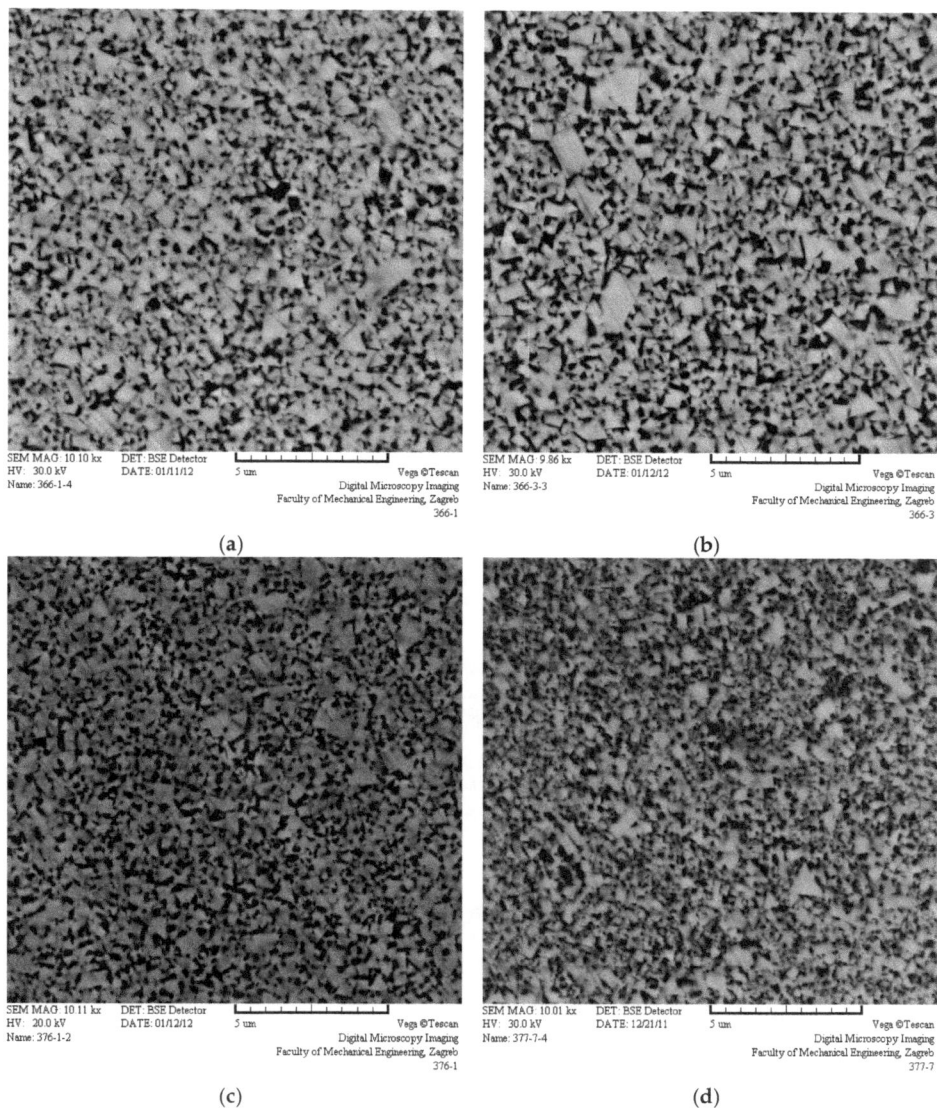

(a)

(b)

(c)

(d)

Figure 3. Microstructures of ultrafine and nanostructured cemented carbides. (**a**) Sample 1-1; (**b**) sample 2-1; (**c**) sample 3-1; (**d**) sample 4-1.

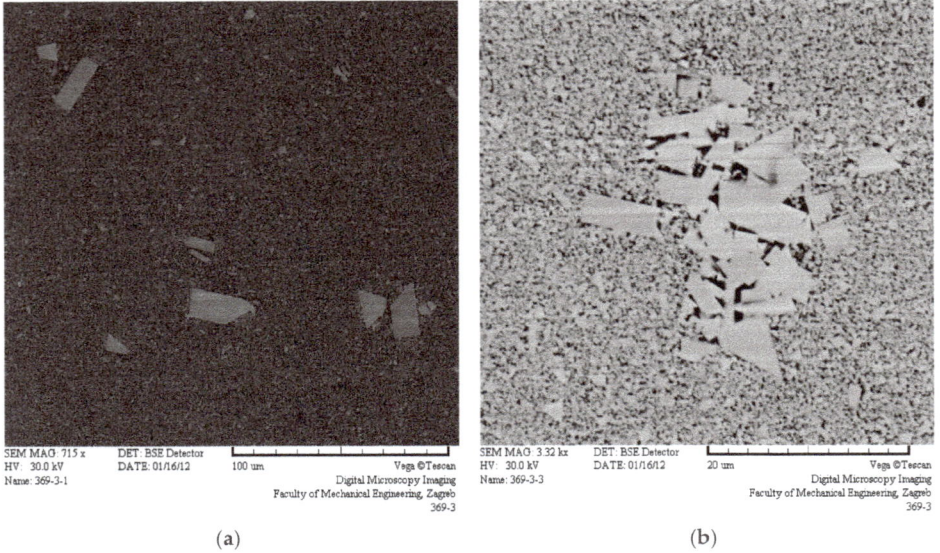

(a) (b)

Figure 4. Microstructural irregularities. (**a**) Discontinuous growth of individual grains on sample 2-3; (**b**) abnormal grain growth of carbide agglomerates on sample 2-3.

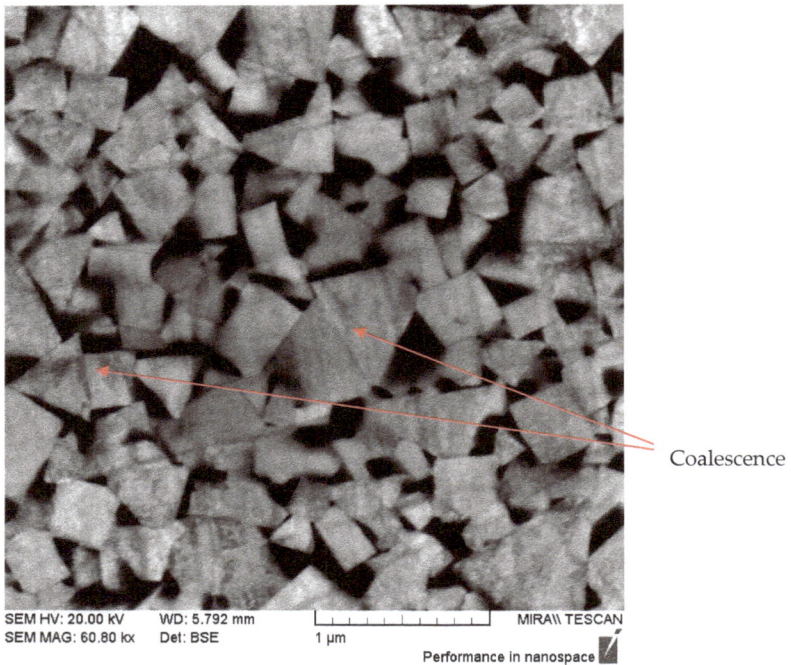

Figure 5. Microstructure with coalescence.

The influence of sintering temperature on WC grain size is illustrated in Figure 6.

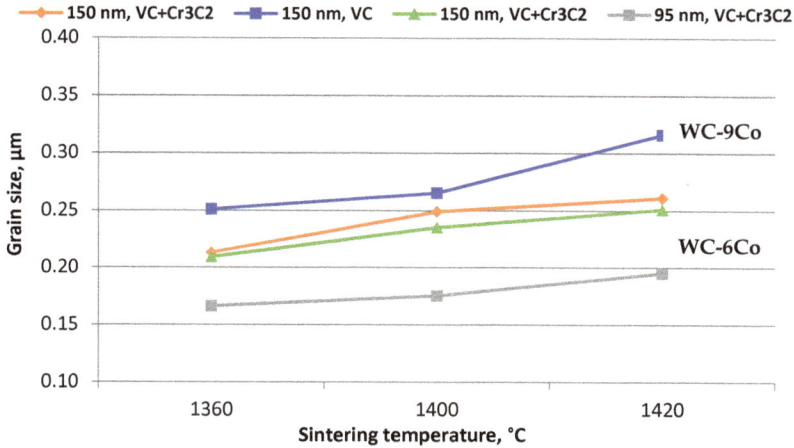

Figure 6. Correlation of WC grain size and sintering temperature.

The sintering temperature contributed significantly to microstructural characteristics of consolidated cemented carbides. The size of WC grains increases with increasing sintering temperatures for all mixtures. The largest increase was recorded for WC-9Co/2; 150 nm and the addition of VC. At the lowest sintering temperature, the measured WC grain size is 0.251 μm, which is an increase of 0.1 μm compared to the starting powder. It is important to note that microstructural analysis showed the existence of abnormal grain growth and grouping of carbides, especially at the highest sintering temperatures. It follows that the addition of 0.27% VC as GGI is too low to retain grain size of the starting powder in the sintered samples. Mixture WC-9Co/1, 150 nm with the addition of Cr_3C_2 and VC, retains a smaller grain size in relation to the starting powder. The grain growth of WC-9Co/1 in general is slightly higher compared to WC-6Co/1 consolidated from the same starting powder but with lower Co content. Co content contributed to the recrystallization and grain growth of near nano- and nanostructured cemented carbides sintered in hydrogen, confirming the role of Co in grain growth, already researched by various scientists [11,13–15]. The smallest grain sizes were measured for WC-6Co/2, 95 nm and the addition of Cr_3C_2 and VC, but in higher content compared to other mixtures. The finer the grain size of the starting powder, the smaller the final grain size. Yet at the highest sintering temperature, the grouping of the carbides and grain growth occurred for all samples. WC grain growth is dependent on the type and amount of GGIs and the grain size of the starting powder.

3.2. Mechanical Properties

The characteristics of starting powders, sintering temperature and achieved microstructure influenced the mechanical properties of consolidated samples. A graphical representation of the correlation between Vickers hardness and sintering temperature is presented in Figure 7.

The measured hardness is substantially influenced by sintering temperature and a consequently developed microstructure. Maximal hardness values were measured at the lowest sintering temperature, where the most uniform, homogenous microstructure was achieved for all mixtures. Furthermore, the characteristics of starting powders, GGIs and WC grain size of the starting powder influenced the measured hardness. The highest hardness was measured for WC-6Co/2 with the smallest grain size of the starting powder and the highest content of GGIs. The lowest hardness was achieved for WC-9Co/2 with the lowest content of GGIs and the biggest grain size. The correlation of Palmqvist toughness and sintering temperatures is presented in Figures 8 and 9.

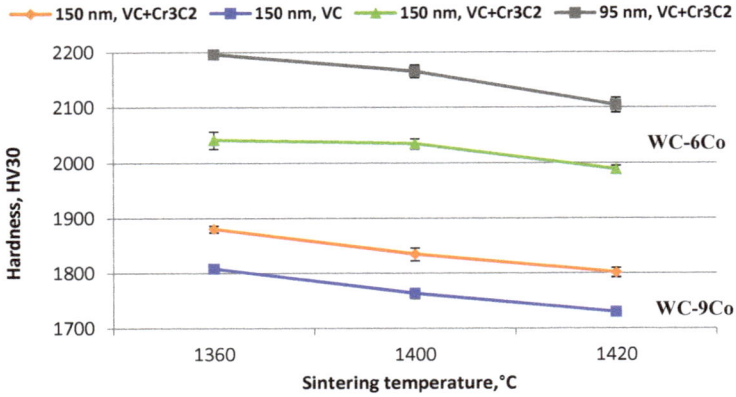

Figure 7. Correlation of Vickers hardness and $T_{sint.}$ for all mixtures.

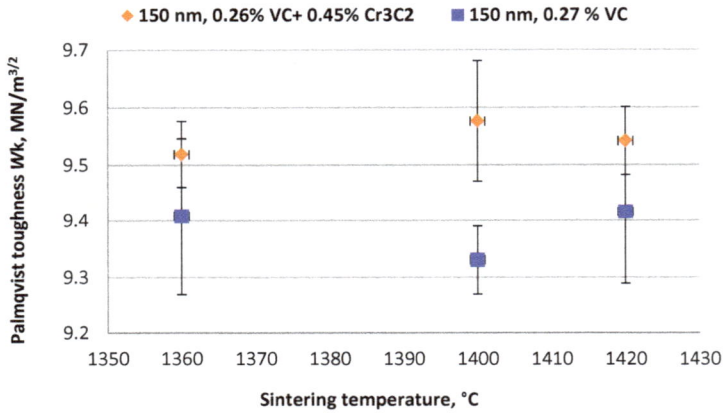

Figure 8. Correlation of Palmqvist toughness and $T_{sint.}$ for WC-9Co mixtures.

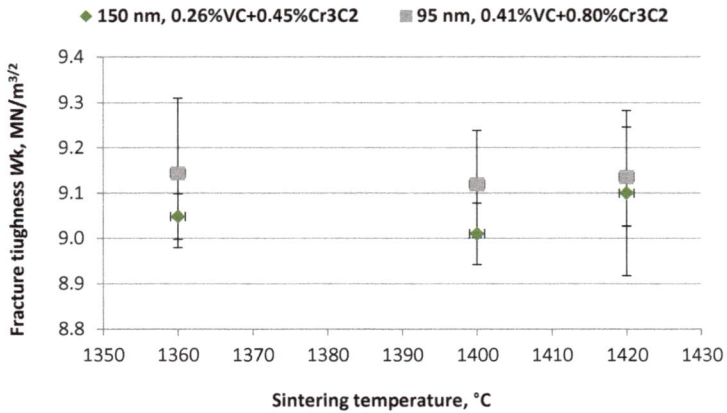

Figure 9. Correlation of Palmqvist toughness and $T_{sint.}$ for WC-6Co mixtures.

The measured Palmqvist toughness of WC-9Co mixtures is the same for all three sintering temperatures. The maximal value of 9.57 MN/m$^{3/2}$ was measured for sample 1-2 sintered at 1400 °C. Higher values were achieved for WC-9Co/1, 150 nm and VC and Cr$_3$C$_2$. Palmqvist toughness does not change with the increase in sintering temperature, which causes microstructural irregularities and consequently a change in hardness.

The sintering temperature did not influence the Palmqvist toughness of WC-6Co mixtures. A slightly higher Palmqvist toughness was achieved for the WC-6Co/2 mixture with a smaller grain size of the starting powder and a higher content of GGI. The correlation of Palmqvist toughness versus Vickers hardness for WC-9Co mixtures with respect to GGIs and the correlation of Palmqvist toughness and Vickers hardness for WC-6Co mixtures with respect to d_{BET} of starting powders and GGI content are presented in Figures 10 and 11.

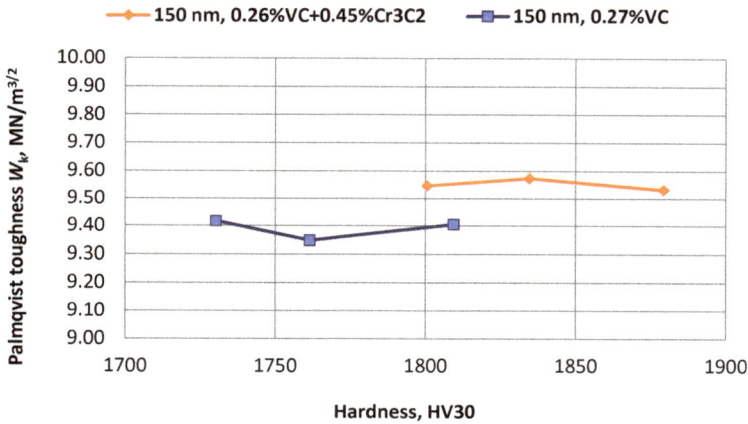

Figure 10. Palmqvist toughness versus hardness for WC-9Co mixtures with respect to GGIs.

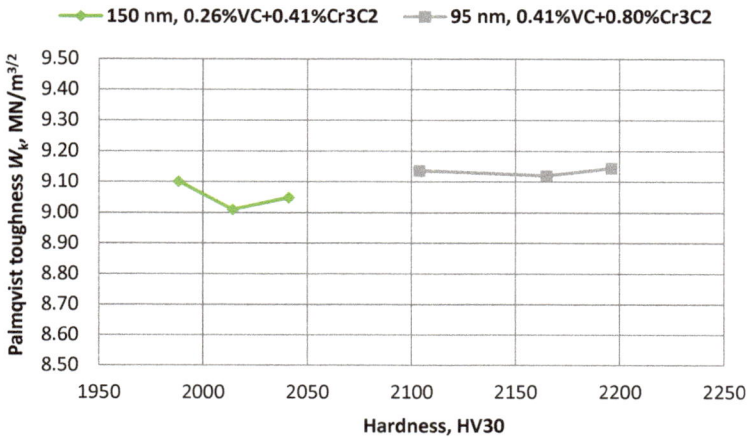

Figure 11. Palmqvist toughness versus hardness for WC-6Co mixtures with respect to d_{BET} of starting powders and GGI content.

A significant difference in measured Vickers hardness was noted comparing WC-9Co mixtures with the same size starting powder, 150 nm, but different GGIs. The addition of 0.45% Cr$_3$C$_2$

increased the Vickers hardness by approximately 70 HV and the fracture toughness by approximately 0.15 MN/m$^{3/2}$ for all sintering temperatures and reduced the hardness decrease with increasing temperature. The correlation between Palmqvist toughness and Vickers hardness is not linear and does not change with the change in hardness.

A significant difference in the measured Vickers hardness was noted comparing WC-6Co mixtures with the same type of GGIs but a different d_{BET} grain size and a specific surface area (BET) of starting WC powders. The reduction in grain size from 150 nm to a real nanosize of 95 nm resulted in significant hardness increase (150 HV) for all sintering temperatures. Nonetheless, in spite of a significant increase in hardness, the measured value of fracture toughness is only slightly higher and more uniform. Fracture toughness stability is not typical of the majority of materials, including conventional WC-Co cemented carbides. It is important to note that GGIs also contributed to this behavior. Grain size reduction may result in significant hardness increase, while there is no decrease in toughness.

4. Conclusions

The following conclusions can be drawn from the conducted research:

(1) In general, higher densities are achieved with higher Co content and greater grain size of the starting powder. Achieved densities of WC-9Co mixtures are close to theoretical densities. Comparing two WC-9Co mixtures with a 150 nm starting powder with different GGIs, it can be concluded that the addition of 0.45 wt. % Cr$_3$C$_2$ did not lead to lower density. The achieved densities are lower compared to densities which would be achieved with a sinter-HIP process or by sintering in vacuum.

(2) The behavior of the 95 nm starting powder, classified as real nanopowder, is completely different compared to the 150 nm staring powder; higher pressures are needed for the consolidation of green parts. Even with higher pressures, a microstructure with a lower densities was obtained. The consolidation of real nanopowders only by sintering in hydrogen without isostatic pressing is not recommended. Sinter-HIP process would be more suitable for the consolidation of real nano powders.

(3) The sintering temperature significantly impacted the microstructure of ultrafine and nanostructured cemented carbides. Increased sintering temperatures lead to microstructural irregularities in a form of discontinuous WC growth, carbide agglomerates and abnormal grain growth, as well as significant hardness decrease, especially for the mixture with only VC as GGIs. The lowest sintering temperature resulted in a fine-grained homogenous microstructure. It is preferred to use a higher amount of GGIs in case of sintering in hydrogen atmosphere.

(4) The characteristic of the starting powder (grain size and amount, as well as type of GGI) significantly influenced the microstructure and mechanical properties. The addition of 0.45% Cr$_3$C$_2$ contributed to microstructure homogeneity, reduced discontinuous and continuous grain growth, increased Vickers hardness by approximately 70 HV and fractured toughness by approximately 0.15 MN/m$^{3/2}$. The reduction of the starting powder grain size from 150 nm to a real nanosize of 95 nm resulted in a significant hardness increase with a simultaneous small increase in fracture toughness.

(5) Fracture toughness is not influenced by microstructural irregularities or the change in Vickers hardness.

Author Contributions: Tamara Aleksandrov Fabijanić performed density, grain size, hardness and fracture toughness measurements, analyzed the data and wrote the paper; Suzana Jakovljević performed microstructural analysis with a scanning electron microscope; Mladen Franz analyzed the data; and Ivan Jeren consolidated the samples.

Conflicts of Interest: The authors declare no conflict of interest.

References

1. Richter, V.; Poetschke, J.; Holke, R.; Michaelis, A. Nanoscaled Cemented carbides—Fiction or Reality? In Proceedings of the 18th International Plansee Seminar, Reutte, Austria, 3–7 June 2013.
2. Fabijanić, T.A.; Alar, Ž.; Ćorić, D. Influence of consolidation process and sintering temperature on microstructure and mechanical properties of near nano and nano-structured WC-Co cemented carbides. *Int. J. Refract. Met. Hard Mater.* **2016**, *54*, 82–89. [CrossRef]
3. Al-Aqeeli, N.; Saheb, N.; Laoui, T.; Mohammad, K. The Synthesis of Nanostructured WC-Based Cemented carbides Using Mechanical Alloying and Their Direct Consolidation. *J. Nanomater.* **2014**. [CrossRef]
4. Weidow, J.; Norgren, S.; Andren, H. Effect of V, Cr and Mn Additions on the Microstructure of WC-Co. *Int. J. Refract. Met. Hard Mater.* **2009**, *27*, 817–822. [CrossRef]
5. Poetschke, J.; Richter, V.; Holke, R. Influence and effectivity of VC and Cr_3C_2 grain growth inhibitors on sintering if binderless tungsten carbide. *Int. J. Refract. Met. Hard Mater.* **2012**, *31*, 218–223. [CrossRef]
6. Fang, Z.Z.; Wang, X.; Ryu, T.; Hwang, K.S.; Sohn, H.Y. Synthesis, sintering, and mechanical properties of nanocrystalline cemented tungsten carbide—A review. *Int. J. Refract. Met. Hard Mater.* **2009**, *27*, 288–299. [CrossRef]
7. Gille, G.; Szesny, B.; Dreyer, K.; van den Berg, H.; Schmidt, J.; Gestrich, T.; Leitner, G. Submicron and ultrafine grained cemented carbides for microdrills and metal cutting inserts. *Int. J. Refract. Met. Hard Mater.* **2002**, *20*, 3–22. [CrossRef]
8. Sun, L.; Jia, C.; Cao, R.; Lin, C. Effects of Cr_3C_2 additions on the densification, grain growth and properties of ultrafine WC-11Co composites by spark plasma sintering. *Int. J. Refract. Met. Hard Mater.* **2008**, *26*, 357–361. [CrossRef]
9. Kishino, J.; Nomura, H.; Shin, S.-G.; Matsubara, H.; Tanase, T. Computational study on grain growth in cemented carbides. In Proceedings of the 15th Internationla Plansee Seminar, Reutte, Austria, May 2001; Kneringer, G., Rodhammer, P., Wildner, H., Eds.; Plansee AG: Reutte, Austria, 2001; Volume 2.
10. Sivaprahasam, D.; Chandrasekar, S.B.; Sundaresan, R. Microstructure and mechanical properties of nanocrystalline WC-12Co consolidated by spark plasma sintering. *Int. J. Refract. Met. Hard Mater.* **2007**, *25*, 144–152. [CrossRef]
11. Mannesson, K. WC Grain Growth during Sintering of Cemented Carbides. Ph.D. Thesis, KTH Industrial Engineering and Management, Stockholm, Sweden, April 2011.
12. Wang, X.; Fang, Z.Z.; Sohn, H.Y. Grain growth during earla stage of sintering of nanosized WC-Co powder. *Int. J. Refract. Met. Hard Mater.* **2008**, *26*, 232–241. [CrossRef]
13. Mannesson, K.; Elfwing, M.; Kusoffsky, A.; Norgeren, S.; Agren, J. Analysis of WC grain growth during sintering using electron backscatter diffraction and image analysis. *Int. J. Refract. Met. Hard Mater.* **2008**, *26*, 449–455. [CrossRef]
14. Da Silva, A.G.P.; Schubert, W.D.; Lux, B. The Role of the Binder Phase in the WC-Co Sintering. *Mat. Res.* **2001**. [CrossRef]
15. Breval, E.; Cheng, J.P.; Agrawal, D.K.; Gigl, P.; Dennis, M.; Roy, R.; Papworthc, A.J. Comparison between microwave and conventional sintering of WC/Co composites. *Mater. Sci. Eng. A* **2005**, *391*, 285–295. [CrossRef]

metals

MDPI

Article

Enhanced Age Strengthening of Mg-Nd-Zn-Zr Alloy via Pre-Stretching

Erjun Guo, Sicong Zhao *, Liping Wang and Tong Wu

School of Materials Science and Engineering, Harbin University of Science and Technology, Harbin 150040, China; guoerjun@126.com (E.G.); lp_wang2003@126.com (L.W.); wu7tong@163.com (T.W.)
* Correspondence: zscwr@163.com; Tel./Fax: +86-451-8639-2517

Academic Editor: Soran Birosca
Received: 2 May 2016; Accepted: 19 August 2016; Published: 24 August 2016

Abstract: Pre-stretching was carried out to modify the microstructure of Mg-Nd-Zn-Zr alloy to enhance its age strengthening. The results indicated that more heterogeneous nucleation sites can be provided by the high density of dislocations caused by the plastic pre-stretching deformation, as well as speeding up the growth rate of precipitates. Comparison of microstructure in non-pre-stretched specimens after artificial aging showed that pre-stretched specimens exhibited a higher number density of precipitates. The fine and coarse plate-shaped precipitates were found in the matrix. Due to an increase in the number density of precipitates, the dislocation slipping during the deformation process is effectively hindered, and the matrix is strengthened. The yield strength stabilizes at 4% pre-stretching condition, and then the evolution is stable within the error bars. The 8% pre-stretched specimens can achieve an ultimate tensile strength of 297 MPa. However, further pre-stretching strains after 8% cannot supply any increase in strength. Tensile fracture surfaces of specimens subjected to pre-stretching strain mainly exhibit a trans-granular cleavage fracture. This work indicated that a small amount of pre-stretching strain can further increase strength of alloy and also effectively enhance the formation of precipitates, which can expand the application fields of this alloy.

Keywords: magnesium alloys; precipitation; pre-stretching; mechanical properties

---

1. Introduction

The addition of rare-earth elements (RE) is an effective way to improve the mechanical properties of magnesium alloys [1–4]. The Mg-Nd-Zn-Zr alloy is one of the most successful Mg-RE alloy systems. This alloy shows a highly desirable combination of relative good room-temperature tensile properties and high-temperature creep resistance, both of which are associated with structure, distribution and number density of precipitates [5]. The precipitation sequence in this alloy during isothermal aging is commonly accepted as: supersaturated solid solution (SSSS) → ordered Guinier-Preston zones (G.P. zones) → β'' → β' → β_1 → β [6–8]. Despite many works regarding the precipitates evolution have been achieved in the past few years. However, further attempts to improve the strength of the alloy via controlling the nucleation and growth of precipitates during aging process are rarely reported. Micro alloying additions of some elements may considerably influence the precipitation during ageing in many magnesium alloy systems. Geng and Buha [9,10] have reported that the addition of Co and Cr to a Mg-Zn alloy can increase number density of precipitates and enhances age-hardening response. Elsayed [11] investigated the influence of Na on age-hardening response of the Mg-9.8Sn-3.0Al-0.5Zn alloy. The Na clusters refine the microstructure by acting as heterogeneous nucleation sites for Mg_2Sn precipitates. However, the recognized micro alloying elements are still less and not applicable for all the magnesium alloy systems, especially for RE-containing or Zr-containing magnesium alloys. For this reason, it is necessary to find other effective methods to enhance the age strengthening for Mg-Nd-Zn-Zr alloy systems.

Another very efficient way to enhance the age strengthening for most magnesium alloy systems may be pre-deformation. The cold pre-deformation after solution treatment, and prior to aging treatment, has been attracted researchers lately due to a tremendous potential for enhancing the nucleation and growth of precipitates. Gazizov [12] has reported pre-deformation is effective in increasing the strength of Al-Cu-Mg-Ag alloy by modifying the normal precipitation sequence. Ozaki [13] studied the effects of pre-compressive strain on the fatigue life of the AZ31 magnesium. These results suggest that the pre-deformation introduces numerous dislocations and twins. The precipitates interact with dislocations during the aging treatment and therefore influence the performance of alloy, which is an area that has been much less explored. Moreover, the pre-deformation by stretching is rarely employed due to poor ductility of many magnesium alloys. In this research, the cold plastic pre-stretching at room temperature (RT) was performed on the Mg-Nd-Zn-Zr alloy system, and the microstructure, mechanical properties and fracture behavior were investigated in detail. It is aimed to further improve the strength of the alloy and, thus, to broaden its application fields.

2. Experimental Section

The nominal compositions of the investigated alloy was Mg-2.7Nd-0.4Zn-0.5Zr (wt. %). Solution treatment of as-cast ingots was performed at 530 °C for 14 h and followed by quenching in water. The specimens with dimensions of 3 mm × 60 mm × 120 mm were cut from the as-quenched ingots by electrical discharge machining (EDM) and cleaned in ethanol. Then the specimens were pre-stretched by 2%, 4%, 6%, 8%, and 10% by electronic material testing machine under a constant strain rate of 1×10^{-3} s^{-1} at room temperature (RT), and subsequently aged at 200 °C for 8 h. The specimens without pre-stretching were also prepared for comparisons. After aging treatment, a series of tensile bars were prepared by EDM and polished to remove the oxide layer from the surface and cleaned with ethanol.

Tensile test was performed on material test machine under a constant strain rate of 1×10^{-3} s^{-1} at RT. Tensile bars with a 10 mm gauge width and 20 mm gauge length parallel to the pre-stretched direction, as illustrated in Figure 1. Metallographic specimens were etched with a solution of 2.5 g picric acid, 2.5 mL acetic acid, 50 mL ethanol and 50 mL water, and observed by an optical microscope (GX71, OLYMPUS, Tokyo, Japan). The transmission electron microscopy (TEM) was used to study the microstructure in the pre-deformed and aged alloys. TEM specimens were prepared by ion-milling (691, Gatan, Pleasanton, CA, USA) and examined in a transmission electron microscope (JEM-2100, JEOL, Tokyo, Japan) operating at 200 kV. The fracture surfaces of specimens after a tensile test were observed by a scanning electron microscope (Quanta 200, FEI, Eindhoven, The Netherlands).

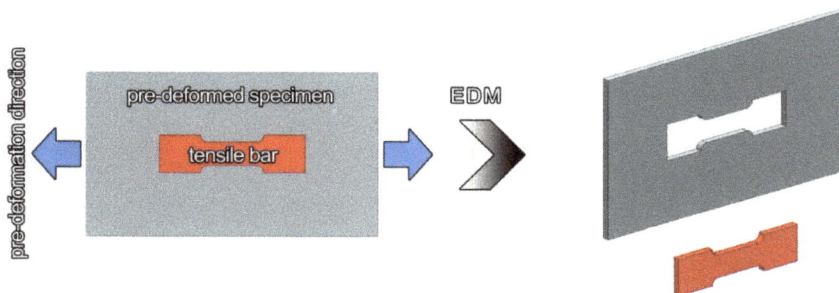

Figure 1. A schematic illustrating the preparation of tensile bar.

3. Results

3.1. Microstructures

Figure 2 shows the microstructure for Mg-2.7Nd-0.4Zn-0.5Zr alloy in as-cast condition. The microstructure of as-cast specimen consists of α-Mg matrix and eutectic compounds. The fine equiaxial α-Mg grains are separated by eutectic compounds. The eutectic compounds mainly gather in grain boundaries, and some small eutectic compound particles also present within grains as is evident in Figure 2a. As previously reported [14], the eutectic compounds for the Mg-Nd system may be β phase, which is tetragonal lattice structure with $a = 1.03$ nm and $c = 0.59$ nm. The composition of the β phase is reported as $Mg_{12}Nd$, which is the main existing phase in the as-cast microstructure. Figure 2b shows a typical bright-field TEM image and corresponding selected area electron diffraction (SAED) patterns of the eutectic compounds on grain boundaries. The beam is parallel to $[102]_{\beta}$ zone axis in this specimen. The eutectic compounds are identified to be the equilibrium β phase, which is irregular in shape and has no orientation relationships with the α-Mg matrix [14]. Figure 3 shows the optical images of the quenched specimens with 0%, 2%, 8%, and 10% pre-stretching. After solution treatment at 530 °C for 14 h and quenching in water, the eutectic compounds completely dissolve into the matrix. It should be noted that many black regions distribute unevenly in a few grains as shown in Figure 3a. These regions contain numerous undissolvable particles, which precipitate during the solution treatment. Another interesting point is the presence of twins after pre-stretching. The number density of twins is varied from grain to grain due to size and orientation difference. The formation of twins is easier in coarse grains during deformation. It is revealed that the stress concentration along the grain boundaries triggers the nucleation of twins and, thus, the nucleation density of twins is sensitive to the grain size. Additionally, the size of twins is limited by grain size, which implies that grain size affects twin growth [15]. Figure 3b,d reveal that the number density and size of twins increase significantly as the strains rise from 2% to 10%. However, the further pre-stretching after 8% cannot supply any visible increase in twins. Figure 4 shows TEM bright-field images of the quenched specimen with 8% pre-stretching. Numerous dislocations distribute evenly in the twins and matrix. Hence, high density dislocations form during pre-stretching plastic deformation.

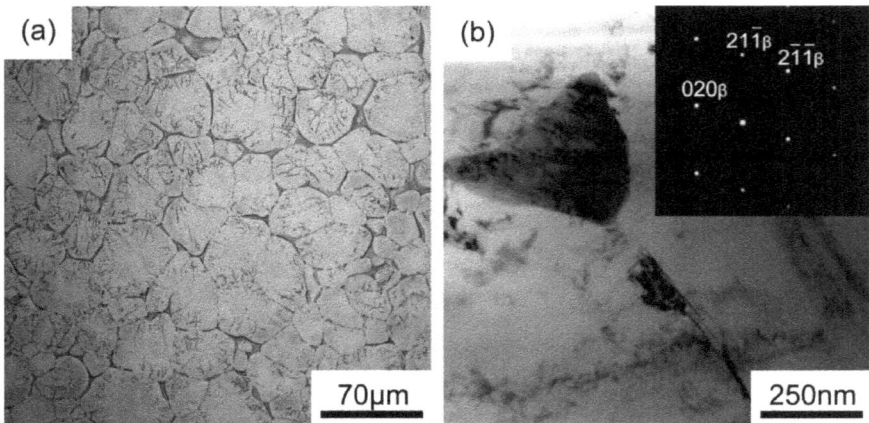

Figure 2. The microstructure of the as-cast specimen: (**a**) optical images and (**b**) TEM bright-field image and corresponding SAED patterns for β phase (B//$[102]_{\beta}$).

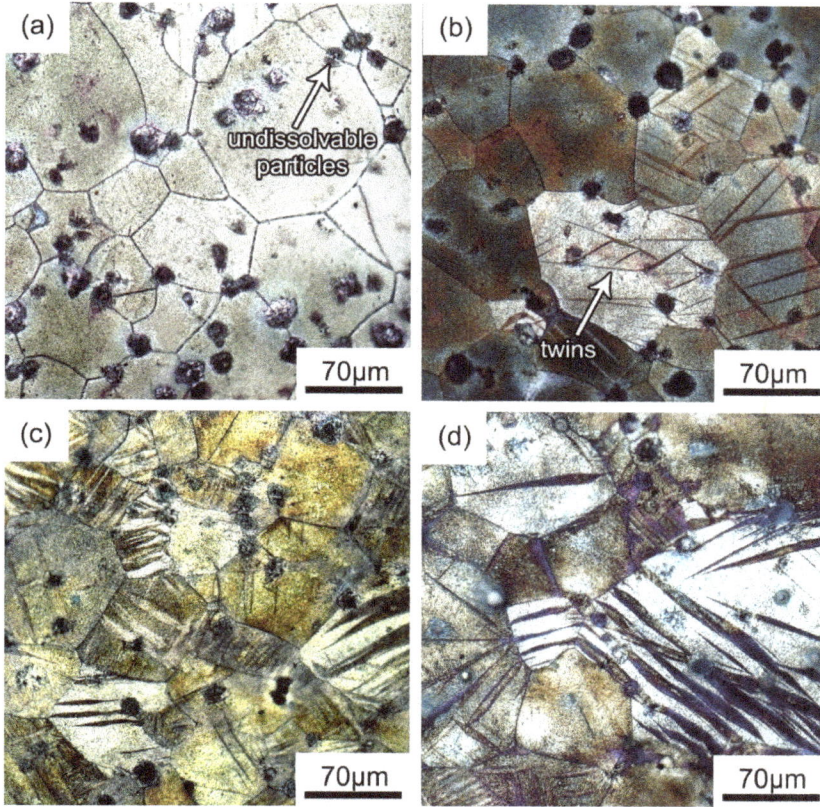

Figure 3. Optical images of quenched specimens with (**a**) 0%, (**b**) 2%, (**c**) 8%, and (**d**) 10% pre-stretching.

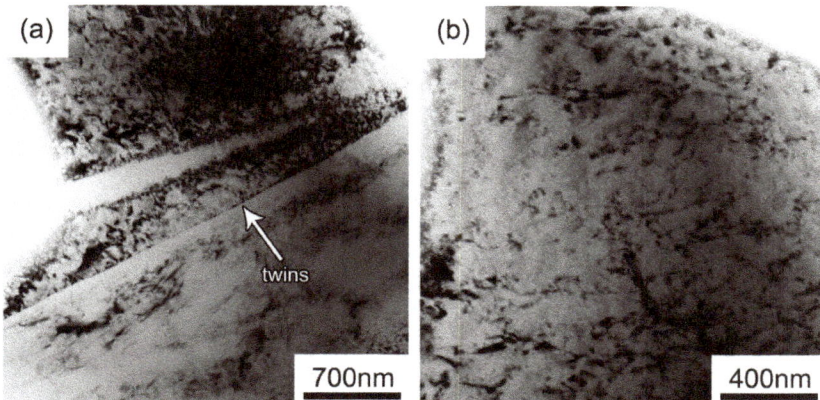

Figure 4. TEM bright-field images of the quenched specimen with 8% pre-stretching: (**a**) twins; (**b**) dislocations.

Figure 5 shows the TEM images of specimens without pre-stretching aged at 200 °C for 8 h. During the aging treatment, the Nd atoms in solid-solution within the matrix decrease due to precipitation. According to bright-field image with beam parallel to $[11\bar{2}0]_\alpha$ direction (Figure 5a), the presence of plate-shaped precipitates distribute evenly in the matrix. The orientation relationship between plate-shaped precipitates and α-Mg is such that $[11\bar{2}0]_{\beta''}//[11\bar{2}0]_\alpha$ and $(01\bar{1}0)_{\beta''}//(01\bar{1}0)_\alpha$. The β'' precipitates in Mg-Nd alloy usually form these kinds of SAED patterns. Moreover, the simulated SAED patterns (right lower quadrant in Figure 5a for this structure are in good agreement with the observed experimental results. Hence, the plate-shaped precipitates are β'' precipitates. Figure 5b shows the high magnification TEM image of the plate-shaped precipitates. The average size and inter-spacing of plate-shaped precipitates are ~30 nm and ~20 nm, respectively, in the non-pre-stretched condition.

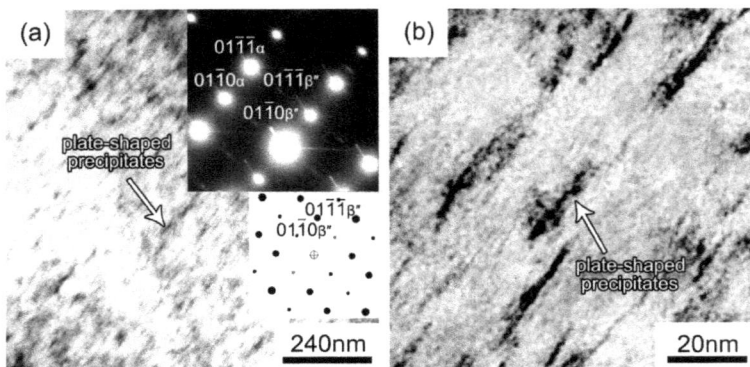

Figure 5. TEM images of specimens without pre-stretching aged at 200 °C for 8 h: (**a**) plate-shaped precipitates in the matrix and corresponding SAED patterns (B//$[11\bar{2}0]_\alpha$); and (**b**) high magnification TEM image of the plate-shaped precipitates.

Figure 6 shows TEM images of specimen with 8% pre-stretching aged at 200 °C for 8 h. Two distinct types of precipitates morphologies are revealed under this condition as seen in Figure 6a. The first type of precipitates is the fine dispersion of plate-shaped β'' precipitates. Compared with the specimen without pre-stretching strain, the size and inter-spacing of plate-shaped precipitates have been significantly reduced when the specimen subjected to pre-stretching strain of 8%. The other precipitates are a large number of coarse precipitates. The average size of this precipitates is ~80 nm in length. In order to determine the type of coarse precipitates, the SAED patterns of coarse precipitates are analyzed. The β'' precipitates in Mg-Nd alloy usually form these kinds of SAED patterns. The orientation relationship between β'' precipitates and α-Mg is such that $[5\bar{1}412]_{\beta''}//[5\bar{1}46]_\alpha$ and $(2\bar{2}0\bar{1})_{\beta''}//(1\bar{1}0\bar{1})_\alpha$. The simulated SAED patterns (right lower quadrant in Figure 6a) for β'' precipitates are in good agreement with the observed experimental results. The dark-field morphology of β'' precipitates are shown in Figure 6b. As it is evident, the fine and coarse precipitates display the same contrast in white. Although they have the different morphologies, they have the same structure. Hence, the coarse precipitates may be also the β'' precipitates. The high magnification TEM image of fine precipitates is shown in Figure 6c, which reveals that the average size and inter-spacing of fine precipitates are ~3 nm and ~5 nm, respectively. Figure 6d shows the distribution of precipitates around grain boundary. After 8% pre-stretching and subsequent aging treatment, there is no visible precipitate-free zones around grain boundary. Figure 6e displays the distribution of precipitates around twins. It exhibits the presence of twins in the pre-stretched specimens even after the aging. Additionally, the distribution and number density of precipitates around the grain boundary and twins is same as the matrix. Figure 6f shows the TEM bright-field morphology of undissolvable rod-shaped particles, which are also present in as-quenched specimens (Figure 3). The projected lengths of the undissolvable rod-shaped particles vary from 100 nm to 900 nm. The energy dispersive X-ray spectrum (EDXS) recorded from an undissolvable rod-shaped particle is shown in the Figure 6f upper right inset. The qualitative results show a prominent Mg K peak, together with significant intensities of Zn K and Zr K peaks. As can be noticed, the ratio of Zn (29.00 at. %) and Zr (40.80 at. %) is nearly two to three. Therefore, the undissolvable rod-shaped particles may be the Zn_2Zr_3 phase similar to what was reported in the literature for this phase [16]. It should be mentioned that the distribution of plate-shaped β'' precipitates is uniform around the Zn_2Zr_3 particles. The Zn_2Zr_3 particles precipitate during the solution treatment. Thus, the distribution of Nd atoms has been slightly affected by Zn_2Zr_3 particles, and the β'' precipitates can precipitate evenly around the Zn_2Zr_3 particles.

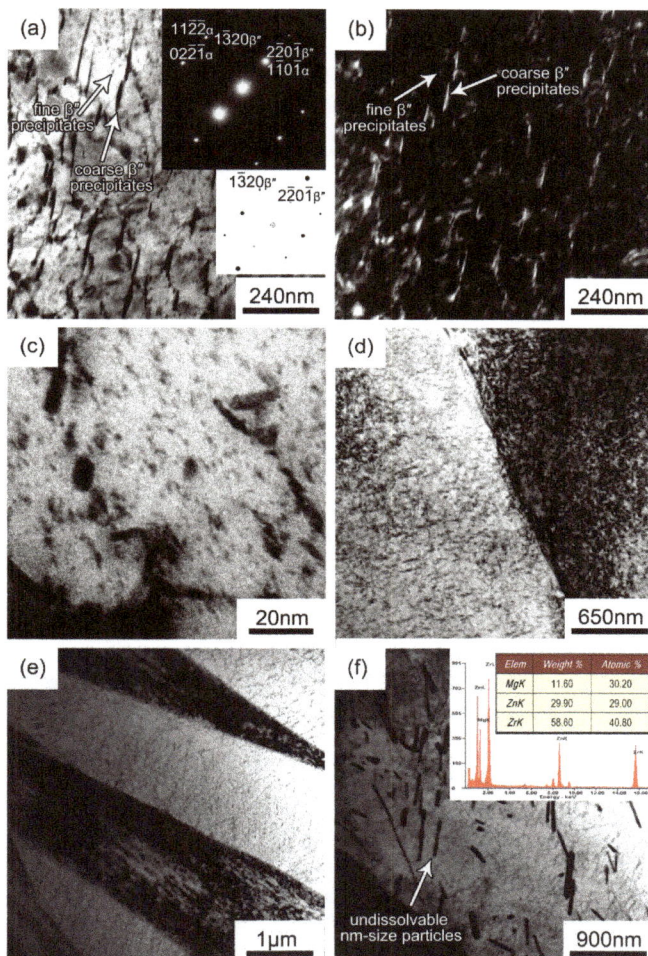

Figure 6. TEM images of specimen subjected to 8% pre-stretching and subsequently aged at 200 °C for 8 h: (**a**) and (**b**) are bright-field and dark-field morphology of plate-shaped precipitates in the matrix $(B//[5\bar{1}\bar{4}6]_\alpha)$, respectively; (**c**) is high magnification TEM image of the fine precipitates; (**d**) and (**e**) are bright-field morphology of grain boundaries and twins, respectively; and (**f**) is bright-field morphology of undissolvable particles and corresponding EDXS recorded from an undissolvable particle.

3.2. Mechanical Properties

Figure 7 shows the evolution of mechanical properties with pre-stretching strains for all specimens aged at 200 °C for 8 h. It clearly shows that the ultimate tensile strength (UTS), yield strength (YS) and elongation change with the changes in the pre-stretching levels. For the case without pre-stretching, the UTS is low at 240 MPa. As pre-stretching rise from 2% to 8%, although the UTS increases from 270 MPa to 297 MPa, the elongation decreased significantly. The YS stabilizes at 4% pre-stretching condition, and then the evolution is stable within the error bars. It is noteworthy that the 8% pre-stretched specimen exhibits the highest UTS, whereas the increment rate of UTS decreases when the pre-stretching after 2%. As the pre-stretching strains increase from 8% to 10%, the average strength tends to slightly reduce, indicating that further pre-stretching strains after 8% cannot supply any increase in strength. Hence, the strength significantly increases after pre-stretching and aging, and ductility largely decreases.

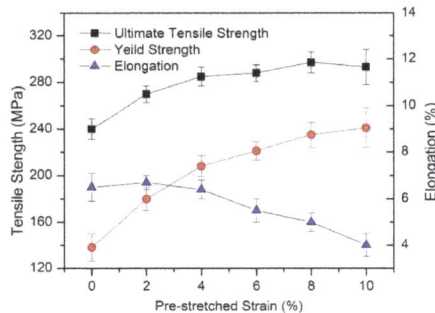

Figure 7. Mechanical properties of specimens subjected to pre-stretch and subsequently aged at 200 °C for 8 h.

3.3. Fractography

Figure 8 shows the SEM images of typical tensile fracture surfaces for specimens with different pre-stretching strains and subsequently aged at 200 °C for 8 h. The examination of the microstructure reveals that alloys have mixed trans-granular and inter-granular fracture. However, the fracture mode is the mainly trans-granular fracture. The fracture surfaces consist of cleavage planes and steps. The size of cleavage planes in the specimen with pre-stretching strain of 8% is smaller than specimen with pre-stretching strain of 0%. Based on the above results, it is inferred that all the tensile fracture surfaces mainly exhibit a trans-granular cleavage fracture, and the cleavage plane size of the pre-stretched alloy is smaller than that of the non-pre-stretched alloy.

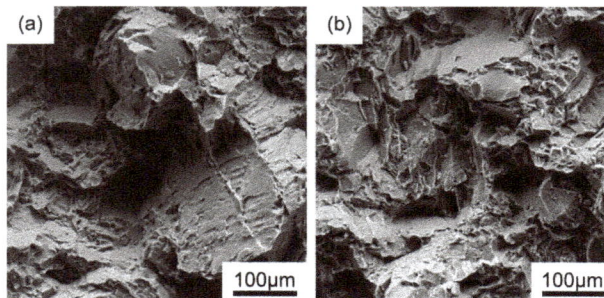

Figure 8. SEM images showing typical fracture surfaces of specimens with (**a**) 0% and (**b**) 8%, respectively, after tensile test at RT.

4. Discussion

4.1. Effect of Pre-Stretching on the Precipitation Evolution

The Mg alloys deformed at RT will promote the activation of twins and basal slip for the former, and the prismatic and pyramidal slip are activated for the later. Moreover, basal plane slip takes a high proportion during cold deformation in random orientated alloys [15,17]. That is why the density of dislocations and twins becomes pronounced when the specimen subjected to plastic pre-stretching deformation. During aging, the morphology, distribution, and number density of precipitates will be modified by dislocations. Compared with the specimen without pre-stretching strain, there are numerous coarse plate-shaped precipitates and fine plate-shaped precipitates exist in the matrix. Moreover, the number density of precipitates is higher than specimens without pre-stretching. This is attributed to heterogeneous nucleation on equilibrium state dislocations [18]. The schematic diagrams of precipitates evolution during aging treatment are shown in Figure 9. For a start, a large number of coarse plate-shaped precipitates heterogeneously nucleate on the dislocations and grow during the aging process. Most of the dislocations and a certain amount of solute Nd are consumed in this process. The continuous dislocations are interrupted and turn into numerous small segments. Subsequently, fine plate-shaped precipitates nucleate and grow on the matrix and the interrupted dislocation segments. It consumes the residual dislocation segments and solute Nd. Due to the growth of precipitates is limited by the residual dislocation size and solute Nd, the size of plate-shaped precipitates formed in the later aging process is smaller. Thus, the potential nucleation sites of precipitates are increased by dislocations, which increased the number density of precipitates. In addition, the length of coarse plate-shaped precipitates is ~80 nm, and the length of the plate-shaped precipitates forming in the non-pre-stretched specimen is ~30 nm. It is noticed that the size of the coarse plate-shaped precipitates is significantly larger than plate-shaped precipitates formed in non-pre-stretched specimen. The reason may be that the dislocations most likely act as fast diffusion pips for the solute Nd. The high density dislocations considerably speed up the growth rate of coarse plate-shaped precipitates. Additionally, non-equilibrium vacancies formed in pre-stretching also increase the kinetics of solute diffusion. Therefore, the pre-stretching strain can greatly enhance the aging kinetics. As noticed in the above discussion, the microstructure includes numerous coarse plate-shaped precipitates, fine plate-shaped precipitates and twins, when the specimens subjected to pre-stretching strain and subsequently aging treatment.

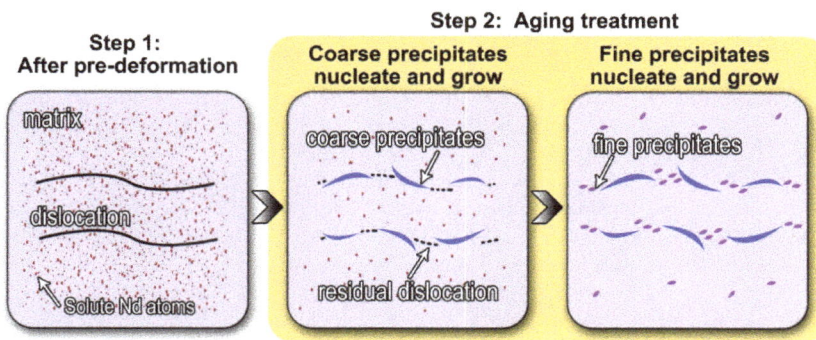

Figure 9. Schematic diagrams of precipitates evolution during aging treatment.

4.2. Effect of Pre-Stretching on the Mechanical Properties

The high strength of the pre-stretched and aged alloy is attributed to the superposition of different strengthening mechanisms. In the case of pre-stretched and aged specimens, the increment in yield strength is determined by work strengthening and precipitation strengthening. The work strengthening is severely weakened by the occurrence of recovering. The density of dislocations gradually reduces during the aging at 200 °C. However, the numerous twins are retained in the matrix after aging treatment. The twin boundaries can effectively act as barriers for dislocation slips and become a source of work hardening. The precipitation strengthening in pre-stretched alloys occurs through the formation of precipitates, which form mainly as coarse and fine plate-shaped particles. It is worth noting that the numerous dislocations increase the number density of precipitates. Additionally, the strength is sensitive to the number density of precipitates, since the more precipitates will shorten the inter spacing of them according to the Orowan law [19]. Then the dislocations are more difficult to pass through them [20]. In addition to the Orowan strength, there is a contribution from the backstress caused by the strain incompatibility and modulus mismatch between precipitates embedded in a sheared matrix [21]. The dislocation motion is impeded simultaneously, rather than sequentially, by both coarse and fine precipitates, which leads to threshold stress higher than for a dislocation interacting with either type of precipitates. In order to illustrate it, a highly simplified geometry is assumed as shown in Figure 10. This situation occurs when the dislocations are pinned at the departure side of the coarse precipitates, while concurrently subjected to the elastic back-stress from 8 neighboring fine precipitates. The overall threshold stress is the sum of the true detachment stress from the precipitates and the back-stress [22]. As it is noticed in the above discussion, the combination of work strengthening and precipitation strengthening provides excellent strength to the alloy after pre-stretching and subsequent aging treatment.

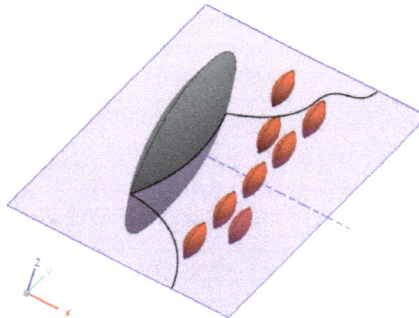

Figure 10. Three-dimensional schematic, showing precipitates contributing to the back-stress on the dislocation.

5. Conclusions

(1) The more heterogeneous nucleation sites can be provided by the high density of dislocations caused by the plastic pre-stretching deformation, as well as speeding up the growth rate of precipitates. The comparisons of microstructure in non-deformed specimens after artificial aging shows that pre-stretched specimens contain not only fine and coarse precipitates but also a higher number density of precipitates.

(2) The dislocation slipping during the deformation process is effectively hindered, and the matrix is strengthened because of an increase in the number density of precipitates and twins. The YS stabilizes at 4% pre-stretching condition, and then the evolution of YS is stable within the error bars.

(3) Tensile fracture surfaces of pre-stretched and subsequently aged specimens mainly exhibit a trans-granular cleavage fracture. The cleavage plane size of the pre-stretched specimens is smaller than that of non-pre-stretched specimen.

Acknowledgments: The authors gratefully acknowledge the financial support from the Heilongjiang Province Natural Science Foundation (No. ZD2016011).

Author Contributions: E.G. and L.W. conceived and designed the experiments; S.Z. and T.W. performed the experiments. T.W. analyzed the data. S.Z. wrote the paper.

Conflicts of Interest: The authors declare no conflict of interest.

References

1. Sravya, T.; Sankaranarayanan, S.; Abdulhakim, A. Mechanical properties of magnesium-rare earth alloy systems: A review. *Metals* **2015**, *5*, 1–39.
2. Ahmad, M.; Mamoun, M. Experimental investigation of the Mg-Nd-Zn isothermal section at 300 °C. *Metals* **2015**, *5*, 84–101.
3. Li, R.; Xin, R.; Chapuis, A.; Liu, Q.; Fu, G.; Zong, L.; Yu, Y.; Guo, B.; Guo, S. Effect of cold rolling on microstructure and mechanical property of extruded Mg-4Sm alloy during aging. *Mater. Charact.* **2016**, *112*, 81–86. [CrossRef]
4. Peng, Q.; Dong, H.; Wang, L.; Wu, Y.; Wang, L. Aging behavior and mechanical properties of Mg-Gd-Ho alloys. *Mater. Charact.* **2008**, *59*, 983–986. [CrossRef]
5. Ma, L.; Mishra, R.K.; Balogh, M.P.; Peng, L.; Luo, A.A.; Sachdev, A.K.; Ding, W. Effect of Zn on the microstructure evolution of extruded Mg-3Nd (-Zn)-Zr (wt. %) alloys. *Mater. Sci. Eng. A* **2012**, *543*, 12–21. [CrossRef]
6. Sanaty-Zadeh, A.; Luo, A.A.; Stone, D.S. Comprehensive study of phase transformation in age-hardening of Mg-3Nd-0.2Zn by means of scanning transmission electron microscopy. *Acta Mater.* **2015**, *94*, 294–306. [CrossRef]
7. Sanaty-Zadeh, A.; Xia, X.; Luo, A.A.; Stone, D.S. Precipitation evolution and kinetics in a magnesium-neodymium-zinc alloy. *J. Alloy. Compd.* **2014**, *583*, 434–440. [CrossRef]
8. Paliwal, M.; Das, S.K.; Kim, J.; Jung, I.H. Diffusion of Nd in hcp Mg and interdiffusion coefficients in Mg-Nd system. *Scr. Mater.* **2015**, *108*, 11–14. [CrossRef]
9. Geng, J.; Gao, X.; Fang, X.Y.; Nie, J.F. Enhanced age-hardening response of Mg-Zn alloys via Co additions. *Scr. Mater.* **2011**, *64*, 506–509. [CrossRef]
10. Buha, J. The effect of micro-alloying addition of Cr on age hardening of an Mg-Zn alloy. *Mater. Sci. Eng. A* **2008**, *492*, 293–299. [CrossRef]
11. Elsayed, F.R.; Sasaki, T.T.; Mendis, C.L.; Ohkubo, T.; Hono, K. Significant enhancement of the age-hardening response in Mg-10Sn-3Al-1Zn alloy by Na microalloying. *Scr. Mater.* **2013**, *68*, 797–800. [CrossRef]
12. Gazizov, M.; Kaibyshev, R. Effect of pre-straining on the aging behavior and mechanical properties of an Al-Cu-Mg-Ag alloy. *Mater. Sci. Eng. A* **2015**, *625*, 119–130. [CrossRef]
13. Ozaki, J.; Yosida, M.; Horibe, S. The effect of pre-compressive strain on the fatigue life of the AZ31 magnesium alloy. *Mater. Sci. Eng. A* **2014**, *604*, 192–195. [CrossRef]
14. Pike, T.J.; Noble, B. The formation and structure of precipitates in a dilute magnesium-neodymium alloy. *Less. Common Met.* **1973**, *30*, 63–74. [CrossRef]
15. Wang, Y.; Choo, H. Influence of texture on Hall-Petch relationships in an Mg alloy. *Acta Mater.* **2014**, *81*, 83–97. [CrossRef]
16. Gao, X.; Muddle, B.C.; Nie, J.F. Transmission electron microscopy of Zr-Zn precipitate rods in magnesium alloys containing Zr and Zn. *Philos. Mag. Lett.* **2009**, *89*, 33–43. [CrossRef]
17. Capolungo, L. Dislocation junction formation and strength in magnesium. *Acta Mater.* **2011**, *59*, 2909–2917. [CrossRef]
18. Zheng, K.Y.; Dong, J.; Zeng, X.Q.; Ding, W.J. Effect of pre-deformation on aging characteristics and mechanical properties of a Mg-Gd-Nd-Zr alloy. *Mater. Sci. Eng. A* **2008**, *491*, 103–109. [CrossRef]
19. Zhang, Z.; Chen, D. Consideration of Orowan strengthening effect in particulate-reinforced metal matrix nanocomposites: A model for predicting their yield strength. *Scr. Mater.* **2006**, *54*, 1321–1326. [CrossRef]
20. Shi, G.L.; Zhang, D.F.; Zhang, H.J.; Zhao, X.B.; Qi, F.G.; Zhang, K. Influence of pre-deformation on age-hardening response and mechanical properties of extruded Mg-6%Zn-1%Mn alloy. *Trans. Nonferr. Met. Soc. China* **2013**, *23*, 586–592. [CrossRef]

21. Robson, J.D.; Paa-Rai, C. The interaction of grain refinement and ageing in magnesium-zinc-zirconium (ZK) alloys. *Acta Mater.* **2015**, *95*, 10–19. [CrossRef]

22. Karnesky, R.A.; Meng, L.; Dunand, D.C. Strengthening mechanisms in aluminum containing coherent Al$_3$Sc precipitates and incoherent Al$_2$O$_3$ dispersoids. *Acta Mater.* **2007**, *55*, 1299–1308. [CrossRef]

Article

Simulation Study on Thermo-Mechanical Controlled Process of 800 MPa-Grade Steel for Hydropower Penstocks

Qingfeng Ding [1], Yuefeng Wang [1], Qingfeng Wang [1,2] and Tiansheng Wang [1,2,*]

[1] State Key Laboratory of Metastable Materials Science and Technology, Yanshan University, Qinhuangdao 066004, China; dingqingfeng@126.com (Q.D.); wangyuefeng@ysu.edu.cn (Y.W.); wqf67@ysu.edu.cn (Q.W.)

[2] National Engineering Research Center for Equipment and Technology of Cold Strip Rolling, Yanshan University, Qinhuangdao 066004, China

* Correspondence: tswang@ysu.edu.cn; Tel.: +86-335-8074631; Fax: +86-335-8074545

Academic Editor: Soran Birosca
Received: 7 July 2016; Accepted: 25 August 2016; Published: 31 August 2016

Abstract: The thermo-mechanical controlled process (TMCP) of 800-MPa-grade non-quenched tempered steel used for penstocks was simulated on a Gleeble-3500 thermo-mechanical simulator. The effect of the finish cooling temperature (FCT) which ranged from 350–550 °C on the microstructure and mechanical properties was studied. The microstructure of TMCP specimens is primarily composed of lath bainite (LB) and granular bainite (GB). The decreased FCT can induce the increase of LB and the decrease of GB in the volume fraction, and the decrease in the amount and the size of Martensite/Austenite (M/A) constituents with a more dispersive distribution. The LB has higher strength and hardness than GB, and the GB with fine and dispersive M/A constituents has excellent impact toughness. The minimum values of the yield strength, tensile strength and hardness, and the maximum value of the impact absorbed energy are obtained for the FCT of 450 °C. For the FCT over 450 °C, the yield strength, tensile strength and hardness are increased slightly, but the impact absorbed energy is rapidly decreased, which is mainly attributed to the formation of block M/A constituents. When the FCT is around 400 °C, the optimal combination of yield strength and impact toughness is obtained, which meets the technical requirements of 800-MPa-grade hydropower penstock steel.

Keywords: hydropower penstocks steel; TMCP; microstructure; mechanical properties

1. Introduction

A higher water head and larger capacities are required for the scale merits of hydropower plants, which make penstocks bear larger pressure. This requires the penstock steel to have a higher strength level, as well as sufficient toughness and excellent weldability. At present, 600-MPa-grade steel used for hydropower penstocks is mostly delivered in a quenched and tempered condition, little is delivered in the thermo-mechanical controlled process (TMCP) or TMCP+ tempering condition. Steel with an 800-MPa-grade used for hydropower penstocks is delivered only in quenched and tempered condition. Therefore, the development of TMCP of an 800-MPa-grade steel sheet used for hydropower penstocks is significant.

TMCP can bring about various advantages, such as the decrease of the weld pre-heating temperature and heat input due to the decrease of the carbon equivalent value, the improvement of the low temperature toughness, the reduction of the manufacturing cost, and the shortening of the delivery time of steel products [1]; otherwise, TMCP can enable high strength and toughness to be realized throughout the thickness through the control of the austenite grain size [2]. As the production process for the non-quenched tempered steel, TMCP includes controlled rolling and controlled cooling. The controlled cooling is the key to governing the mechanical properties. The cooling rate and finish cooling temperature (FCT) are two important parameters of controlled cooling which have been investigated by some researchers. Ai et al. (2005) [3] studied the TMCP of a 60Si2MnA spring steel rod, showing that increasing the cooling rate can reduce the interlamellar spacing of pearlite and increase the strength. Rasouli et al. (2008) [4] indicated that a faster cooling rate results in a higher strength and lower elongation in medium-carbon non-quenched tempered steel. Yi et al. (2014) [5] reported that interphase precipitation and diffusion precipitation can be observed at different FCTs in Nb-Ti micro-alloyed steel. A higher FCT induces the interphase precipitation, while a lower FCT induces the diffusion precipitation. The interphase precipitation has a larger contribution to enhancing the yield strength than the diffusion precipitation does.

Some studies [6,7] indicated that the yield strength and low temperature toughness can be improved by adding nickel above 1.4 wt %, which increases the cost. For thick plates and higher yield strength grades, a tempering process was usually used after controlled cooling [8], which reduces the production efficiency and increases the costs. Xie et al. (2014) [9] reported that a novel 1000-MPa-grade low-carbon microalloyed steel plate (0.08–0.11 wt % C) can be produced by TMCP at a high cooling rate and low FCT in the absence of a tempering process. The thickness of the TMCP steel plate is only 6 mm because of the large rolling reduction, and the Charpy V-notch impact absorbed energy at $-20\,^{\circ}\mathrm{C}$ tested using one-fourth-size specimens ($2.5 \times 10 \times 55\ \mathrm{mm}^3$) is 24 J. This is suggested that if the strength is moderately depressed by decreasing the cooling rate, the impact absorbed energy will be further increased. Therefore, the Charpy V-notch impact absorbed energy at $-20\,^{\circ}\mathrm{C}$ for the standard specimen could be inferred to be more than 47 J which is required for the 800-MPa-grade hydropower penstocks.

In this study, the TMCP of the 800-MPa-grade non-quenched tempered steel used for hydropower penstocks was simulated on a Gleeble-3500 thermo-mechanical simulator. The aim is to study the influence of a wider range of FCTs on the microstructure and mechanical properties to optimize the FCT for the TMCP of the steel.

2. Materials and Methods

The hot-rolled steel plate with thickness of 60 mm was used in the present study. The chemical composition is listed in Table 1.

Table 1. Chemical compositions of test steel (wt %).

C	Mn	Si	Cu	Ni	Cr	Mo	Nb	V	Ti	B	Ceq
0.09	1.50	0.30	0.20	0.50	0.40	0.25	0.025	0.045	0.013	0.0011	0.52

Note: Ceq = C + Mn/6 + (Cr + Mo + V)/5 + (Ni + Cu)/15(%).

TMCP simulation was performed on Gleeble-3500 thermo-mechanical simulator (DSI, Saint Paul, MN, USA) using round bar specimens with a length of 75 mm and diameter of 15 mm sampled from the one-fourth-thickness of the hot-rolled steel plate. The TMCP regimes are given in Figure 1. The compressive deformation includes two steps, "roughing" and "finishing", those were respectively performed at 1020 °C (γ recrystallization region) and 830 °C (γ non-recrystallization region). Accelerated cooling (ACC) was started at 780 °C with a cooling rate of 15 °C/s and interrupted at different FCTs, and then the specimen was reheated with 2 °C/s for 15 s to simulate self-tempering schedule, and then cooled with 1 °C/s to simulate air cooling.

The microstructure of the simulated TMCP specimens was examined at the central area in the longitudinal section by optical microscopy (OM, Axiover-200MAT, Zeiss, Heidenheim, Germany) and transmission electron microscopy (TEM, JEM-2010, JEOL, Musashino, Japan). Specimens for OM were mechanically ground, polished and etched with 4% nital. For TEM examination, ~0.6 mm-thick foils were cut via wire electro-discharging and ground to ~30 μm in thickness using waterproof abrasive paper, and then thinned to perforation on a TenuPol-5 twinjet electro-polishing device (Struers, Ballerup, Denmark) using an electrolyte composed of 7% perchloric acid and 93% glacial acetic acid solution at room temperature and a voltage of 28 V.

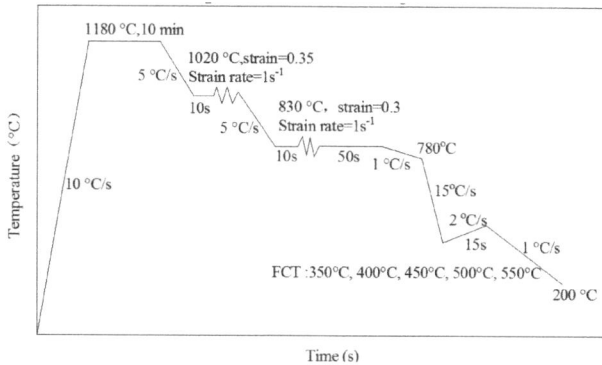

Figure 1. Process curve of simulated TMCP.

For the simulated TMCP specimens, the tensile property and hardness were measured at room-temperature, and the Charpy V-notch impact toughness was evaluated at −20 °C. Because of the limitation of the simulated specimen size, the sub-size tensile specimens was used to measure the tensile properties. Chu et al. (2000) [10] have experimentally demonstrated that there is no much difference in strength between sub-size tensile specimen and standard specimen. The dimensions of the sub-size specimens used in this work are shown in Figure 2a, which were given by Fan et al. (2014) [11]. The tensile specimens were wire-cut from the TMCP simulated specimens (Figure 2b). The tensile property was measured on a tensile testing machine (INSPEK TABLE100, Hegewald&Peschke, Nossen, Germany) with a cross-head speed of 0.25 mm/min. The impact absorbed energy was measured at −20 °C on an impact testing machine (JBN-300B 150 J, Shidaizhifeng, Beijing, China) using standard Charpy V-notch specimens with sizes of $10 \times 10 \times 55$ mm^3. Because the incompletely symmetrical deformation was induced by compressive deformation in TMCP simulation due to the active force applied on one side of the specimen, the notch of the impact specimens were machined at the maximal diameter of simulated TMCP specimens. Additionally, the impact fracture surface was examined on a scanning electron microscope (SEM, Hitachi S-3400, Hitachi, Tokyo, Japan). The Vickers hardness of the specimens was measured on a micro-Vickers hardness tester (FM-ARS9000, Future Tech, Kanagawa, Japan) with a load of 9.8 N.

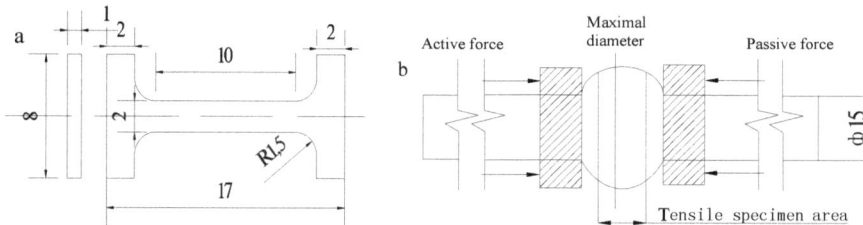

Figure 2. (**a**) Dimensions of tensile specimens cut from the center of the compressive deformation specimens along transversal direction; (**b**) Dimensions of specimens after simulated TMCP.

3. Results and Discussion

3.1. Effect of FCT on Mechanical Properties

Figure 3a gives the influence of the FCT on the mechanical properties of the simulated TMCP specimens, and Figure 3b shown the stress-strain curves of the specimens under tensile testing. It is indicated that with the FCT increasing, both the strength and hardness decrease first and then slightly increase, and the minimum is obtained at a FCT of 450 °C. When the FCTs are 350 °C and 400 °C, the yield strength exceeds 800 MPa. The effect of the FCT on the impact absorbed energy (−20 °C) is opposed to that on the strength and hardness (Figure 3a); the impact absorbed energy increases first and then decreases with the FCT increasing. The maximum impact absorbed energy occurs at a FCT of 450 °C, and the impact absorbed energy rapidly decreases when the FCT is above 450 °C. When the FCT is 400 °C and 450 °C, the measured values of the impact absorbed energy are much more than 47 J. In addition, from Figure 3b one can see that the tensile elongation is below 15% only at the FCT of 550 °C. Therefore, the optimized FCT for 800-MPa-grade hydropower penstock steel could be around 400 °C, which results in the room-temperature yield strength of 820–830 MPa and the −20 °C impact absorbed energy of 52–66 J.

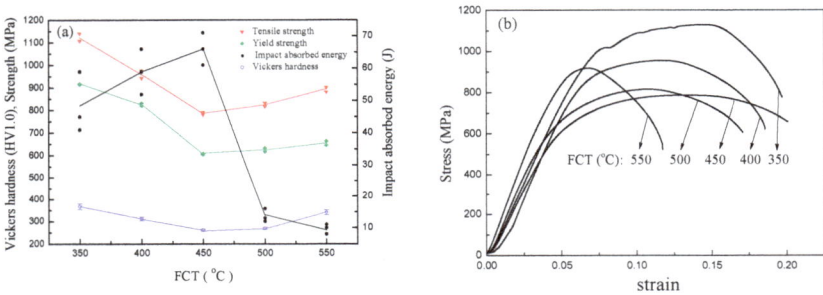

Figure 3. (**a**) Relationships between mechanical properties (tensile strength, yield strength, impact absorbed energy, and hardness) and FCT; (**b**) Engineering stress-strain curves.

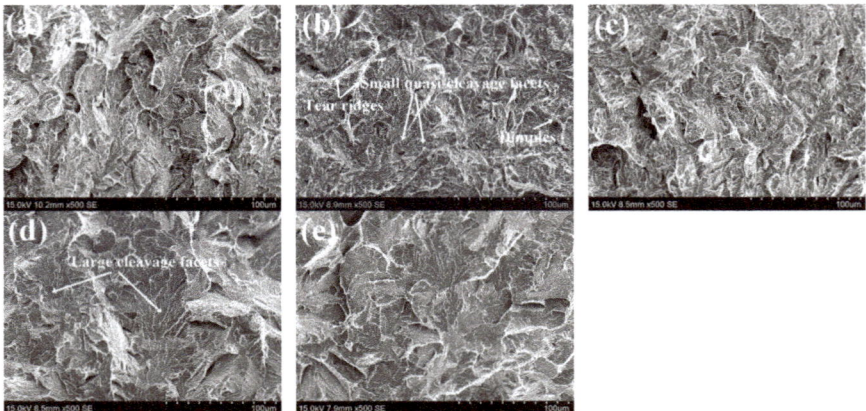

Figure 4. SEM morphologies of impact fracture of specimens after TMCP with different FCTs: (**a**) 350 °C; (**b**) 400 °C; (**c**) 450 °C; (**d**) 500 °C; (**e**) 550 °C.

The SEM morphologies of the impact fracture surfaces of the simulated TMCP specimens are shown in Figure 4. For the specimens with a FCT not above 450 °C, one can observe small cleavage facets, tear ridges, and some dimples in the fracture surface (Figure 4a–c), which absorb more impact energy. However, for the specimens with a FCT above 450 °C (such as 500 °C and 550 °C), many large cleavage facets were observed in the fracture (Figure 4d,e), which absorb less impact energy.

3.2. Relationship between Microstructure and Mechanical Properties

Optical microscopy (OM) microstructures in specimens after simulated TMCP with FCTs of 350 °C, 400 °C, 450 °C, 500 °C, and 550 °C are illustrated in Figure 5a–e, respectively. It can be seen that the microstructures of all TMCP specimens are primarily composed of lath bainite (LB) and granular bainite (GB). The LB is a fine lamella structure, and the GB is some island constituents distributed in the matrix. The volume fractions of LB and GB were roughly evaluated, as shown in Table 2, indicating that as the FCT increases, the fraction of GB increases, but LB decreases. Usually, a slower cooling rate promotes the formation of GB, while a faster cooling rate promotes the formation of LB. The increased FCT is equivalent to the decrease in the undercooling degree or cooling rate, which promotes the formation of GB.

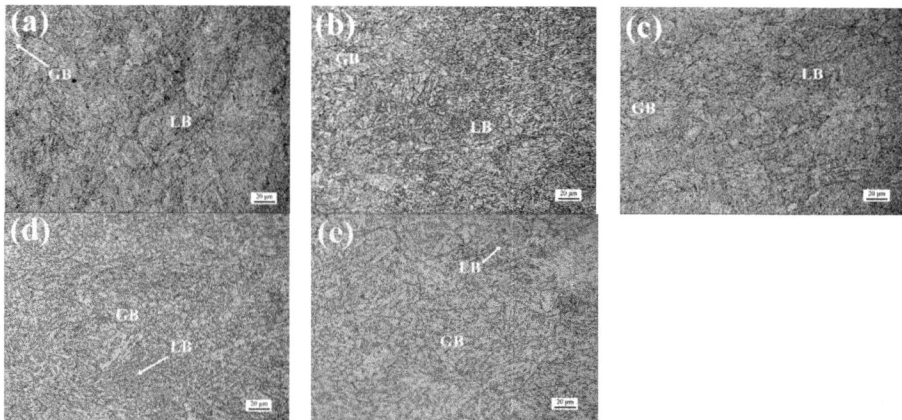

Figure 5. OM micrographs of specimens after TMCP with different FCTs: (**a**) 350 °C; (**b**) 400 °C; (**c**) 450 °C; (**d**) 500 °C; (**e**) 550 °C.

Table 2. Volume fractions of LB and GB in TMCP specimens with different FCTs.

FCT (°C)	350	400	450	500	550
LB(vol %)	92 ± 6	60 ± 5	19 ± 5	8 ± 4	6 ± 3
GB(vol %)	8 ± 6	40 ± 5	81 ± 5	92 ± 4	94 ± 3

In order to observe the details of the microstructures, TEM examinations were performed for two TMCP specimens with FCTs of 550 °C and 350 °C, the former mainly composed of GB and the latter mainly composed of LB (see Table 2). Figure 6a–d show the TEM micrographs of GB in the specimen after simulated TMCP with a FCT of 550 °C, displaying the lath-like morphology. This is similar to what was observed by Luo et al. (2010) [12] and Zhao et al. (2016) [13] through TEM; that is, the GB was composed of bainitic ferrite lath and islands of austenite and martensite (M/A constituents) which are transformed from austenite during the accelerated cooling processes. Figure 6a,b show two kinds of GB, i.e., bainitic ferrite laths with clear and unclear boundaries. The unclear lath boundaries could be caused by the recovery of GB formed at a higher temperature, which can be indirectly proved by the fact that the fine rod-like carbides precipitated inside laths can be detected (Figure 6b). Additionally, one can also observe M/A constituents with micro-twins in martensite (Figure 6c) and the blocky M/A constituents in prior austenite boundaries (Figure 6d).

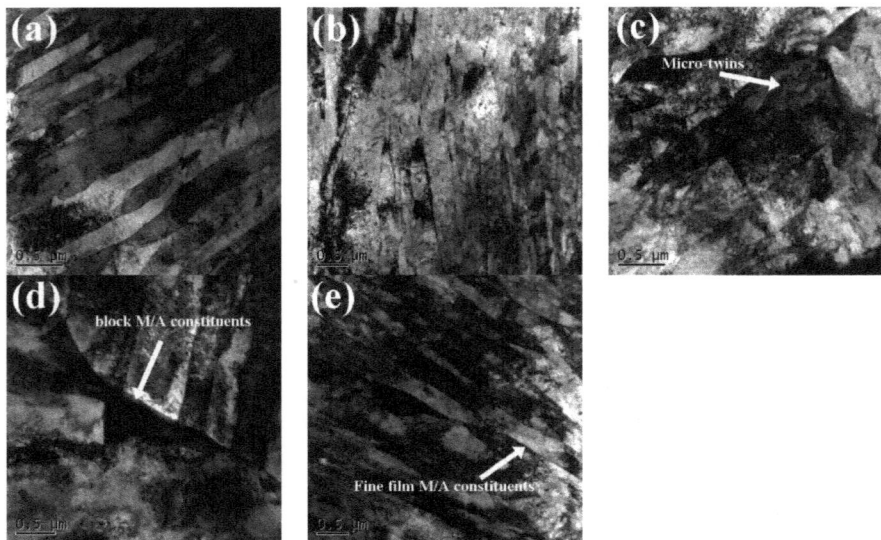

Figure 6. TEM micrographs of specimens after TMCP with FCT of 550 °C (**a–d**) and 350 °C (**e**).

Figure 6e shows the TEM micrographs of LB in the specimen after the simulated TMCP with a FCT of 350 °C, displaying the lath-like morphology, i.e., LB. The dislocation density in LB is higher than that in GB; meanwhile, film-like M/A constituents are located in between the laths, which could be attributed to the relatively faster cooling rate that promotes the formation of LB. These differences between LB and GB can result in the difference in mechanical properties. It is well-known that the LB can make a greater contribution to strength and hardness than GB does. Therefore, the highest strength and hardness were obtained in the specimen after TMCP with a FCT of 350 °C. The increase in the FCT results in the increase in the volume fraction of GB, which leads to the decrease in the strength and hardness. When the FCT increases to 450 °C, the microstructure is composed of 81 vol % GB and 19 vol % LB (Table 2), obtaining the lowest strength and hardness. A further increase in the FCT induces the slight increase in strength and hardness, which could result from the dispersion strengthening of fine precipitates (Figure 6b) and the large-sized M/A constituents (Figure 6c,d).

When the FCTs are high (500 °C and 550 °C), the impact absorbed energy is only about 10 J, which is attributed to the block of the M/A constituents (Figure 6c,d) in the microstructure. The brittle block M/A constituents distributed on the grain boundaries are usually regarded as the site of the crack initiation and deteriorate the impact toughness [14]. When the FCT decreases to 450 °C, the impact absorbed energy increases rapidly, which results from the lower FCT (equivalent to a faster cooling rate) making the amount and the size of M/A constituents decrease and its form changes to dot-like; meanwhile, the M/A constituents distribute more dispersively [14,15]. The fine and dispersive M/A constituents can effectively prevent crack propagation [15,16]. When the FCT decreases from 450 °C to 350 °C, the amount of M/A constituents further decreases and its form changes to film-like (Figure 6e), and the microstructure is mainly composed of LB (see Table 2). Although LB with a low-angle grain boundary has little effect for inhibiting crack propagation [14], the film-like M/A constituents between bainitic laths strongly prevent dislocation motion and crack propagation [17], and thus the impact absorbed energy does not decrease markedly.

4. Conclusions

As the FCT decreases from 550 °C to 350 °C, mechanical properties of simulated TMCP specimens are non-monotonic functions of the FCT. The minimum values of the yield and tensile strength and hardness and the maximum value of the impact absorbed energy are obtained at the FCT of 450 °C. The microstructures in the simulated TMCP specimens are primarily composed of LB and GB. The decrease in the FCT can result in the increase in the volume fraction of LB and the decrease in the volume fraction of GB, and the decrease in the amount and the size of M/A constituents with more dispersive distribution. The optimum FCT of the TMCP for the steel is around 400 °C, which results in the room-temperature yield strength of 820–830 MPa and the −20 °C impact absorbed energy of 52–66 J, which meets the technical requirements of 800-MPa-grade hydropower penstock steel.

Acknowledgments: This work was supported by the National Natural Science Foundation of China (Grant No. 51471147).

Author Contributions: Tiansheng Wang designed the experiments and wrote the paper; Qingfeng Ding and Yuefeng Wang performed the experiments, analyzed the data and wrote the paper, Qingfeng Wang performed the experiments and analyzed the data.

Conflicts of Interest: The authors declare no conflict of interest.

References

1. Chiaki, O. Development of Steel Plates by Intensive Use of TMCP and Direct Quenching Processes. *ISIJ Int.* **2001**, *41*, 542–553.
2. Nishiokaand, K.; Ichikawa, K. Progress in thermo-mechanical control of steel plates and their commercialization. *Sci. Technol. Adv. Mater.* **2012**, *13*, 1–20.
3. Ai, J.H.; Zhao, T.C.; Gao, H.J.; Hu, Y.H.; Xie, X.S. Effect of controlled rolling and cooling on the microstructure and mechanical properties of 60Si2MnA spring steel rod. *J. Mater. Process. Technol.* **2005**, *160*, 390–395. [CrossRef]
4. Rasouli, D.; Asl, S.K.; Akbarzadeh, A.; Daneshi, G.H. Effect of cooling rate on the microstructure and mechanical properties of microalloyed forging steel. *J. Mater. Process. Technol.* **2008**, *206*, 92–98. [CrossRef]
5. Yi, H.L.; Yang, X.; Sun, M.X.; Liu, Z.Y.; Wang, G.D. Influence of finishing cooling temperature and holding time on nanometer-size carbide of Nb-Ti microalloyed steel. *J. Iron Steel Res. Int.* **2014**, *21*, 433–438. [CrossRef]
6. Onishi, K.; Katsumoto, H.; Kamo, T.; Sakaibori, H.; Kawabata, T.; FujiwAra, K.; Okaguchi, S. Study on application of advanced TMCP to high tensile strength steel plates for penstock. *Weld World* **2008**, *52*, 523–530.
7. Tsuzuki, T.; Kawabata, N.; Okushima, M.; Tokunoh, K.; Okamura, Y. Development and Application of 950MPa and 780MPa Class High Strength Steel for Penstock. In Proceedings of the High Strength Steels for Hydropower Plants, Takasaki, Japan, 20–22 July 2009.
8. Falko, S. Structural steel for the application in offshore, wind and hydro energy production: Comparison of application and welding properties of frequently used materials. *Int. J. Microstruct. Mater. Prop.* **2011**, *6*, 4–19.
9. Xie, H.; Du, L.X.; Hu, J.; Misra, R.D.K. Microstructure and mechanical properties of a novel 1000 MPa grade TMCP low carbon microalloyed steel with combination of high strength and excellent toughness. *Mater. Sci. Eng. A* **2014**, *612*, 123–130. [CrossRef]
10. Chu, R.Q.; Duan, Z.Q.; Dong, H.; Yang, K.; Li, S.X.; Wang, Z.G. Design of a kind of mini-tensile-specimen and its application in research of the high-performance pipe line steel. *Acta Metall. Sin.* **2000**, *36*, 626–629.
11. Fan, L.; Wang, T.L.; Fu, Z.B.; Zhang, S.M.; Wang, Q.F. Effect of heat-treatment on-line process temperature on the microstructure and tensile properties of a low carbon Nb-microalloyed steel. *Mater. Sci. Eng. A* **2014**, *607*, 559–568. [CrossRef]
12. Luo, Y.; Peng, J.M.; Wang, H.B.; Wu, X.C. Effect of tempering on microstructure and mechanical properties of a non-quenched bainitic steel. *Mater. Sci. Eng. A* **2010**, *527*, 3433–3437. [CrossRef]
13. Zhao, Z.P.; Qiao, G.Y.; Tang, L.; Zhu, H.W.; Liao, B.; Xiao, F.R. Fatigue properties of X80 pipeline steels with ferrite/bainite dual-phase microstructure. *Mater. Sci. Eng. A* **2016**, *657*, 96–103. [CrossRef]

14. Chen, X.W.; Qiao, G.Y.; Han, X.L.; Wang, X.; Xiao, F.R.; Liao, B. Effects of Mo, Cr and Nb on microstructure and mechanical properties of heat affected zone for Nb-bearing X80 pipeline steels. *Mater. Des.* **2014**, *53*, 888–901. [CrossRef]

15. Wang, C.M.; Wu, X.F.; Liu, J.; Xu, N.A. Transmission electron microscopy of martensite/austenite islands in pipeline steel X70. *Mater. Sci. Eng. A* **2006**, *438–440*, 267–271. [CrossRef]

16. Sung, H.K.; Lee, S.H.; Shin, S.Y. Effects of Start and Finish Cooling Temperatures on Microstructure and Mechanical Properties of Low-Carbon High-Strength and Low-Yield Ratio Bainitic Steels. *Metall. Mater. Trans. A* **2013**, *45*, 2004–2013. [CrossRef]

17. Zhong, Y.; Xiao, F.R.; Zhang, J.W.; Shan, Y.Y.; Wang, W.; Yang, K. In situ TEM study of the effect of M/A films at grain boundaries on crack propagation in an ultra-fine acicular ferrite pipeline steel. *Acta Mater.* **2006**, *54*, 435–443. [CrossRef]

metals

MDPI

Article

Effects of Current Stressing on the Grain Structure and Mechanical Properties of Ag-Alloy Bonding Wires with Various Pd and Au Contents

Chien-Hsun Chuang [1,*], Chih-Hsin Tsai [1], Yan-Cheng Lin [2] and Hsin-Jung Lin [2]

[1] Wire Technology Co. LTD., Taichung 432, Taiwan; kkmanspace@wiretech.com.tw
[2] Institute of Materials Science and Engineering, National Taiwan University, Taipei 106, Taiwan;
     D02527013@ntu.edu.tw (Y.-C.L.); hsinjung@itri.org.tw (H.-J.L.)
*   Correspondence: josh604@hotmail.com; Tel.: +886-2-2392-9635

Academic Editor: Soran Birosca
Received: 9 May 2016; Accepted: 1 August 2016; Published: 4 August 2016

Abstract: Ag-alloy bonding wires containing various Pd and Au elements and traditional 4 N Au and Pd-coated 4 N Cu bonding wires were stressed with a current density of 1.23×10^5 A/cm^2 in air. The amounts of annealing twins in the Ag-alloy wires were much higher than those in Au and Pd-coated Cu wires. The percentages of twinned grains in these Ag-alloy wires increased obviously with current stressing. However, the grains in Ag-3Pd and Ag-15Au-3Pd grew moderately under current stressing, in contrast to the dramatic grain growth in the other bonding wires. In addition, the breaking loads and elongations of the various Ag-alloy wires changed slightly, similar to the case of Au wire. The results implied that degradation of the mechanical properties of these annealing twinned Ag-alloy wires due to electromigration was limited. Pd-coated Cu wire was severely oxidized after current stressing for only 1 h in air, which drastically degraded both the breaking load and elongation.

Keywords: Ag-alloy bonding wires; current stressing; grain growth; mechanical properties

---

1. Introduction

Since the invention of wire bonding technology by Bell Laboratories in 1957, Au has become the most popular material for this packaging process. In recent years, the rising cost of Au has created a demand for new bonding wires to replace Au wire. For this purpose, Cu wire and Pd-coated Cu wire are under consideration due to their low cost, high tensile strength, and good electrical and thermal conductivity. However, the inherent tendency of Cu to oxidize restricts its application to wire bonding, for it raises serious concerns about the reliability of electronic products. This concern cannot be completely eliminated, not even by coating Cu wire with Pd, Au, or Pt thin films. Furthermore, in contrast to the overgrowth of intermetallic compounds that occurs at the interface between Au wire and Al pad, the risk of using Cu wire is that an insufficient intermetallics layer may form in the joints, leading to a poor bonding effect [1,2]. In addition, the high hardness of Cu wire requires a large bonding force. The use of greater force can cause under-pad chip cratering, especially in the packaging of certain integrated circuit (IC) chips with low K substrates [3]. It has also been recognized that Cu wire is unsuitable for certain advanced wire bonding technologies, such as the stacking die, bonding ball on stitch (BBOS), or bonding stitch on ball (BSOB) methods.

Recently, Pd- and Au-doped Ag-alloy wires with large amounts of annealing twins have been developed to replace Au and Cu bonding wires in IC and LED packaging [4,5]. The addition of Pd and Au into Ag-alloy wire both increases the mechanical strength and improves its oxidation and corrosion resistance. Adding Pd also reduces the electromigration of the pure Ag conductor and inhibits the overgrowth of intermetallic compounds at the interface between the Ag-bonding wire and Al pad. In addition, a new method of multiple drawings interspersed with multiple annealing treatments can produce a high percentage of annealing twinned grains in this Ag-alloy wire. It is known that the low energy of twin boundaries yields many beneficial effects on the mechanical properties of structural materials [6–8]. In previous studies, it was verified that annealing twinned Ag-8Au-3Pd and Ag-4Pd alloy wires retain the thermal stability of the grain structure and material properties at elevated temperature for longer than conventional grained wires with fewer twinned grains [9,10]. After a further wire bonding process in an electronic package, the high thermal stability of the grain structure in annealing twinned Ag-8Au-3Pd and Ag-4Pd bonding wires results in a smaller heat affected zone (HAZ) near the ball bond than that in conventional grained wires, ensuring the applicability of these Ag-alloy wires to ultra-low loop wire bonding [11].

During the operation of IC devices, current stressing in the interconnections of packages can cause electromigration, leading to the early wear-out of electronic products. Such a failure mode under current stressing has attracted attention in recent years due to the current trend of minimizing interconnections. Although electromigration has been known to the IC industry for several decades [12], it has only recently been observed in electronic packages, such as flip chip solder bumps [13] and ball grid array (BGA) solder joints [14]. However, few researchers have reported on the effects of current stressing on the grain structure and material properties of bonding wires. Tse and Lach reported a failure case of Al bonding wires in an octal buffer inverter device after a few years of field usage [15]. They observed a phenomenon of thinning and thickening of the diameter that accompanied the breakage and melting of wire material, which implied that a mass redistribution of Al had occurred due to current stressing. In addition, grain growth of the wire material—even to the development of a special "bamboo" structure—was also observed. Orchard et al. [16] reported that current stressing indeed accelerated the failure of wire bonding joints. Zin et al. further indicated that the damage caused by electrical current was greater than that caused by thermal aging [17]. They also found that the growth of intermetallic compounds at the Au/Al interface was more rapid when electrons flowed from Au wire to Al pad than when they flowed in the opposite direction. In contrast, Schepper et al. [18] and Krabbenborg [19] reported that current stressing negligibly affected the contact failure of a package with Au wire bonding on Al pad.

It is known that doping Ag-alloy wires with different amounts of Pd and Au can change the electrical conductivity, which affects the temperature of wire materials due to the Joule effect. In addition, the atomic diffusion of Ag-alloy wires with various Pd and Au contents under current stressing can also influence the grain boundary migration. This study therefore examined the effects of the alloy concentrations of Ag-Au-Pd bonding wires under current stressing on their grain structures and the resultant degradation of mechanical properties.

2. Experimental Section

A series of Ag-alloy wires doped with various contents of Au and Pd were melted in a vacuum furnace of 10^{-3} Torr, followed by continuous casting to form a thick wire with a diameter of 6 mm. The thick wire was then sequentially drawn to a fine wire of 20 μm diameter. The drawing processes were interspersed with multiple annealing treatments at 530 °C for 2 s. Finally, the wire materials were drawn to 17.6 μm and annealed at 570 °C for 5 s. For comparison, commercial 4 N Au and Pd-coated 4 N Cu wires with the same diameter were employed. For the experiments, the grain structures and material properties of these bonding wires were assessed through stressing with a current of 0.3 A in air, which corresponded to a current density of 1.23×10^5 A/cm^2. Such a current density is typical for the operation of electronic devices. For the observations of grain morphology, the wire materials were bombarded by a focused ion beam (FIB, FEI Helios 600i, Hillsboro, OR, USA) along the longitudinal cross-sections using a tilt angle of 52°, as shown in Figure 1a. After the ion-cutting, the wire specimen (Figure 1b) was rotated 180° and tilted −6° to an angle vertical to the ion beam, as shown in Figure 1c. The micrographs were taken from the secondary electron images using the ion beam incidence in this FIB equipment. Because the cutting terminal of the ion beam can deviate from the central line of the specimen, the widths of wire images in certain micrographs will be smaller than the actual diameter of the wire (17.6 µm). The grain size was taken from the FIB images, from the mean intercept length of grain boundaries in random measurements in lines of 15 µm. The percentage of twinned grains as a ratio of the total number of grains was directly counted in the micrographs. The breaking loads and elongations of various specimens were measured with tensile tests using an MTS-Tytron 250 microforce tester (MTS Systems, Eden Prairie, MN, USA) at a crosshead speed of 10 mm/min.

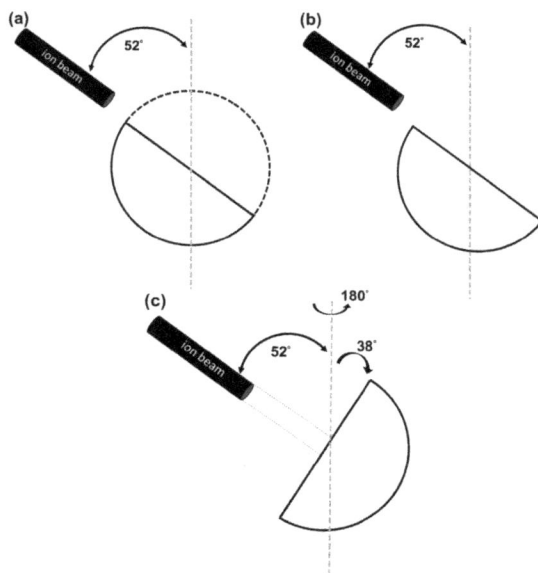

Figure 1. Preparation of micrograph using a focused ion beam: (**a**) ion bombardment along the longitudinal cross section of wire specimen; (**b**) after ion cutting; (**c**) secondary electron image taken from the ion beam.

3. Results

Figures 2–7 show the grain structures in longitudinal cross-sections of the bonding wires after stressing at a current density of 1.23×10^5 A/cm^2 for various time periods. It can be seen in Figures 2a and 3a that the original bonding wires with compositions of Ag and Ag-0.5Pd comprised equiaxial grains with coarse grain sizes of 6.5 and 9.7 µm, respectively. In Figure 6a, many equiaxial fine grains of about 1.6 µm with a texture-oriented structure can be observed in the original pure Au wire. In contrast, slender grains appeared in the central parts of Ag-3Pd, Ag-15Au-3Pd, and Pd-coated Cu wires, as shown in Figures 4a, 5a and 7a. Surrounding the central slender grains in these alloy wires were equiaxial fine grains of 6.2, 2.5, and 4.8 µm in size, respectively, with a mass of annealing twins. The finer grain sizes in Ag and Ag-15Au-3Pd, as compared to those of the other bonding wires, indicated that the addition of Au to Ag alloy wires can obstruct recrystallization during the final annealing treatment after the drawing process in the manufacturing of these bonding wires.

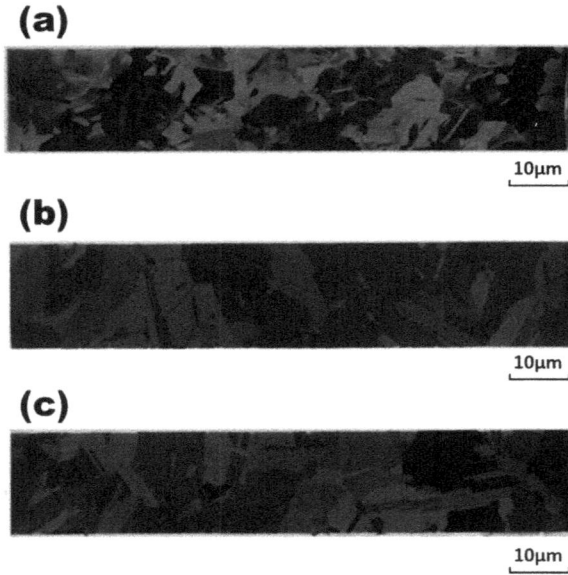

Figure 2. FIB image shows the grain structures in longitudinal cross-sections of the Ag wire after current stressing at 1.23×10^5 A/cm^2 for various times: (**a**) original; (**b**) 1 h; (**c**) 20 h.

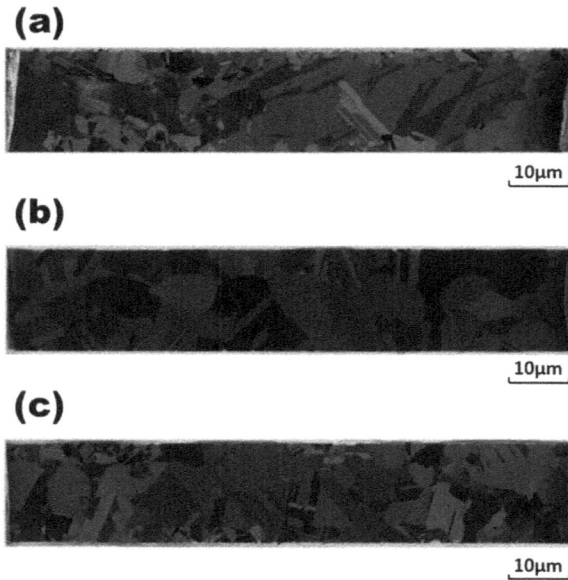

Figure 3. FIB image shows the grain structures in longitudinal cross-sections of the Ag-0.5Pd wire after current stressing at 1.23×10^5 A/cm^2 for various times: (**a**) original; (**b**) 1 h; (**c**) 20 h.

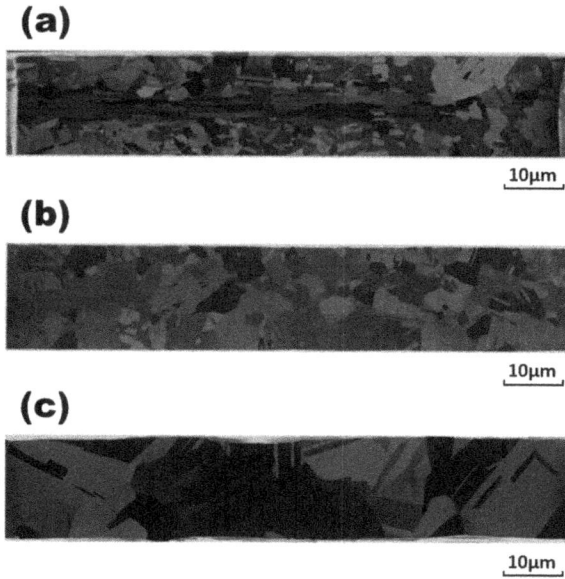

Figure 4. FIB image shows the grain structures in longitudinal cross-sections of the Ag-3Pd wire after current stressing at 1.23×10^5 A/cm^2 for various times: (**a**) original; (**b**) 1 h; (**c**) 20 h.

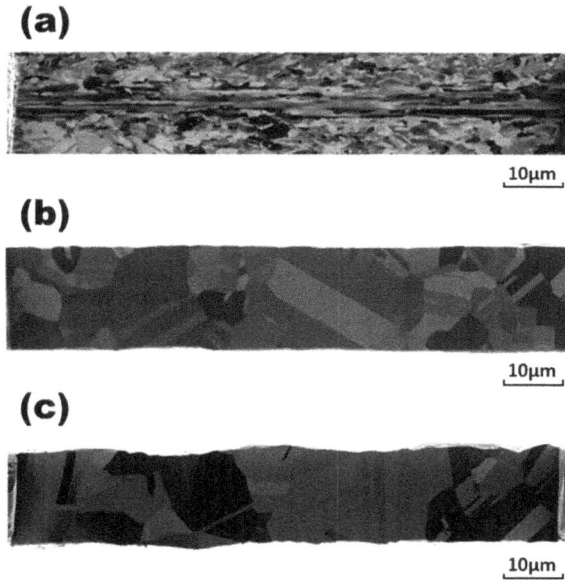

Figure 5. FIB image shows the grain structures in longitudinal cross-sections of the Ag-15Au-3Pd wire after current stressing at 1.23×10^5 A/cm^2 for various times: (**a**) original; (**b**) 1 h; (**c**) 20 h.

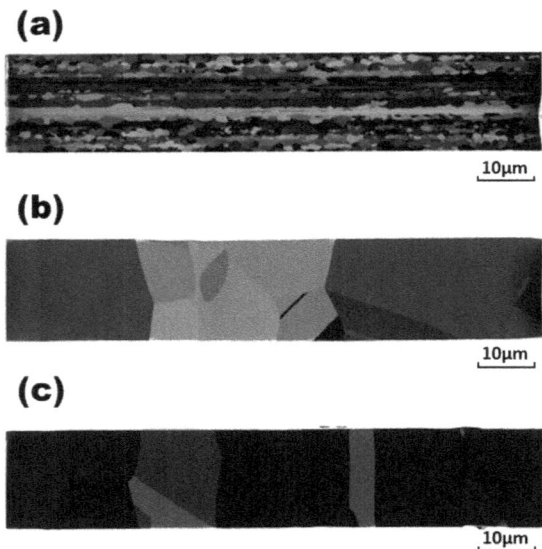

Figure 6. FIB image shows the grain structures in longitudinal cross-sections of the pure Au wire after current stressing at 1.23×10^5 A/cm^2 for various times: (**a**) original; (**b**) 1 h; (**c**) 20 h.

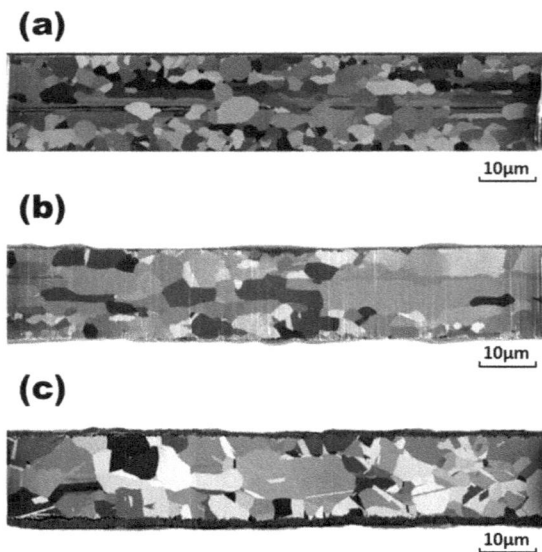

Figure 7. FIB image shows the grain structures in longitudinal cross-sections of the Pd-coated Cu wire after current stressing at 1.23×10^5 A/cm^2 for various times: (**a**) original; (**b**) 30 min; (**c**) 1 h.

During electrical stressing at a current density of 1.23×10^5 A/cm^2 for 1 h, the central slender grains in Ag-3Pd, Ag-15Au-3Pd, and Pd-coated Cu wires disappeared, and the equiaxial grains were redistributed throughout the wires, as shown in Figures 4b, 5b and 7b. It can also be seen in Figures 4 and 7 that the equiaxial grains in the original Ag-3Pd wire grew from 6.2 μm to 11.9 μm. In contrast to the moderate grain growth in Ag-3Pd wire, the growth of the grains in the pure Ag, Ag-0.5Pd, pure Au, and Pd-coated Cu wires was far more severe, as shown in Figures 2, 3, 6 and 7. It is obvious that the addition of 3 wt. % Pd in Ag-alloy wires retarded further grain growth during current stressing. However, the grains in Ag-15Au-3Pd wire also grew drastically, from 2.5 μm to 13.7 μm, as evidenced in Figure 5a,b, similar to the rapid grain growth in pure Au wire.

It should be noted here that, during current stressing, the wire temperature increases due to the Joule effect. However, direct measurement of the temperature of wire specimens with a diameter of 17.6 μm (such as those in this study) is quite difficult and prone to inaccuracy. An alternate method for indirectly determining the wire temperature—by measuring its electrical resistivity—was proposed in our previous study [20]. It was also shown in that report that the temperatures of Ag-15Au-3Pd and pure Au wires were higher than that of Ag-4Pd (similar to the Ag-3Pd), leading to the larger grain structure of the former Au-containing wires than that of the Ag-3Pd. In this study, increasing the stressing time to 20 h caused the grains of all the bonding wires to grow further. However, the growth rates became much slower than those in the early stage of stressing for 1 hr. However, as shown in Figures 5c and 6c, the grains in Ag-15Au-3Pd and pure Au even grew into a "bamboo" structure, similar to the phenomenon observed by Tse and Lach [15] in Al bonding wire. It seems that Au can exaggerate the electromigration of bonding wires and enhance their grain growth effects.

From the observations of grain structures in these bonding wires, it is clear that the electrical current caused recrystallization and grain growth. Although electrical stressing at a current density of 1.23×10^5 A/cm^2 can increase the wire temperature to about 100 °C (as reported in our previous study [20]), it was shown in another of our studies that the grain structure of a wire specimen remained almost unchanged during heating at 200 °C for 50 h in an air furnace without electrical current stressing [21]. Therefore, it is suggested that the grain growth in the current stressed wires was predominantly caused by the grain boundary motion driven by the electron flow, rather than by recrystallization driven by the thermal effect. Figure 7b indicates that, for the Pd-coated Cu wire, electrical stressing for 30 min caused an oxide layer of about 1.5 μm to form on its outer surface, and the grain size of this bonding wire increased from 4.8 μm to 13.2 μm. The oxide layer thickened rapidly to about 5.3 μm, as shown in Figure 7c, and this thickening was accompanied by grain growth to 15.3 μm in the remaining Cu wire when the current stressing time was lengthened to 1 h. The Pd-coated Cu wire was oxidized completely after current stressing for 3 h.

The grain sizes of various bonding wires after current stressing for different time periods are summarized in Figure 8. The results indicate that grain growth was much more rapid in pure Au and high Au-content Ag-15Au-3Pd wire than in the Ag and Ag-alloy wires. In the Ag-alloy wires, the addition of Pd retarded grain growth. Among all the bonding wires in this study, Ag-3Pd had the finest grains. In addition, the curves in Figure 8 reveal that the grains of all bonding wires first grew rapidly and then more slowly during current stressing. This kinetic behavior implies that the grain growth rate varies inversely with the grain size, which leads to the relation:

$$\frac{dD}{dt} = \frac{K_1}{D} \tag{1}$$

where D is the mean grain diameter after stressing time (t) and K_1 is constant, depending on wire composition. When the initial grain diameter (D_0) is much smaller than the grain diameter after various current stressing times (D_t), the grain growth Equation (1) can be integrated as:

$$D_t - D_0 = K_2 \times t^n \quad n = 0.5 \tag{2}$$

where n is the grain growth exponent. However, the relation of $n = 0.5$ in Equation (2) only exists for a high purity metal and the aging temperatures near its melting point, corresponding to the soap cells model [22]. In general, the n values depend on the purity of metals, the aging temperatures, and the texture in this metal [23]. For the calculation of the grain growth exponent (n), Equation (2) is expressed in a logarithmic form:

$$\log (D_t - D_0) = \log K_2 + n \log t \tag{3}$$

which is plotted in Figure 9 for various bonding wires. The grain growth exponents (n) can be calculated from the slopes of the straight lines in Figure 9, and they are summarized in Table 1. For the determination of the n value of pure Au wire, it was noticed that after current stressing for 20 h, the grains of pure Au wire had grown to 29.4 μm, which was much larger than the wire diameter (17.6 μm), leading to the appearance of the bamboo morphology of the grain structure. In this case, the grain boundary was anchored by the wire surface, and the grain growth was not uniform throughout the whole wire during further current stressing. Therefore, the data on the Au wire in Figure 10 for a stressing time of over 20 h were omitted from the calculation of its grain growth exponent.

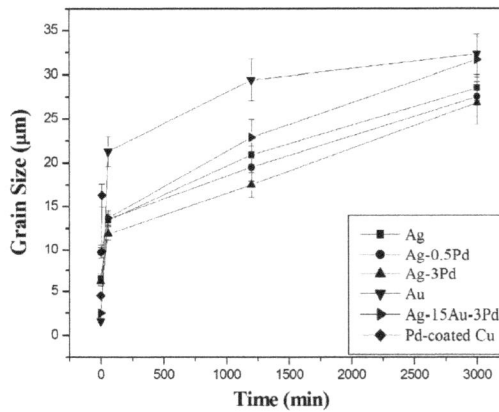

Figure 8. Variations of the grain sizes of bonding wires after current stressing at 1.23×10^5 A/cm^2 for various times.

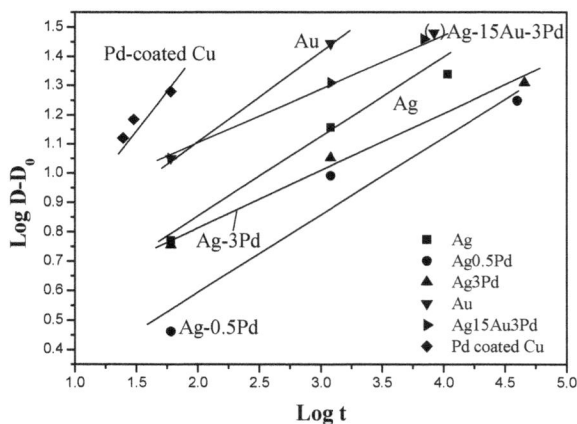

Figure 9. Log plots for the grain growth ($D_t - D_0$) of bonding wires after current stressing at 1.23×10^5 A/cm^2 for various times (t).

Table 1. Grain growth exponents (*n*) and constants (*k*) for various bonding wires during stressing at a current density of $1.23 \times 10^5 \text{ A/cm}^2$.

Bonding Wires	Log *k*	*k*	*n*
Ag	0.33	2.14	0.25
Ag-0.5Pd	0.3	1.99	0.28
Ag-3Pd	0.43	2.69	0.19
Ag-15Au-3Pd	0.70	5.01	0.20
Au	0.60	3.98	0.30
Pd-coated Cu	0.51	3.23	0.38

The results in Table 1 indicate that the grain growth exponents (*n*) of all bonding wires were larger than the ideal *n* value of a pure metal (0.5) according to the soap cells model [22]. This discrepancy might be attributed to the restriction of grain boundary mobility by the wire surface due to the fine diameter of the bonding wires in this study. In addition, the enrichment of annealing twins in these bonding wires can also play a role in dragging the grain boundaries during grain growth. In Table 1, it can be found that the grain growth exponents of Ag-alloy and pure Ag wires ranged from 0.19 to 0.28; this range is lower than those of pure Au wire (0.30) and Pd-coated Cu wire (0.38). The results were correlated with the twinned grain percentages in the pure Ag and Ag-alloy wires, which were much higher than those in the conventional Au wire and Pd-coated Cu wire, as shown in Figure 10. In addition, the similar grain growth exponents of Ag-3Pd and Ag-15Au-3Pd wires imply that the solid solution of Au atoms in Ag matrix has a minor effect on its grain growth kinetics. Furthermore, the pure Ag and near-pure Ag-0.5Pd wires have similar grain growth exponents, which are obviously larger than those of Ag-3Pd and Ag-15Au-3Pd alloy wires. This finding can be attributed to the solute drag effect of 3% Pd element on the grain boundary migration of Ag- and Ag-Au- alloy wires [22,23].

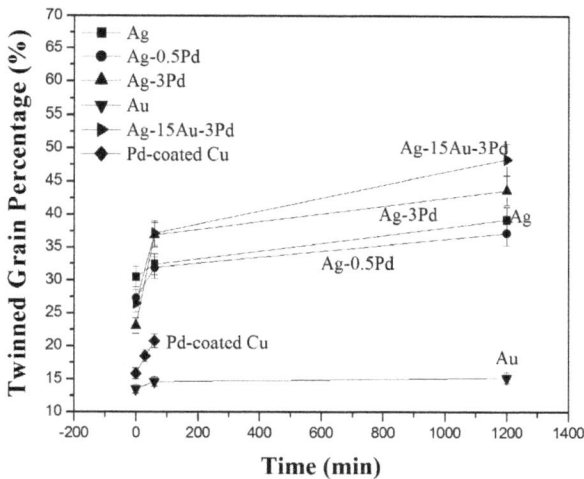

Figure 10. Percentages of annealing twins in all bonding wires in this study after current stressing at $1.23 \times 10^5 \text{ A/cm}^2$ for various times.

The enrichment of annealing twinned grains is attributed to the lower stacking energy after cold working and annealing treatments, as reported by Dillamore et al. [24]. In addition, the manufacturing method of multiple drawing interspersed with annealing treatments further enhanced the formation of annealing twins [5]. The abundant twin boundaries with low interfacial energy anchored the high angle grain boundaries, retarding grain growth in these annealing twinned Ag-alloy wires. Furthermore, it is notable in Figure 10 that the twinned grain percentages in all bonding wires increased with the current stressing. Especially, for the Ag-alloy wires containing various Pd and Au contents, the percentages of annealing twins in the original wire grains—ranging from 23% to 30%—rose to between 37% and 48% after 20 h of stressing. It is obvious that the electron flow can also promote the formation of annealing twins, which inhibit the grain growth. This result seems to be contradictory to the phenomenon that the current stressing can be a driving force for grain growth. This contradiction is explained by the fact that the grains without annealing twins have a higher tendency to recrystallize and grow during current stressing, so the total number of grains decreases. Meanwhile, new twins are created in these recrystallized and grown grains through the electron flow, so the number of twinned grains increases. The two effects combine to increase the twinned grain percentage, as shown in Figure 10.

Figures 11 and 12 show the results of tensile tests of all bonding wires after electrical stressing at 1.23×10^5 A/cm^2 for various times. The results in Figure 11 indicate that the breaking loads decreased with the lengthening of stressing time, which can be attributed to the grain growth in the wires during the current stressing. In addition, the concave notches on the wire surfaces visible in the micrographs of Figure 2c, Figure 3c, Figure 4c, Figure 5c, Figure 6c, Figure 7c were created by the mass transportation during current stressing due to the equilibrium of interfacial energies at the intersection of the grain boundary and the wire surface. These notches can also lower their breaking loads. Figure 11 also shows that the breaking loads of all of the Ag-alloy wires were higher than those of the pure Au and pure Ag wires under current stressing, with a sequence of Ag-15Au-3Pd > Ag-3Pd > Ag-0.5Pd > Au > Ag. It is obvious that the addition of Pd and Au strengthened the wire materials. In addition, the enrichment of annealing twins in Ag-alloy wires can also increase their breaking loads. Although the original Pd-coated Cu wire had a high breaking load of 5.3 g, it dropped drastically to a low value of 3.0 g after current stressing at 1.23×10^5 A/cm^2 for only 1 h due to severe oxidation on its outer surface.

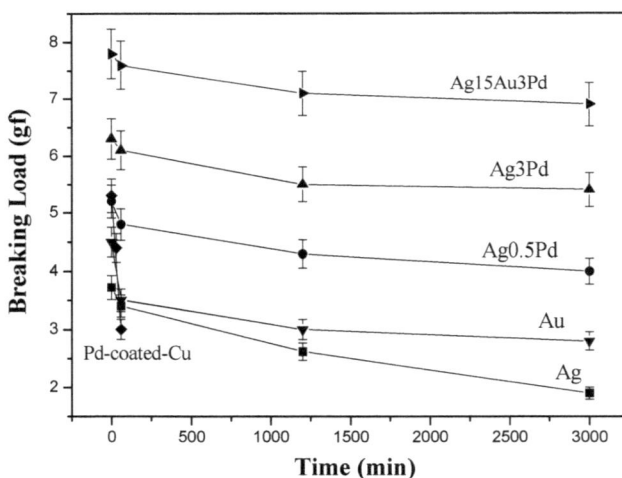

Figure 11. Variations in the breaking loads of bonding wires after current stressing at 1.23×10^5 A/cm^2 for various times.

The original pure Ag and Pd-coated Cu wires had very high elongations, 15.6% and 12.2%, respectively. After electrical stressing for 1 h, the elongation of pure Ag wire increased due to the recrystallization and grain growth, as shown in Figure 12. Although further current stressing decreased its elongation moderately, it was still higher than those of the other wires. In contrast, severe oxidation on the surface of Pd-coated Cu wire reduced its elongation to a low value of 4.3% after only 1 h of current stressing at 1.23×10^5 A/cm^2. It is also evidenced in Figure 12 that the elongations of pure Ag and Ag-alloy wires were higher than that of pure Au wire under electrical stressing, with a sequence of Ag > Ag-15Au-3Pd > Ag-3Pd > Ag-0.5Pd > Au. Similar to that of pure Ag wire, the elongations of pure Au and Ag-alloy wires also increased after initial current stressing for 1 h and then dropped slightly during further stressing. However, the decline was smaller in Ag-alloy wires than in pure Au wire. The surface morphology and mass transportation in Ag-alloy wires after electrical stressing seem to have insignificant influences on the variations in elongation due to the existence of annealing twins.

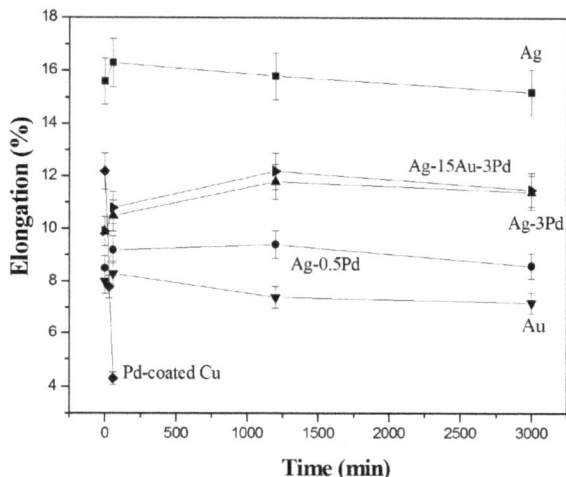

Figure 12. Variations of the elongations of bonding wires after current stressing at 1.23×10^5 A/cm^2 for various times.

4. Conclusions

Pd and Au-doped Ag-alloy bonding wires containing a large amount of annealing twins have been developed as replacements for traditional 4 N Au and Pd-coated 4 N Cu wires in electronic packaging. In contrast to the rapid grain growth during electrical stressing that occurs in the pure Au wire, these annealing twinned Ag-alloy wires have a highly stable grain structure. The *n*-values in the grain growth kinetics ($D_t - D_0 = K_2 \times t^n$) range from 0.19 to 0.28 for Ag-alloy and pure Ag wires, which are lower than those of the pure Au wire (0.30) and Pd-coated Cu wire (0.38). The slow grain growth in these annealing twinned Ag-alloy wires during current stressing results in mechanical properties that are superior to those of pure Au wire, with the sequence of breaking loads being Ag-15Au-3Pd > Ag-3Pd > Ag-0.5Pd > Au > Ag, and the sequence of elongations being Ag > Ag-15Au-3Pd ~Ag-3Pd > Ag-0.5Pd > Au. Although the original Pd-coated Cu wire had a high breaking load and high elongation, it dropped drastically to very low values after only 1 h of current stressing at 1.23×10^5 A/cm^2 in air due to severe oxidation on its outer surface. The experimental results also indicate that the degradation of the tensile strengths and elongations of these Ag-alloy wires under the effect of electromigration is limited.

Acknowledgments: This study was sponsored by the Industrial Technology Development Program (ITDP) of the Ministry of Economic Affairs (MOEA), Taiwan, under Grant No. 102-EC-17-A-08-11-0068, and by the industrial and academic cooperation program of Wire Technology Co. and the Ministry of Science and Technology, Taiwan, under Grant No. MOST 103-2622-E-002-012-CC2.

Author Contributions: C.H. Chuang conceived and designed the current stressing experiments for bonding wires, analyzed the results and wrote the paper. C.H. Tsai produced the Ag-alloy wires with various Pd and Au contents. Y.C. Lin and H.J. Lin performed the current stressing tests and measured the materials properties.

Conflicts of Interest: The authors declare no conflict of interest.

References

1. Noolu, N.; Murdeshwar, N.; Ely, K.; Lippold, J.; Baeslack, W., III. Phase transformations in thermally exposed Au-Al ball bonds. *J. Electron. Mater.* **2004**, *33*, 340–352. [CrossRef]
2. Chuang, T.H.; Chang, C.C.; Chuang, C.H.; Lee, J.D.; Tsai, H.H. Formation and growth of intermetallisc in an annealing-twinned Ag-8Au-3Pd wire bonding package during reliability tests. *IEEE Trans. Compon. Packag. Manuf. Technol.* **2013**, *3*, 3–9. [CrossRef]
3. Chauhan, P.; Zhong, Z.W.; Pecht, M. Copper wire bonding concerns and best practices. *J. Electron. Mater.* **2013**, *42*, 2415–2434. [CrossRef]
4. Chuang, T.H.; Wang, H.C.; Tsai, C.H.; Chang, C.C.; Chuang, C.H.; Lee, J.D.; Tsai, H.H. Thermal stability of grain structure and material properties in an annealing-twinned Ag-8Au-3Pd alloy wire. *Scr. Mater.* **2012**, *67*, 605–608. [CrossRef]
5. Lee, J.D.; Tsai, H.H.; Chuang, T.H. Alloy wire and methods for manufacturing the same. US patent 0171470A1, JP 279489, KR 10-1328863, TW I384082, 2013.
6. Souai, N.; Bozzolo, N.; Naze, L.; Chastel, Y.; Loge, R. About the possibility of grain boundary engineering via hot-working in a nickel-base superalloy. *Scr. Mater.* **2010**, *62*, 851–854. [CrossRef]
7. Lee, S.Y.; Chun, Y.B.; Hua, J.W.; Hwang, S.K. Effect of thermomechanical processing on grain boundary characteristics in two-phase brass. *Mater. Sci. Eng. A* **2003**, *363*, 307–315. [CrossRef]
8. Pande, C.; Rath, B.; Imam, M. Effect of annealing twins on Hall-Petch relation in polycrystalline materials. *Mater. Sci. Eng. A* **2004**, *367*, 171–175. [CrossRef]
9. Chuang, T.H.; Tsai, C.H.; Wang, H.C.; Chang, C.C.; Chuang, C.H.; Lee, J.D.; Tsai, H.H. Effects of annealing twins on the grain growth and mechanical properties of Ag-8Au-3Pd bonding wires. *J. Electron. Mater.* **2012**, *41*, 3215–3222. [CrossRef]
10. Chuang, T.H.; Lin, H.J.; Chuang, C.H.; Shiue, Y.Y.; Shieu, F.S.; Huang, Y.L.; Hsu, P.C.; Lee, J.D.; Tsai, H.H. Thermal Stability of Grain Structure and Material Properties in an Annealing Twinned Ag-4Pd Alloy Wire. *J. Alloy. Compd.* **2014**, *615*, 891–898. [CrossRef]
11. Tsai, C.H.; Chuang, C.H.; Tsai, H.H.; Lee, J.D.; Chang, D.; Lin, H.J.; Chuang, T.H. Materials Characteristics of Ag-Alloy Wires and Their Applications in Advanced Packages. *IEEE Trans. Compon. Packag. Manuf. Technol.* **2016**, *6*, 298–305. [CrossRef]
12. Black, J.R. Electromigration failure modes in aluminum metallization for semiconductor devices. *Proc. IEEE* **1969**, *57*, 1587–1594. [CrossRef]
13. Liu, C.Y.; Chen, C.; Tu, K.N. Electromigration in Sn-Pb solder strips as a function of alloy composition. *J. Appl. Phys.* **2000**, *88*, 5703–5709. [CrossRef]
14. Lin, H.J.; Lin, J.S.; Chuang, T.H. Electromigration of Sn-3Ag-0.5Cu and Sn-3Ag-0.5Cu-0.5Ce-0.2Zn solder joints with Au/Ni(P)/Cu and Ag/Cu pads. *J. Alloy. Compd.* **2009**, *487*, 458–465. [CrossRef]
15. Tse, P.K.; Lach, T.M. Aluminum electromigration of 1-mil bond wire in octal inverter integrated circuits. In Proceedings of 45th IEEE ECTC, Las Vegas, NV, USA, 21–24 May 1995; pp. 900–905.
16. Orchard, H.T.; Greer, A.L. Electromigration effects on intermetallic growth at wire-bond interfaces. *J. Electron. Mater.* **2006**, *35*, 1961–1968. [CrossRef]
17. Zin, E.; Michael, N.; Kang, S.H.; Oh, K.H.; Chul, U.; Cho, J.S.; Moon, J.T.; Kim, C.U. Mechanism of Electromigration in Au/Al Wirebond and Its Effects. In Proceedings of IEEE Electronic Components and Technology Conference, San Diego, CA, USA, 26–29 May 2009; pp. 943–947.
18. Schepper, L.D.; Ceuninck, W.D.; Leken, G.; Stals, L.; Vanhecke, B.; Roggen, J.; Beyne, E.; Tielemans, L. Accelerated ageing with in situ electrical testing: A powerful tool for the building-in approach to quality and reliability in electronics. *Qual. Reliab. Eng. Int.* **1994**, *10*, 15–26. [CrossRef]
19. Krabbenborg, B. High current bond design rules based on bond pad degradation and fusing of the wire. *Microelectron. Reliab.* **1999**, *39*, 77–88. [CrossRef]
20. Chuang, T.H.; Lin, H.J.; Wang, H.C.; Chuang, C.H.; Tsai, C.H. Mechanism of electromigration in Ag-alloy bonding wires with different Pd and Au content. *J. Electron. Mater.* **2015**, *44*, 623–629. [CrossRef]

21. Chuang, T.H.; Lin, H.J.; Chuang, C.H.; Tsai, C.H.; Lee, J.D.; Tsai, H.H. Durability to Electromigration of an Annealing-Twinned Ag-4Pd Alloy Wire Under Current Stressing. *Metall. Mater. Trans. A* **2014**, *45*, 5574–5583. [CrossRef]

22. Fullman, R.L. Metal Interfaces. In Proceedings of ASM Seminar, Cleveland, OH, USA, 13 October 1952; p. 179.

23. Brickenkamp, W. Grain growth, secondary and ternary recrystallization. In *Rekristallisation Metallischer WERKSTOFFE*; Gottstein, G., Deutscher Gesellschaft für Metallkunde, E.V., Eds.; DGM-Informationsgesellschaft mbH: Oberursel, Germany, 1984; pp. 83–97.

24. Dillamore, I.L.; Smallman, R.E.; Roberts, W.T. A determination of the stacking-fault energy of some pure F.C.C. metals. *Philo. Mag.* **1964**, *9*, 517–526. [CrossRef]

metals

MDPI

Article

Effect of Mo Content on Microstructure and Property of Low-Carbon Bainitic Steels

Haijiang Hu, Guang Xu *, Mingxing Zhou and Qing Yuan

The State Key Laboratory of Refractories and Metallurgy, Key Laboratory for Ferrous Metallurgy and Resources Utilization of Ministry of Education, Wuhan University of Science and Technology, Wuhan 430081, China; hhjsunny@sina.com (H.H.); kdmingxing@163.com (M.Z.); 15994235997@163.com (Q.Y.)
* Correspondence: xuguang@wust.edu.cn; Tel.: +86-27-6886-2813

Academic Editor: Soran Birosca
Received: 17 June 2016; Accepted: 19 July 2016; Published: 23 July 2016

Abstract: In this work, three low-carbon bainitic steels, with different Mo contents, were designed to investigate the effects of Mo addition on microstructure and mechanical properties. Two-step cooling, i.e., initial accelerated cooling and subsequent slow cooling, was used to obtain the desired bainite microstructure. The results show that the product of strength and elongation first increases and then shows no significant change with increasing Mo. Compared with Mo-free steel, bainite in the Mo-containing steel tends to have a lath-like morphology due to a decrease in the bainitic transformation temperature. More martensite transformation occurs with the increasing Mo, resulting in greater hardness of the steel. Both the strength and elongation of the steel can be enhanced by Mo addition; however, the elongation may decrease with a further increase in Mo. From a practical viewpoint, the content of Mo could be ~0.14 wt. % for the composition design of low-carbon bainitic steels in the present work. To be noted, an optimal scheme may need to consider other situations such as the role of sheet thickness, toughness behavior and so on, which could require changes in the chemistry. Nevertheless, these results provide a reference for the composition design and processing method of low-carbon bainitic steels.

Keywords: bainitic transformation; low carbon; microstructure; property

---

1. Introduction

Low-carbon bainitic steels are used widely in many industrial fields due to their favorable combination of strength and toughness. For the composition design of low-carbon bainitic steels, several alloying elements such as B, Mo, and Cr are usually added to achieve sufficient hardenability [1–4]. In doing so, the desired microstructure of bainitic ferrite is obtained during continuous cooling in industrial production processes.

Since B-Mo bainitic steel was developed by Pickering and Irvine [5], the function of Mo in steels has been widely investigated by many researchers. It is generally accepted that Mo addition can separate the bainitic transformation zone to obtain the desired bainitic microstructure over a wide range of cooling rates [6,7]. Mo is added to decrease the diffusion coefficient of carbon, resulting in the retardation of ferrite and pearlite transformation. Khare et al. [8] stated that the most dramatic effect of Mo is to hinder high temperature transformation, but 0.25 wt. % Mo had no significant influence on bainitic transformation kinetics in a 0.32 wt. % C bainite steel. Sourmail et al. [9] reported that the effect of Mo on bainite transformation was not clear with negligible or no retardation influence on the bainite formation kinetics, although the calculated effect on the driving force led to an expected acceleration. However, Kong et al. [10] claimed that bainitic transformation in a low-carbon alloyed steel containing 0.40 wt. % Mo was slowed down when the cooling rate was below 15 °C/s. Chen et al. [11] reported that it is necessary to add suitable Mo to improve the toughness and strength of high Nb-bearing

X80 pipeline steels (0.26 wt. % Mo, 0.07 wt. % Nb). They considered that Mo addition can suppress pearlite and ferrite (PF) transformation and decrease the transformation temperature, resulting in refined transformed products, so the mechanical properties are improved.

Although the effects of Mo on transformation, microstructure and properties of low-carbon bainitic steels have been widely investigated by several studies, an optimal scheme on the amount of Mo to be added is not fully understood. In most works, the amount of Mo added to low-carbon bainitic steels is designed to be $\geqslant$0.25 wt. % [12–16], which could be considered an expensive addition in the modern context. It is necessary to optimize the amount of Mo added for commercial scenarios. So far, how the comprehensive mechanical properties, especially the product of strength and elongation (PSE), change with the amount of Mo added has seldom been investigated.

Therefore, in the present work, three low-carbon bainitic steels were designed to investigate the effect of Mo content on the PSE. The evolution of the bainitic microstructure with Mo addition was also analyzed. The results provide a theoretical reference for the composition design of low-carbon bainitic steels.

2. Materials and Methods

Three low-carbon bainitic steels with the chemical compositions given in Table 1 were refined using a 100 kg vacuum furnace. Steel #1 was free from Mo and acted as a standard with which the Mo-bearing steels could be compared. The cast ingots were heated to 1250 °C before hot-rolling to 12-mm-thick plates by seven passes. The starting temperatures of the rough roll and finishing roll processes were 1070 °C and 880 °C, respectively. After hot rolling, the plates were initially fast cooled to 500 °C at ~30 °C/s, then air cooled to room temperature. This experimental procedure is basically consistent with the industrial technology. The bainite starting temperature (Bs) and martensite starting temperature (Ms) of three tested steels were calculated according to the following equations [17]:

$$B_S = 839 - \sum_i P_i x_i - 270 \times [1 - \exp(-1.33x_c)] \tag{1}$$

$$M_S = 565 - \sum_i K_i x_i - 600 \times [1 - \exp(-0.96x_c)] \tag{2}$$

where i = Mn, Si, Cr, Ni, and Mo, and the concentration x in wt. %.

$$\sum_i P_i x_i = 86x_{Mn} + 23x_{Si} + 67x_{Cr} + 67x_{Ni} + 75x_{Mo} \tag{3}$$

$$\sum_i K_i x_i = 31x_{Mn} + 13x_{Si} + 10x_{Cr} + 18x_{Ni} + 12x_{Mo} \tag{4}$$

Table 1. Chemical compositions of steels (wt. %).

Steel	C	Si	Mn	Mo	P	S	Bs	Ms
#1	0.223	1.523	2.187	0	<0.01	<0.003	546	357
#2	0.219	1.504	2.095	0.134	<0.01	<0.003	543	360
#3	0.225	1.519	2.034	0.273	<0.01	<0.003	538	358

The aim of calculating Bs and Ms is to design the finishing temperature point of the fast cooling stage and the results are given in Table 1. The specimens were mechanically polished and etched with a 4% nital solution for microstructure examination using optical microscopy (OM) and scanning electron microscopy (SEM). The morphology of bainitic ferrite was examined using a Nova 400 Nano field emission scanning electron microscope (FEI, Hillsboro, OR, USA) operated at an accelerating voltage of 20 kV. In addition, the fine microstructure was observed using a JEM-2100F transmission electron microscope (TEM, JEOL, Tokyo, Japan). The volume fractions of retained austenite (RA) in different samples were determined using an X'Pert diffractometer (Panalytical, Almelo, The Netherlands) with Co Kα radiation under the following conditions: acceleration voltage, 40 kV; current, 150 mA; and step, 0.06°. Tensile tests were performed using a UTM-5305 electronic universal tensile machine (Instron, Norwood, MA, USA) at room temperature. Tensile specimens were prepared according to ASTM standards and the strain rate was ~4 × 10⁻³/s. Vickers hardness tests were performed with a HV1000A micro-hardness tester (0.2 kg-1960 mN, Matsuzawa, Tokyo, Japan). The average value of at least ten individual measurements was calculated, including several martensite bands and bainite blocks in the microstructure.

3. Results and Discussions

3.1. Characteristics of Microstructure

Optical micrographs of the specimens are presented in Figure 1. It can be seen that bainite was obtained in all samples with or without Mo addition. In steel #1 (Figure 1a), the microstructure mainly consists of bainite with a very small amount of martensite/austenite (M/A) islands. The microstructure is fine due to the initial fast cooling rate of ~30 °C/s. Martensite in bands (darker regions in Figure 1b,c) appears in Mo-added steels. Actually, microstructural bands are caused by the segregation of the substitutional alloying element of manganese, which is abundant in these materials. Compared to steel #1, the microstructure of steels #2 and #3 is finer.

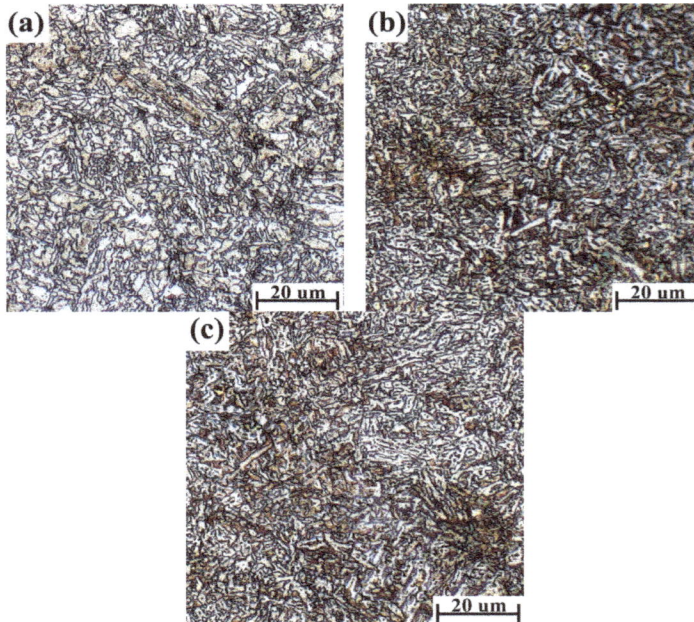

Figure 1. Optical micrographs of steels with different Mo content: (**a**) steel #1, Mo-free; (**b**) steel #2, 0.134 wt. % Mo; (**c**) steel #3, 0.273 wt. % Mo.

In order to further clarify the changes in the microstructure, SEM micrographs of steels with different Mo additions are given in Figure 2. The microstructure in the Mo-free steel consists of granular bainite (GB) and M/A islands, shown by the arrow in Figure 2a. With a 0.134 wt. % Mo addition, lath bainite (LB), GB and M/A islands are observed (Figure 2b); the amount of GB decreases and the M/A islands become finer. From Figure 2c, it can be seen that the amount of LB increases with increasing Mo. The results indicate that Mo can decrease the amount of GB transformed by undercooled austenite, and contributes to obtaining LB and fine M/A islands during the continuous cooling process.

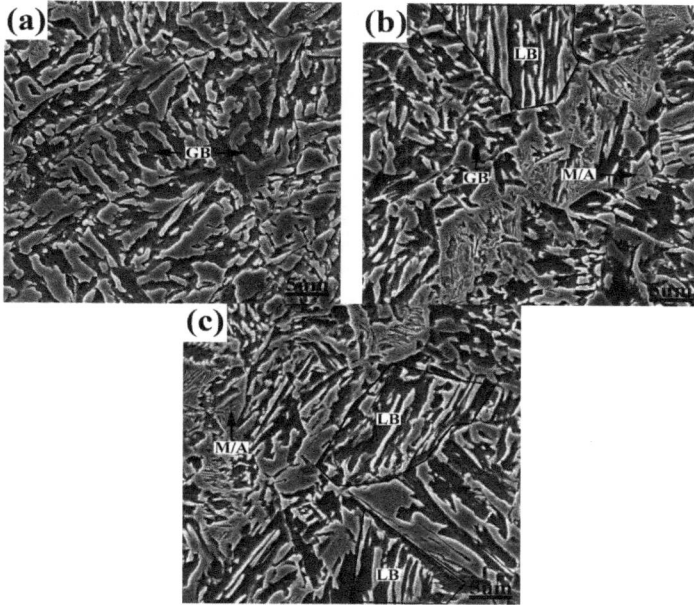

Figure 2. SEM micrographs of steels with different Mo contents: (**a**) steel #1, Mo-free; (**b**) steel #2, 0.134 wt. % Mo; (**c**) steel #3, 0.273 wt. % Mo.

The appearance of LB and M/A islands demonstrates that bainite transformation in Mo-containing steel occurs at a lower temperature range than in Mo-free steel. As the strong carbide-forming element Mo is added to the steel, the carbon diffusion activation energy in austenite increases and the carbon diffusion coefficient decreases [10]. Bainite transformation is closely related to carbon diffusion. Therefore, bainite transformation starts at a lower temperature range when Mo is added. As the Bs point decreases, the parent austenite transforms into LB at a low transformation temperature. In addition, a phenomenon called "incomplete transformation" usually occurs because bainite transformation stops prematurely before the equilibrium amount is attained [18,19]. Thus, residual austenite after bainite transformation partially transforms to martensite.

Figure 3 shows TEM micrographs of the microstructure of the three tested steels. The morphologies of GB and LB are clearly presented, as shown in Figure 3a–c. The bainite in steel #1 (Mo-free) is almost GB with little LB, whereas the microstructure in steel #2 (0.134 wt. % Mo) consists of GB and LB (Figure 3b). With the increase of Mo, the amount of LB increases and the width of the bainitic laths decreases (Figure 3c). In addition, net-like dislocation lines are observed on the surface of LB, as shown in Figure 3d. When adding Mo to steel, the morphology of bainite changes from GB to LB and dislocation begins to occur. This is due to the decrease of the transformation temperature, which changes the mechanism of bainite transformation. The relatively high dislocation density associated with bainitic ferrite is often attributed to the shape deformation caused by displacive transformation [20]. Caballero et al. [21] investigated the influence of the transformation temperature on the dislocation density and the corresponding bainitic ferrite thickness. They found that the dislocation density increased with a decrease in the transformation temperature and the corresponding bainitic ferrite thickness decreased.

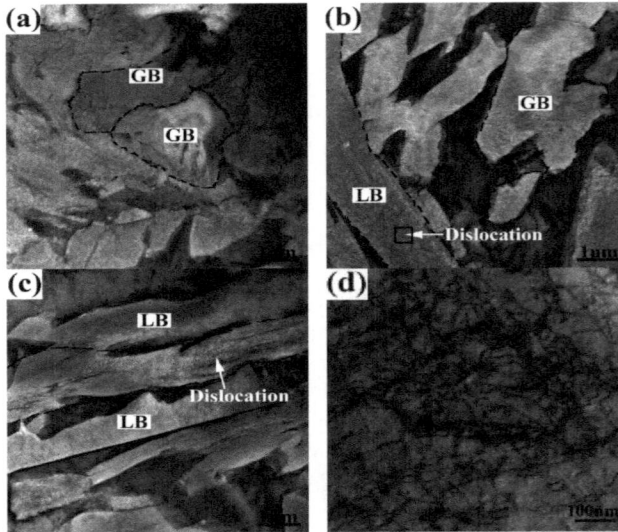

Figure 3. TEM micrographs showing the morphology of bainite and dislocation: (**a**) steel #1, Mo-free; (**b**) steel #2, 0.134 wt. % Mo; (**c**) steel #3, 0.273 wt. % Mo; (**d**) micrograph in high magnification of the square zone in (**b**).

3.2. Quantitative Analysis

The calculation of RA is based on integrated intensities of the $(200)\alpha$, $(211)\alpha$, $(200)\gamma$, $(220)\gamma$ and $(311)\gamma$ diffraction peaks obtained by X-ray diffraction (XRD). The volume fractions of RA were calculated according to Equation (5) and the value is the average of V_i [22].

$$V_i = \frac{1}{1 + G(I_\alpha/I_\gamma)} \tag{5}$$

where V_i is the volume fraction of austenite for each peak, I_α and I_γ are the integrated intensities of ferrite and austenite peaks, respectively, and G is a certain value for each peak. As bainite and martensite have a similar crystalline structure, it is difficult to distinguish them in the diffraction diagrams. The volume fraction of bainite was evaluated using the method in Reference [23]. The results of the quantitative analysis are given in Table 2. In order to minimize error, an average value of each case was obtained from at least 10 individual measurements. The volume fraction of GB decreased with Mo content, whereas the volume fractions of LB and M increased. The Mo-containing steels had more RA than the Mo-free steel, demonstrating that Mo addition contributes to the stabilization of austenite.

Table 2. Quantitative data on the microstructure of different steels.

Sample	V_{GB}	V_{LB}	V_M	V_γ
#1 (Mo-free)	0.69 ± 0.04	0.05 ± 0.01	0.19 ± 0.02	0.07 ± 0.01
#2 (0.134% Mo)	0.39 ± 0.03	0.25 ± 0.02	0.25 ± 0.03	0.11 ± 0.01
#3 (0.273% Mo)	0.06 ± 0.01	0.47 ± 0.03	0.37 ± 0.03	0.10 ± 0.01

Notes: V_{GB}, volume fraction of granular bainite; V_{LB}, volume fraction of lath bainite; V_M, volume fraction of martensite; V_γ volume fraction of RA (determined by electron micrographs and XRD).

3.3. Mechanical Properties

The tensile test results of the three steels are given in Table 3. The results are the average values of four tests. The yield strength (YS) and ultimate tensile strength (UTS) are improved by the addition of Mo. Meanwhile, the total elongation (TE) first increases and then decreases within a small range. The percentage reduction of area (Z) also changes a little with varying Mo content. As mentioned previously, the addition of Mo decreases the transformation temperature of bainite, resulting in a finer bainitic microstructure and a certain amount of martensite. Moreover, LB provides a higher phase strengthening effect than GB. These two reasons lead to the increase in strength of the Mo-containing steel compared to the Mo-free steel. It should be noted that the increment of UTS (112 MPa, Mo from 0 to 0.134 wt. %; 56 MPa, Mo from 0.134 wt. % to 0.273 wt. %) reduces with the increment of the Mo content. For the industrial production of low-carbon bainitic steels, comprehensive mechanical properties are normally considered. Thus, the value of PSE, i.e., UTS × TE (GPa %), versus Mo addition is plotted in Figure 4. It can be observed that the PSE showed no significant change when the Mo content went beyond 0.134 wt. %. This is due to the slow increase in strength and decrease in elongation. In many current studies, the amount of Mo added to low-carbon bainitic steels is normally designed to be ⩾0.25 wt. % [12–16]. However, in the present study, it is demonstrated that it is more suitable to add Mo at ~0.134 wt. % considering the production cost. It should be noted that an optimal scheme sometime needs to consider other factors such as the role of sheet thickness, toughness behavior and so on which could require changes in the chemistry. Nevertheless, the result of the present work provides a reference for the composition and processing design of low-carbon bainitic steels.

Table 3. Mechanical properties of samples with different compositions.

Steel	YS (MPa)	UTS (MPa)	TE (%)	Z (%)	PSE (GPa %)
#1	561	1015	15.6	41	15.8
#2	583	1127	18.3	37	20.6
#3	610	1173	16.8	38	19.7

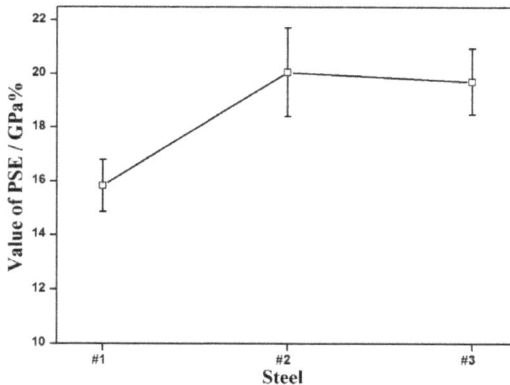

Figure 4. The product of strength and elongation (PSE) in steels with different Mo contents.

Figure 5 shows the results of the Vickers micro-hardness tests. The average value of the micro-hardness in steel #1 (4.83 GPa) is the lowest of all the steels. The highest hardness values of steels #2 and #3 reach ~5.5 GPa, which corresponds to the M/A constituents in the microstructure. In addition, the ratio of the high hardness value in steel #3 is larger than that in steel #2, which is due to the larger amount of martensite. Moreover, the bainite matrix is harder in the Mo-containing steel than in the Mo-free one. The occurrence of LB contributes to the increase of the hardness of the matrix. The results of the hardness tests are closely related to the microstructure (Figures 1 and 2).

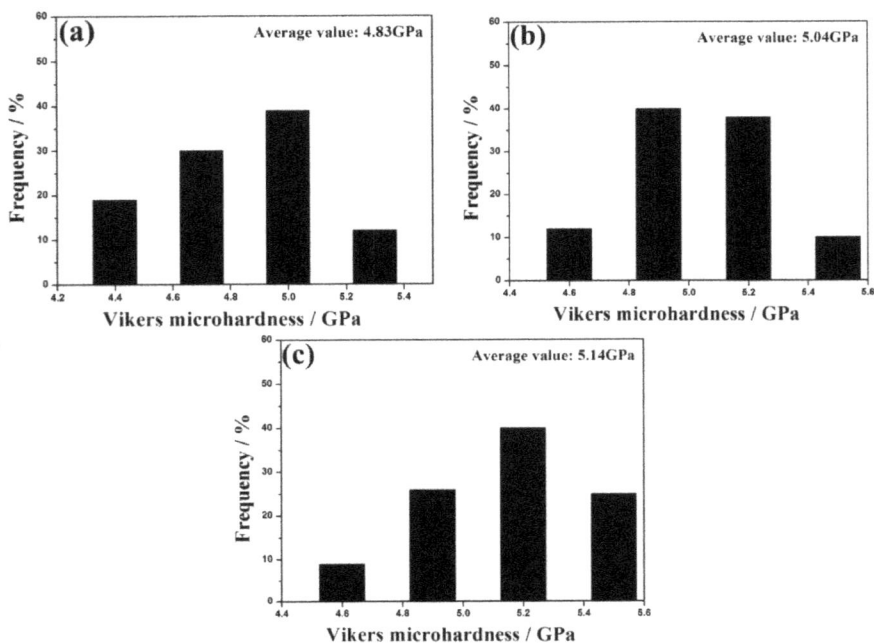

Figure 5. Vickers micro-hardness values of samples with different Mo content: (**a**) steel #1, Mo-free (**b**) steel 2#, 0.134 wt. % Mo, (**c**) steel #3, 0.273 wt. % Mo.

3.4. Design of Processing

As discussed above, the constitution and morphology of the microstructure are changed by adding Mo into steels. It is inferred that the transformation kinetics are affected by the addition of Mo. Time-temperature-transformation (TTT) diagrams are often used as references for the design of the composition and processing technology of low-carbon bainitic steels [24,25]. Figure 6 shows the TTT diagrams calculated based on the compositions of tested steels by MUCG83 [26,27]. Compared to the Mo-free steel, both steels #2 and #3 have an obvious bay between the two C-curves (one at high temperature representing reconstructive transformation and the other at low temperature representing displacive transformation [27]). This is due to Mo enhancing the hardenability of the steel, causing a separation of the bainite C-curve. In addition, when the Mo content is increased from 0.134 wt. % to 0.273 wt. %, the C-curves show no significant change. Kong et al. [9] reported that 0.40 wt. % Mo can further reduce the Bs when the cooling rate is low, but the difference is not obvious when the cooling rate is above 30 °C/s. The calculated TTT diagrams suggest that a two-step cooling regime is a promising method for producing an optimal bainite microstructure. The results in the present work provide a practical reference for the design of the composition and processing routes of low-carbon bainitic steels.

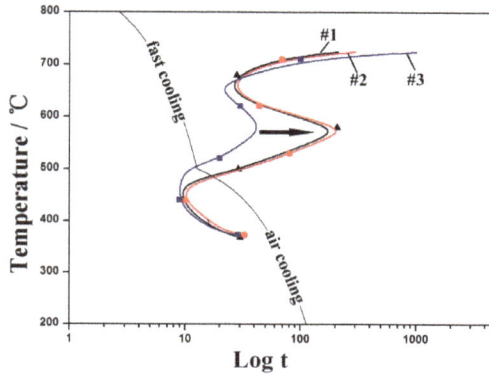

Figure 6. Calculated TTT diagrams for the Mo-free steel (#1) and the Mo-containing steels (#2 and #3).

4. Conclusions

Three low-carbon bainitic steels were designed to investigate the effects of Mo addition on the microstructure and mechanical properties. The results show that the microstructure in Mo-free steel consists of GB + M/A with little LB, whereas the microstructure in Mo-added steels consists of GB + LB + M/A. With increasing the Mo, the amount of LB increases and the bainite laths become finer. The product of strength and elongation first increases and then shows no significant change with further Mo addition. Lath-like bainite and M/A islands in the Mo-bearing steels contribute to the increase in strength and the improvement in elongation compared to the Mo-free steel. However, the elongation may decrease with increasing Mo. From a practical viewpoint, the content of Mo could be ~0.14 wt. % for the composition design of low-carbon bainitic steels in the present work. It should be noted that an optimal scheme sometime needs to consider other factors such as the role of sheet thickness, toughness behavior and so on which could require changes in the chemistry. Nevertheless, the result of the present work provides a reference for the composition and processing design of low-carbon bainitic steels.

Acknowledgments: The authors are grateful for the financial supports from the National Natural Science Foundation of China (NSFC) (No. 51274154), the State Key Laboratory of Development and Application Technology of Automotive Steels (Baosteel Group) (No. 20121101).

Author Contributions: Guang Xu and Haijiang Hu conceived and designed the experiments; Haijiang Hu performed the experiments; Haijiang Hu and Mingxing Zhou analyzed the data; Qing Yuan contributed materials and tools; Haijiang Hu wrote the paper.

Abbreviations

The following abbreviations are used in this manuscript:

PSE	product of strength and elongation
LSCM	high temperature laser scanning confocal microscopy
Bs	bainite starting temperature
Ms	martensite starting temperature
OM	optical microscopy
SEM	scanning electron microscopy
TEM	transmission electron microscope
XRD	X-ray diffractogram
M/A	martensite/austenite
GB	granular bainite

LB	lath bainite
RA	retained austenite
YS	yield strength
UTS	ultimate tensile strength
TE	total elongation
TTT	time-temperature-transformation

References

1. Soliman, M.; Palkowski, H. Development of the low temperature bainite. *Arch. Civ. Mech. Eng.* **2016**, *16*, 403–412. [CrossRef]
2. Furuhara, T.; Tsuzumi, K.; Miyamoto, G.; Amino, T.; Shigesato, G. Characterization of Transformation Stasis in Low-Carbon Steels Microalloyed with B and Mo. *Metall. Mater. Trans. A* **2014**, *45A*, 5990–5996. [CrossRef]
3. Bhadeshia, H.K.D.H. Computational design of advanced steels. *Scr. Mater.* **2014**, *70*, 12–17. [CrossRef]
4. Hu, Z.W.; Xu, G.; Yang, H.L.; Zhang, C.; Yu, R. The Effects of Cooling Mode on Precipitation and Mechanical Properties of a Ti-Nb Microalloyed Steel. *J. Mater. Eng. Perform.* **2014**, *23*, 4216–4222. [CrossRef]
5. Ivine, K.J.; Pickering, F.B. Low Carbon Bainitic Steels. *J. Iron Steel Inst.* **1957**, *187*, 292–309.
6. Grange, R.A. Estimating the hardenability of carbon steels. *Metall. Trans.* B **1973**, *4*, 2231–2244. [CrossRef]
7. Pickering, F.B. *Physical Metallurgy and the Design of Steels*; Applied Science Publishers Ltd.: London, UK, 1978.
8. Khare, S.; Lee, K.; Bhadeshia, H.K.D.H. Relative effects of Mo and B on ferrite and bainite kinetics in strong steels. *Int. J. Mater. Res.* **2009**, *100*, 1513–1520. [CrossRef]
9. Sourmail, T.; Smanio, V. Low temperature kinetics of bainite formation in high carbon steels. *Acta Mater.* **2013**, *61*, 2639–2648. [CrossRef]
10. Kong, J.H.; Xie, C.S. Effect of molybdenum on continuous cooling bainite transformation of low-carbon microalloyed steel. *Mater. Des.* **2006**, *27*, 1169–1173. [CrossRef]
11. Chen, X.W.; Qiao, G.Y.; Han, X.L.; Wang, X.; Xiao, F.R.; Liao, B. Effects of Mo, Cr, and Nb on microstructure and mechanical properties of heat affected zone for Nb-bearing X80 pipeline steels. *Mater. Des.* **2014**, *53*, 888–901. [CrossRef]
12. Soliman, M.; Mostafa, H.; El-Sabbagh, A.S.; Palkowski, H. Low temperature bainite in steel with 0.26 wt. % C. *Mater. Sci. Eng. A* **2010**, *527*, 7706–7713. [CrossRef]
13. Qian, L.; Zhou, Q.; Zhang, F.; Meng, J.; Zhang, M.; Tian, Y. Microstructure and mechanical properties of a low carbon carbide-free bainitic steel co-alloyed with Al and Si. *Mater. Des.* **2012**, *39*, 264–268. [CrossRef]
14. Wang, Y.H.; Zhang, F.C.; Wang, T.S. A novel bainitic steel comparable to maraging steel in mechanical properties. *Scr. Mater.* **2013**, *68*, 763–766. [CrossRef]
15. Long, X.Y.; Zhang, F.C.; Kang, J.; Lv, B.; Shi, X.B. Low-temperature bainite in low-carbon steel. *Mater. Sci. Eng. A* **2014**, *594*, 344–351. [CrossRef]
16. Wang, X.L.; Wu, K.M.; Hu, F.; Yu, L.; Wan, X.L. Multi-step isothermal bainitic transformation in medium-carbon steel. *Scr. Mater.* **2014**, *74*, 56–59. [CrossRef]
17. Bohemen, S.M.C. Bainite and martensite start temperature calculated with exponential carbon dependence. *Mater. Sci. Technol.* **2012**, *28*, 487–495. [CrossRef]
18. Aaronson, H.I.; Reynolds, W.T.; Purdy, G.R. The incomplete transformation phenomenon in steel. *Metall. Mater. Trans. A* **2006**, *37A*, 1731–1745. [CrossRef]
19. Xia, Y.; Miyamoto, G.; Yang, Z.G.; Zhang, C.; Furuhara, T. Direct measurement of carbon enrichment in the incomplete bainite transformation in Mo added low carbon steels. *Acta Mater.* **2015**, *91*, 10–18. [CrossRef]
20. Bhadeshia, H.K.D.H.; Christian, J.W. Bainite in steels. *Metall. Trans.* **1990**, *21A*, 767–769. [CrossRef]
21. Caballero, F.G.; Garcia-Mateo, C.; Santofimia, M.J.; Miller, M.K.; de Andrés, C.G. New experimental evidence on the incomplete transformation phenomenon in steel. *Acta Mater.* **2009**, *57*, 8–17. [CrossRef]
22. Wang, C.Y.; Shi, J.; Cao, W.Q.; Dong, H. Characterization of microstructure obtained by quenching and partitioning process in low alloy martensic steel. *Mater. Sci. Eng. A* **2010**, *527*, 3442–3449. [CrossRef]
23. Hu, H.J.; Xu, G.; Wang, L.; Xue, Z.L.; Zhang, Y.L.; Liu, G.H. The Effects of Nb and Mo Addition on Transformation and Properties in Low Carbon Bainitic Steels. *Mater. Des.* **2015**, *84*, 95–99. [CrossRef]
24. Caballero, F.G.; Santofimia, M.J.; Capdevila, C.; García-Mateo, C.; de Andrés, C.G. Design of advanced bainitic steels by optimization of TTT diagrams and T_0 curves. *ISIJ Int.* **2006**, *46*, 1479–1488. [CrossRef]

25. Caballero, F.G.; Santofimia, M.J.; García-Mateo, C.; Chao, J.; de Andrés, C.G. Theoretical design and advanced microstructure in super high strength steels. *Mater. Des.* **2009**, *30*, 2077–2083. [CrossRef]
26. Bhadeshia, H.K.D.H. A thermodynamic analysis of isothermal transformation diagrams. *Met. Sci.* **1982**, *16*, 159–165. [CrossRef]
27. Bhadeshia, H.K.D.H. *Bainite in Steels*; Institute of Materials: London, UK, 1992; pp. 1–450.

![metals logo] *metals*

MDPI

Article

Effect of Titanium on the Microstructure and Mechanical Properties of High-Carbon Martensitic Stainless Steel 8Cr13MoV

Wen-Tao Yu *, Jing Li, Cheng-Bin Shi and Qin-Tian Zhu

State Key Laboratory of Advanced Metallurgy, University of Science and Technology Beijing,
Beijing 1000083, China; lijing@ustb.edu.cn (J.L.); shicb09@163.com (C.-B.S.); zqtustb@163.com (Q.-T.Z.)
* Correspondence: ywt82@163.com; Tel.:+86-188-1304-8785

Academic Editor: Soran Birosca
Received: 20 May 2016; Accepted: 17 August 2016; Published: 22 August 2016

Abstract: The effect of titanium on the carbides and mechanical properties of martensitic stainless steel 8Cr13MoV was studied. The results showed that TiCs not only acted as nucleation sites for δ-Fe and eutectic carbides, leading to the refinement of the microstructure, but also inhibited the formation of eutectic carbides M_7C_3. The addition of titanium in steel also promoted the transformation of M_7C_3-type to $M_{23}C_6$-type carbides, and consequently more carbides could be dissolved into the matrix during hot processing as demonstrated by the determination of extracted carbides from the steel matrix. Meanwhile, titanium suppressed the precipitation of secondary carbides during annealing. The appropriate amount of titanium addition decreased the size and fraction of primary carbides in the as-cast ingot, and improved the mechanical properties of the annealed steel.

Keywords: carbide; mechanical property; high-carbon martensitic stainless steel; titanium

---

1. Introduction

High-carbon martensitic stainless steels have high hardness, high strength, and good corrosion and wear resistance. They have been widely used to produce knives and scissors, valves, structural components [1–4]. The high-carbon martensitic stainless steel 8Cr13MoV is currently used to produce high-grade knives and scissors [4].

Carbon atoms can combine with alloying elements to form carbides, which can improve the strength and wear resistance of the steel. For martensitic stainless steel, carbides precipitate in the matrix through heat treatment in the final process, which can achieve secondary hardening [4–6]. However, high carbon content can also result in some undesired problems, especially for the HMSS. Because of the segregation of the alloying elements, many large eutectic carbides precipitate in the cooling process [7]. These carbides can hardly be removed in the following hot working and heat treatment [8,9]. In the working process of steel, large carbides can induce a stress concentration [8], causing the generation of cracks in steel [10,11]. Edges crazing is one of the most common failure modes during the cold rolling of high-carbon martensitic stainless steels used for high-grade knives and scissors. Eutectic carbides should be responsible for the edges crazing, which also occurs in tool steel and die steel production [12–14]. Therefore, before cold rolling, high ductility and low hardness of steel are desired. These mechanical properties of steel could be improved by optimizing the microstructure and carbides in the steel. In the practical production process, the segregation is inevitable so measures are taken such as decreasing or refining the eutectic carbides, increasing nucleation or inhibiting the growth of eutectic carbides. However, methods such as accelerating the cooling rate [15,16], adopting low superheat [17], adding an alloy element [18–21] or rare earth for modification treatment [22–25] bring a very limited improvement to the mechanical properties of steel. The addition of alloying elements in high-carbon tool steels is one of the countermeasures in controlling carbides in these steels. However, the effect of alloying elements on carbides in high-carbon martensitic stainless steels has not been studied yet.

Titanium is a strong carbide-forming element, and its ability to combine with carbon is higher than that of chromium [26]. Formation of TiC could partially substitute M_7C_3 in high-carbon high-Cr tool steel, consequently improving the crack resistance of the steel [14]. In this study, the effect of titanium on the carbides and mechanical properties of martensitic stainless steel 8Cr13MoV was studied. The transformation of microstructure and carbides in the steel after electroslag remelting (ESR), forging and spheroidizing annealing were analyzed. The effect of titanium on the microstructure and mechanical properties of high-carbon martensitic steel 8Cr13MoV was revealed.

2. Experimental

2.1. Experimental Materials and Experimental Process

The 8Cr13MoV steel was melted in a vacuum induction furnace. Different amounts of ferro-titanium were added into liquid steel to adjust composition. The liquid steel was cast in a mold, and then forged into electrode with a diameter of 100 mm. The electrode was electroslag remelted to produce as-cast ingot of 160 mm in diameter. The electric current was 2400 A and the voltage was 40 V in the ESR process. The remelting process was conducted in the argon atmosphere. The chemical composition of samples after ESR was confirmed by inductively coupled plasma optical emission spectrometer, which is shown in Table 1. The experiment process also included forging and spheroidizing annealing. The as-cast ESR ingots were forged after held at 1200 °C for 120 min. The forging finish temperature was no less than 800 °C. For spheroidizing annealing, the specimens that were taken from the forged steel were heat treated in an electric resistance furnace, consisting of two holding process and a cooling process. The thermal history in heat treatment is shown in Figure 1.

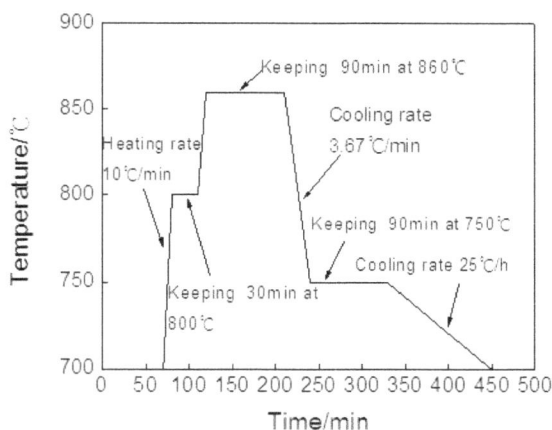

Figure 1. The flow chart of spheroidizing annealing.

Table 1. Chemical composition of 8Cr13MoV steel (mass %).

Sample No.	C	Cr	Mo	Mn	Si	V	Ni	P	S	Ti	Fe
No.1	0.72	13.6	0.43	0.5	0.32	0.14	0.16	0.03	0.004	0.043	Bal.
No.2	0.71	13.44	0.41	0.44	0.33	0.16	0.16	0.031	0.004	0.771	Bal.
No.3	0.73	14.05	0.42	0.42	0.35	0.16	0.16	0.029	0.003	1.22	Bal.

2.2. Microstructure and Carbides

Firstly, ESR ingot was forged into 100 mm × 100 mm × 30 mm steel plate. Secondly, the steel plate was spheroidizing annealed. A steel sample of 12 mm × 12 mm × 5 mm was taken from the edge of ESR ingot, forged steel plate and the spheroidizing annealed steel plate. Finally, the metallographic samples were analyzed by SEM (FEI MLA250, FEI, Hillsboro, OR, USA) after grinding, polishing and etching with $FeCl_3$ hydrochloric acid alcohol ($FeCl_3$ 5 g, 38% HCl 25 mL, C_2H_5OH 25 mL). Meanwhile, the samples of the ESR ingot were corroded by the potassium permanganate solution ($KMnO_4$ 20 g, 0.5 mol/L NaOH 300 mL, H_2O 200 mL), this corrodent can only corrode carbides and it was of no effect on the matrix. The carbides were observed using OM (Leica DM4M Leica, Wetzlar, Germany) and the number and size of carbides were analyzed by Image Analysis Software-Image Pro Plus (IPP).

2.3. Carbides Collection Using Electrolytic Extraction Technique

The samples taken from ESR ingot and annealed steel plate were machined to Ø15 mm × 80 mm round bar. The carbides were extracted from steel matrix in organic solution (methanol, tetramethylammonium chloride, glycerin, diethanol amine) by electrolysis. Part of the carbides were analyzed by XRD (Rigaku D_{max}-RB, Rigaku, Tokyo, Japan) to confirm the type, part of carbides were observed by SEM for the morphology.

2.4. Mechanical Properties

The ESR ingot and spheroidizing annealed steel plates were sampled and machined to Ø8 mm × 80 mm tensile samples based on the Chinese national standard GBT228-2002. The mechanical properties were tested by universal material testing machine (Tinius Olsen H50K, Tinius Olsen, Shanghai, China).

3. Results

3.1. Influence of Titanium on Microstructure

3.1.1. Influence of Titanium on Microstructure of As-Cast ESR Ingot

The SEM images of the ESR ingot microstructure are shown in Figure 2a–c. The grain sizes of samples No. 1 and No. 2 were similar, while the grain size of No. 3 was smaller. Five SEM-BES images were selected from each sample, and the grain size and volume of martensite were calculated using the Count/Size module of IPP. The grain size of sample No. 1 and No. 2 was similar, and the average grain diameter of sample No. 1 is 369.11 ± 46.33 μm, while the grain size of sample No. 3 was much smaller with the average diameter of 233.05 ± 25.74 μm. The amount of martensite was larger and the acicular structure of martensite was much finer. Results showed that the volume fraction of martensite in sample No. 1 is 32.61% ± 3.22%, while that in samples No. 2 and No. 3 was doubled. Carbides in sample No. 1 were distributed on the grain boundaries, which were typical eutectic carbides with a large size.

These carbides presented two kinds of morphology in sample No. 2, granular carbides with small size and some irregular carbides, as shown in Figure 2e. According to EDS analysis, the complex carbides were confirmed to be primary carbides of TiC as a nucleus of eutectic carbides M_7C_3. The EDS results for Figure 2d–g are shown in Table 2. Carbides in sample No. 3 have three kinds of morphologies, which included single granular and irregular flake as shown in Figure 2f, and rod-like with an interior tiny laminar structure as shown in Figure 2g. The eutectic carbides were separated.

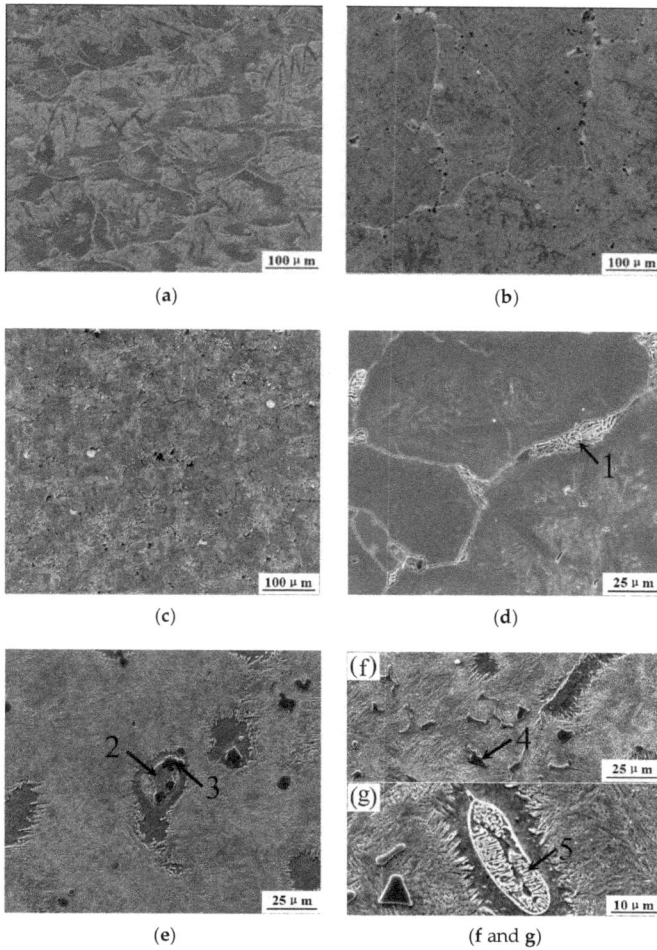

Figure 2. SEM images of as-cast ESR ingot microstructure: (**a**) and (**d**) sample No. 1; (**b**) and (**e**) sample No. 2; (**c**) and (**f**) and (**g**) sample No. 3.

Table 2. The EDS results for carbides (mass %).

Point	Element				
	C	**Cr**	**Fe**	**Ti**	**Mo**
1	13.22	51.89	31.25	—	1.45
2	13.81	53.38	26.94	1.05	2.48
3	23.86	4.43	8.89	51.76	7.79
4	20.28	2.91	12.2	61.47	3.14
5	8.68	42.53	47.90	—	1.40

3.1.2. Influence of Titanium on Forged Microstructure and Annealed Organization

The SEM images of the forged ESR ingot microstructure were shown in Figure 3a–c. The eutectic carbides in sample No. 1 were broken. Meanwhile, the small secondary carbides in sample No. 1

precipitated around the eutectic carbides. No small secondary carbides precipitated around the broken eutectic carbides as shown in sample No. 2 and No. 3. It was found that secondary carbides precipitated along the grain boundaries in these three samples. These secondary carbides were much less after titanium was added to the steel. The precipitated carbides in sample No. 3 along the grain boundaries were granule-shaped. The martensite structure of sample No. 2 was the finest, followed by that in sample No. 3.

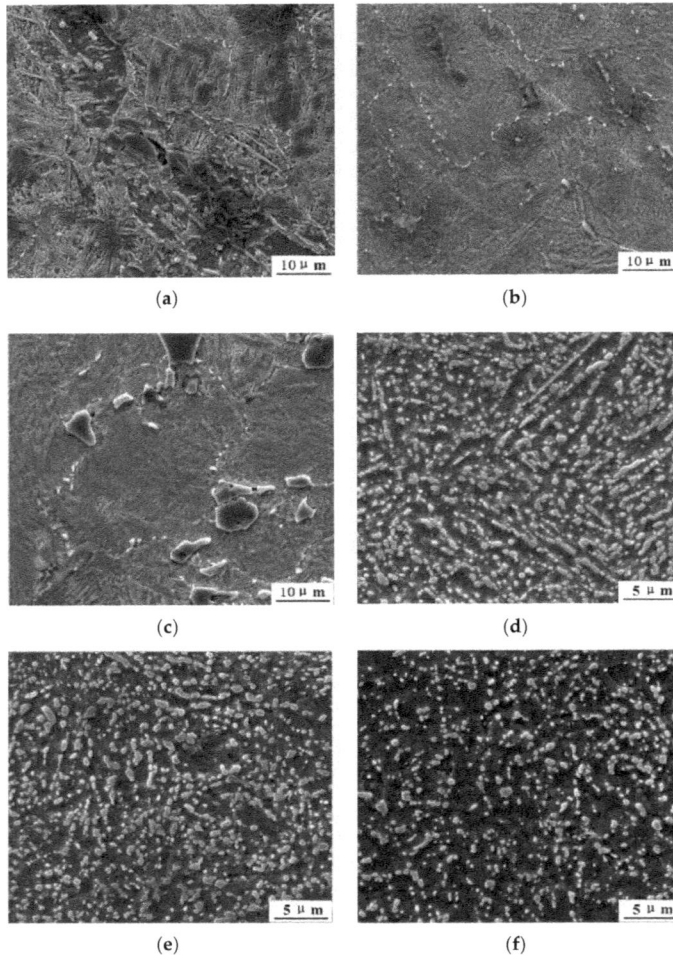

Figure 3. SEM images of specimens before and after spheroidizing annealing: (**a**–**c**) samples No. 1, No. 2, No. 3 before spheroidizing annealing; (**d**–**f**) samples No. 1, No. 2, No. 3 after spheroidizing annealing.

The SEM images of the spheroidizing annealed microstructure were shown in Figure 3d–f. The amount of carbides decreased from sample No. 1 to No. 3, which was 1.27 ± 0.02 particles per μm^2, 1.06 ± 0.05 particles per μm^2 and 0.93 ± 0.04 particles per μm^2, respectively. The average diameter of carbides in samples No. 1, No. 2 and No. 3 is 0.62 ± 0.03 μm, 0.52 ± 0.05 μm and 0.44 ± 0.03 μm, in turn. The martensite structure was different before annealing so that the most carbides of sample No. 1 were chains and had certain directivity. During annealing, the carbides would precipitate along the acicular martensite and become a long strip shape [27]. The precipitated carbides have certain directivity after being annealed if separated incompletely during annealing. In summary, the addition of titanium could inhibit the precipitation of the secondary carbides during heat treatment, and this effect increases with the increasing titanium content.

3.2. Influence of Titanium on Carbides

3.2.1. Carbides Amount

The number of carbides was analyzed by the Count/Size module in IPP. Six images for each sample were analyzed, and the results are shown in Figures 4 and 5. According to the statistical results, the volume fraction of carbides decreased first and then increased with the increasing titanium content. The number of carbides in sample No. 1 is the smallest among these three samples. The volume fraction of carbides in sample No. 2 was the smallest among the three samples while the total amount of carbides was 1.7 times larger than that in sample No. 1. Obviously, the volume fraction of carbides in sample No. 2 was the smallest and the carbides distributed diffusely, which is favorable for improving the mechanical properties of steel.

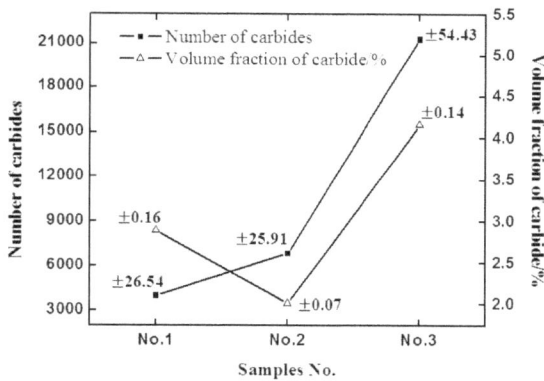

Figure 4. Statistical result of carbides in three samples with different titanium content: values above each point were the standard deviation for each statistic.

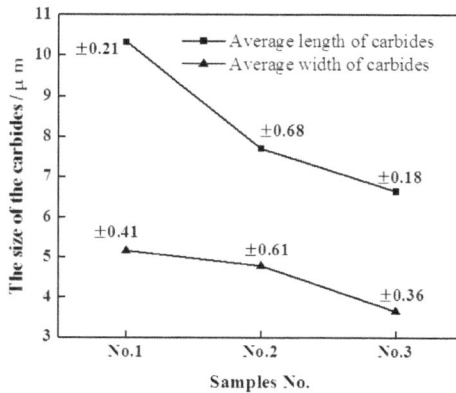

Figure 5. The size of carbides in cast-ingot: values above each point were the standard deviation for each statistic.

3.2.2. The Type of Carbides

Carbides extracted from the steel matrix were analyzed by XRD and the results are shown in Figure 6. Carbides in sample No. 1 were confirmed to be M_7C_3. The type of carbides changed after the addition of titanium because titanium combined carbon and nitrogen to form TiC and Ti(C, N). Meanwhile, M_7C_3 changed to $M_{23}C_6$ after the addition of titanium. This result indicated that titanium could enhance the complete transition from M_7C_3 to $M_{23}C_6$.

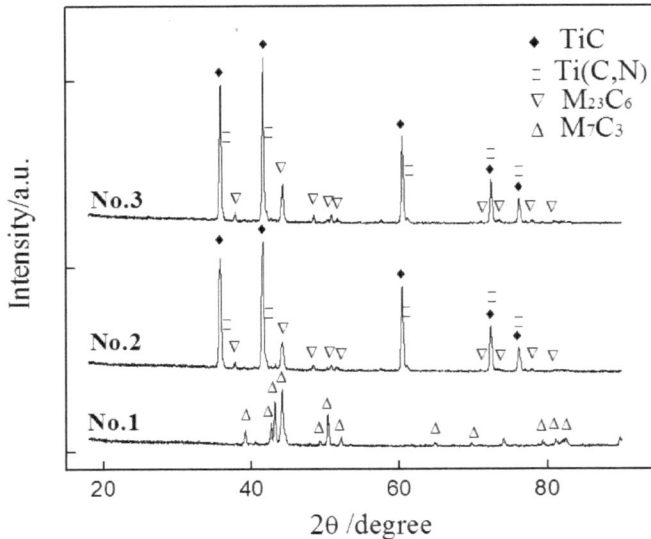

Figure 6. XRD results for three samples with different titanium content: (**a**) sample No. 3; (**b**) sample No. 2; (**c**) sample No. 1.

3.2.3. The Morphology of Carbides

The SEM images of carbides extracted from steel are shown in Figure 7. In sample No. 1, the size of the carbides was large, and the maximum size was over 50 μm. The carbides were typical eutectic carbides and their morphology was a skeleton shape made of a cluster of long-strip carbide, as shown in Figure 7a,d.

In sample No. 2, the size of the carbides was relatively small. Most of the carbides were the complex structure of TiC combined with $M_{23}C_6$. Others were pure TiC, as shown in Figure 7b. The size of the complex carbides was large, and their morphology was similar with that of eutectic carbides, but different from that in sample No. 1, as shown in Figure 7e. It was deduced from the morphology that the large carbides $M_{23}C_6$ were transformed from eutectic carbides M_7C_3 rather than precipitated directly from austenite.

In sample No. 3, the morphology of most carbides was small, irregular flake TiC and Ti(C, N), as shown in Figure 7c. Only a small amount of carbides $M_{23}C_6$ were regular shapes as shown in Figure 7e, which corresponded to the rod-like carbides shown in Figure 2g. The small carbides $M_{23}C_6$ in sample No. 3 were different from those in sample No. 2. These carbides were transformed from M_7C_3 precipitated from austenite. There were no eutectic carbides among the extracted carbides, which indicated that the addition of titanium inhibited the formation of eutectic carbides.

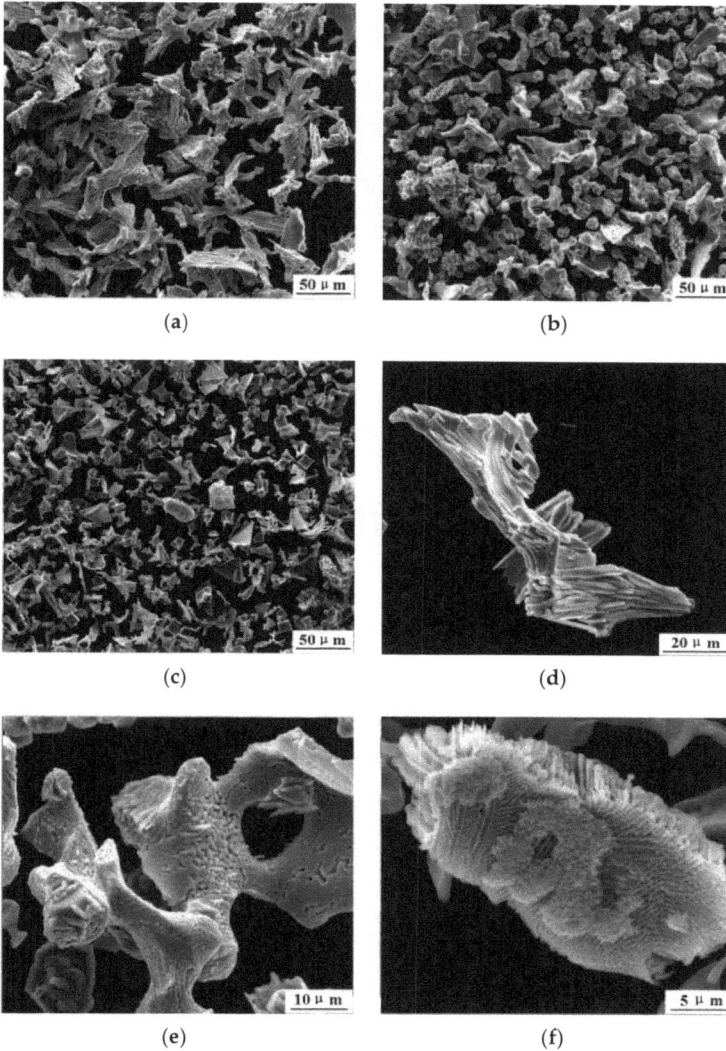

Figure 7. SEM images of carbides power: (**a**) and (**d**) sample No. 1; (**b**) and (**e**) sample No. 2; (**c**) and (**f**) sample No. 3.

3.2.4. The Composition of Carbides

In sample No. 1, there existed a little titanium. It was found by EDS, as shown in Figure 8a, that there was an enrichment of titanium, vanadium and nitrogen in the area where (Ti, V)N precipitated. Thus, (Ti, V)N has a high melting point and acts as the core of the eutectic carbides after precipitating from liquid steel. It was proved that titanium was prior to combine with nitrogen when the content of titanium was low. The chromium content of carbides $M_{23}C_6$ in samples No. 2 and No. 3 was less than that of sample No. 1 as shown in Table 3. In sample No. 2, it was clear that nitrogen, vanadium, molybdenum and titanium are enriched in the same area, confirming that molybdenum had the tendency to dissolve into TiC. The EDS result for the compound containing titanium in Figure 8 is shown in Table 4.

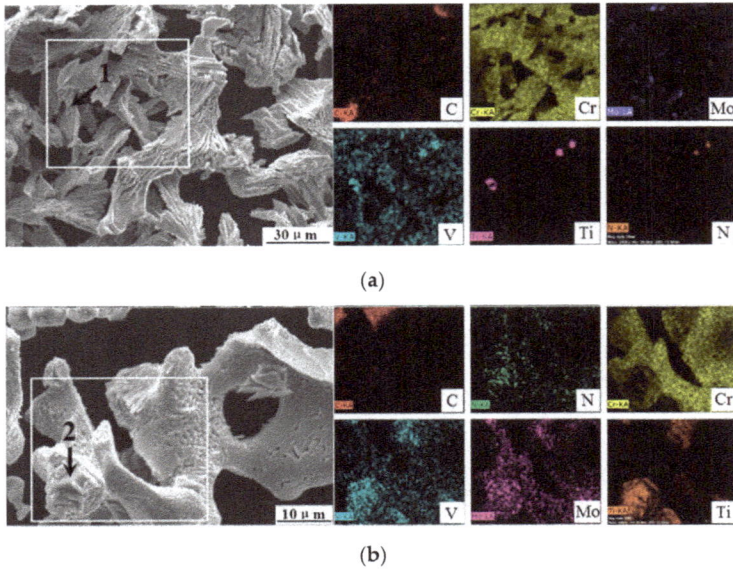

(a)

(b)

Figure 8. EDS element mappings of carbides: (**a**) sample No. 1; (**b**) sample No. 2.

Table 3. EDS-analyzed results of carbides containing chromium (mass %).

Sample No.	Element					
	C	Cr	Fe	Mo	V	Ti
No. 1	14.59	53.90	26.25	2.33	2.93	0.0
No. 2	13.10	37.53	40.19	2.79	1.69	2.6
No. 3	11.08	43.98	28.07	2.67	1.41	2.8

Table 4. EDS-analyzed results of compound containing titanium (mass %).

Point	Element						
	C	N	Cr	Fe	Ti	Mo	V
1	15.20	15.57	1.32	1.28	64.44	0.98	1.21
2	26.66	——	2.76	1.21	62.76	5.46	1.15

3.2.5. Morphology of Carbides after Forging and Heat Treatment

The electrolysis method was employed to extract the carbides in samples after spheroidizing annealing, and the result of the SEM analysis is shown in Figure 9. A large amount of fine carbides precipitated after being spheroidized. As shown in Figure 9b, there are no large complex carbides in sample No. 2. This indicated that most of carbides $M_{23}C_6$ surrounding TiC dissolved, and a small number of carbides $M_{23}C_6$ were spheroidized as shown in Figure 9e. The eutectic carbides in sample No. 1 were spheroidized to some degree, but the size was still large, as shown in Figure 9d. The size of carbides in sample No. 2 was similar to that in sample No. 3.

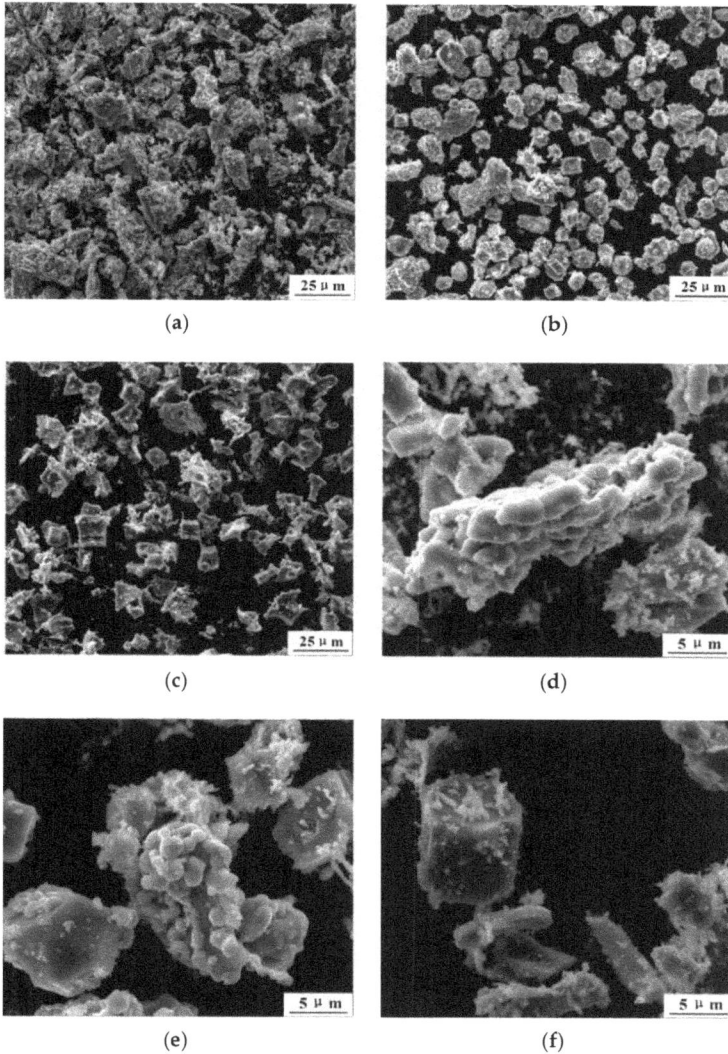

Figure 9. SEM images of carbides power after spheroidizing annealing: (**a**) and (**d**) No. 1; (**b**) and (**e**) No. 2; (**c**) and (**f**) No. 3.

3.3. Influence of Titanium on Mechanical Property of Steel before and after Heat Treatment

The hardness and tensile strength of the ESR ingot were measured, and the results are shown in Table 5. The hardness and tensile strength of cast alloy increased greatly when titanium was added in steel. The tensile strength of sample No. 3 was 1315.66 MPa. Titanium could refine the grain, increase the hardness and tensile strength, and combine with carbon to form TiC to inhibit the formation of carbides containing chromium, resulting in the dissolution of chromium in the matrix and increasing the hardenability and the quantity of martensite.

Table 5. Mechanical properties of steel samples before and after heat treatment

Sample No.	Hardness		Tensile Strength/MPa		Elongation after Fracture/%	
	Before/HRC	After /HRB	Before	After	Before	After
No. 1	52.9 ± 0.21	95.2 ± 0.14	635.95	739.53	1	17.85
No. 2	56.7 ± 0.15	93.6 ± 0.29	974.49	714.01	1	20.95
No. 3	57.6 ± 0.31	94.1 ± 0.52	1315.66	694.39	1	20.41

The morphology of carbides in samples No. 2 and No. 3 was apparently different. In sample No. 2, complex carbides formed by TiC and $M_{23}C_6$ were large in size, and most single TiCs were granular. In sample No. 3, most TiCs had a laminated structure with a multi-direction which was beneficial to matrix strengthening in comparison with the granular structure.

The tensile fracture after spheroidizing annealing is shown in Figure 10. There were large carbides at the fracture surface of sample No. 1, which are labeled in Figure 10a. There are no large carbides in the fracture of sample No. 2. TiC particles distributed on the fracture surface of samples No. 2 and No. 3 uniformly. The number of TiC in sample No. 3 was larger than that in sample No. 2. Due to the existence of a large amount of TiC, the tensile strength and elongation after fracture decreased. It suggested that a moderate addition of titanium was in favor of improving the property of steel, while the excessive addition of titanium would increase the volume fraction of TiC and deteriorate the mechanical property of steel.

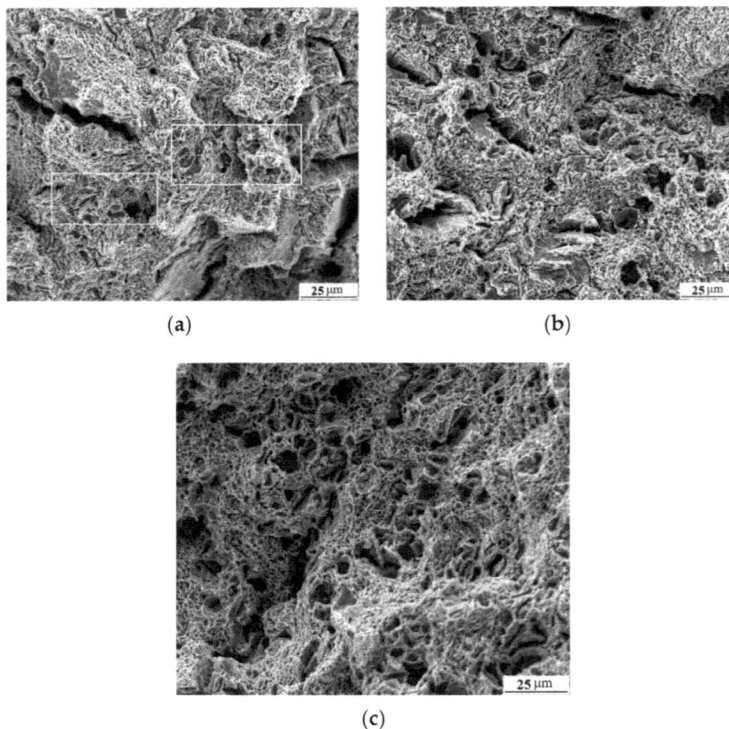

(a)

(b)

(c)

Figure 10. SEM images of tensile fracture after spheroidizing annealing: (a) sample No. 1; (b) sample No. 2; (c) sample No. 3.

4. Discussions

4.1. Effect of Ti on the Solidification Microstructure

The effect of titanium on the equilibrium phase during the cooling process of 8Cr13MoV was calculated by Thermo-Calc. The calculated results are shown in Figure 11. The precipitation temperature of TiC increased with increasing the titanium content. When the titanium content reached up to 0.5%, TiC precipitated at 1600 °C from liquid steel. It is the precipitation of TiC from liquid steel that provides an essential condition for the heterogeneous nucleation of other precipitates and the grain refinement of steel.

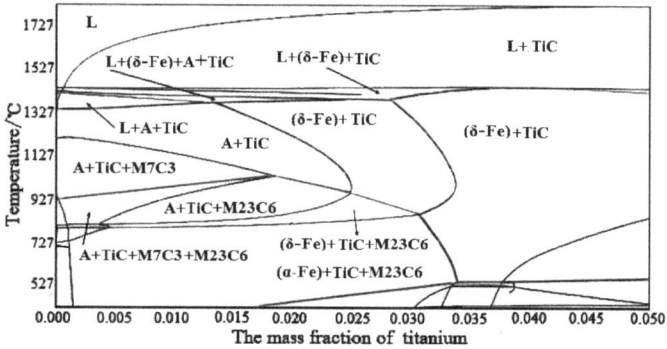

Figure 11. Equilibrium phase formation of 8Cr13MoV steel calculated using Thermo-Calc: L-Liquid, (δ-Fe)-(δ-Ferrite), A-Austenite, (α-Fe)-(α-Ferrite).

According to the Bramfitt theory [28], when the disregistry between the lattice parameters of the substrate and the nucleating phase is less than 6%, the nucleating agent is very effective for heterogeneous nucleation. If the disregistry equals to 6%–12%, the nucleating agent is moderately effective. If the disregistry is greater than 12%, the potency is poor. According to the calculation by Bramfitt [28], the disregistry between TiC and δ-Fe is 5.9%, and the promoting effect of TiC for the nucleation sites of liquid iron was proved by his experiment.

As shown in Figure 11, the microstructure of 8Cr13MoV steel was refined with the addition of titanium in the steel. This is attributed to the precipitation of TiCs, which served as heterogeneous nucleation sites for liquid steel. The present result is consistent with the findings that the addition of titanium in high-carbon high-chromium steel or cast iron could refine the eutectic carbides M_7C_3 and the grain size of the steel as reported by other researchers [26,27,29].

In the 8Cr13MoV steel with the Ti addition, eutectic carbides M_7C_3 were found to nucleate on TiC as shown in Figures 7e and 8b. The disregistry concept proposed by Bramfitt [28] was employed to predict the nucleation between TiC and M_7C_3, as expressed in the following equation:

$$\delta_n^s = \sum_{i=1}^{3} \frac{\left| d_{[uvw]_s^i} \cos\theta - d_{[uvw]_n^i} \right|}{d_{[uvw]_n^i}} \times 100 \qquad (1)$$

$(hkl)_s$ = a low-index plane of the substrate;
$[uvw]_s$ = a low-index direction in $(hkl)_s$;
$(hkl)_n$ = a low-index plane in the nucleated solid;
$[uvw]_n$ = a low-index direction in $(hkl)_n$;
$d[uvw]_n$ = the interatomic spacing along $[uvw]_n$;

d[uvw]$_s$ = the interatomic spacing along [uvw]$_s$;

θ = the angle between the [uvw]$_s$ and [uvw]$_n$.

The crystal structure of TiC and M_7C_3 was face-centered cubic and hexagonal, respectively. The lattice plane (0001) of M_7C_3 is superimposed on the lattice plane (001) of TiC. The lattice parameters for the corresponding crystal faces are 4.31 Å and 4.45 Å. The corresponding angles of crystal orientation are 15° and 30°, respectively. According to above equation, the disregistry between TiC and M_7C_3 was calculated to be 4.7%, which proved that TiC could act as nucleation site for M_7C_3. Therefore, it was concluded that TiC played an important role in the refinement of M_7C_3.

4.2. Eutectic Carbides M_7C_3 and Secondary Carbides $M_{23}C_6$

All carbides in as-cast 8Cr13MoV steel containing 0.043% titanium are M_7C_3-type. The chromium-containing carbides in samples No. 2 and No. 3 are $M_{23}C_6$-type, instead of M_7C_3-type. As shown in Figures 2e and 8b, a large amount of carbides in sample No. 2 are complex carbides consisting of TiC and $M_{23}C_6$. Carbides $M_{23}C_6$ formed around TiC. It can be seen, from the morphology and size, that carbides $M_{23}C_6$ exhibit the characteristics of eutectic carbides.

The Scheil module in Thermo-Calc software was employed to calculate the non-equilibrium solidification of liquid steel, as shown in Figure 12. As shown in Figure 12, M_7C_3 started precipitating from liquid steel when the mole fraction of the solid phase reached 90%. Prior to the precipitation of M_7C_3, the solid phase consists of austenite and TiC. This indicated that carbides $M_{23}C_6$ arose from the transformation of eutectic carbides M_7C_3.

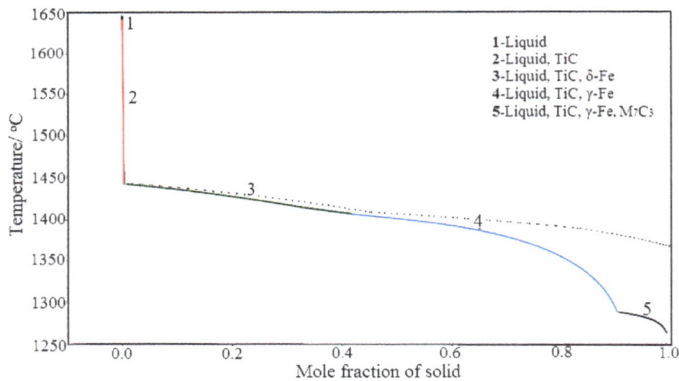

Figure 12. Mole fraction of solid phase as a function of temperature during non-equilibrium solidification.

During solidification under the equilibrium condition, carbides M_7C_3 precipitated from austenite, and then transformed to $M_{23}C_6$. As shown in Figure 11, the precipitation temperature range of M_7C_3 shrinks with the increasing titanium content, which indicated that titanium suppressed the precipitation of M_7C_3. Meanwhile, the transformation temperature of M_7C_3 to $M_{23}C_6$ increased with the increase of the titanium content, which indicated that titanium can promote the transformation process of M_7C_3 to $M_{23}C_6$. Although M_7C_3 still precipitated from liquid steel containing 0.71% Ti (see Figure 8), they would transform to $M_{23}C_6$ eventually because of the effect of titanium during solidification and the cooling process of ESR casting.

Compared with carbides M_7C_3, carbides $M_{23}C_6$ are easier to dissolve in the steel matrix during hot working and the heat treatment process. Hence, the transformation of M_7C_3 to $M_{23}C_6$ was expected. The results shown in Figures 9 and 10 indicated that large carbides $M_{23}C_6$ in sample No. 2 did dissolve in the steel matrix. The volume fraction of carbides in the as-cast No. 2 sample was the lowest. After hot processing, a large amount of large $M_{23}C_6$ dissolved, which is favorable for the subsequent process and can improve the mechanical property of steel. This has been confirmed by the mechanical property measurements, which show the highest elongation after fracture and the lowest hardness of sample No. 2 after forging and spheroidizing annealing.

5. Conclusions

1. TiCs as the first phase formed from liquid steel could act as nucleation sites to refine the grain and the eutectic carbides M_7C_3. The amount and size of eutectic carbides M_7C_3 decreased gradually with the increasing titanium content in 8Cr13MoV steel. The fraction of carbides in the as-cast ESR ingot decreased first and then increased with the increase of the titanium addition.

2. Titanium could promote the transformation from carbides M_7C_3 to carbides $M_{23}C_6$ during solidification and the cooling process. This is favorable for the dissolution of large carbides during hot processing, as demonstrated by the determination of extracted carbides from the steel matrix.

3. The 0.771% titanium addition decreased the fraction and the size of the primary carbides, and improved the distribution of these carbides in 8Cr13MoV steel, contributing to the increase of the mechanical properties of annealed steel in terms of hardness and elongation after fracture. However, excessive titanium addition led to the significant increase in the volume fraction of TiC. In addition, the titanium addition could inhibit the precipitation of the secondary carbides in the spheroidizing annealing process, and this effect increases with the increasing titanium content. These two factors above are adverse for the mechanical properties of annealed steel.

Acknowledgments: This work was financially supported by the National Natural Science Foundation of China (Grant No. 51574025).

Author Contributions: For research articles with several authors, a short paragraph specifying their individual contributions must be provided. The following statements should be used "W.-T.Y. and J.L. conceived and designed the experiments; W.-T.Y. and Q.-T.Z. performed the experiments; W.-T.Y. and C.-B.S. analyzed the data; J.L. contributed reagents/materials/analysis tools; W.-T.Y. wrote the paper." Authorship must be limited to those who have contributed substantially to the work reported.

Conflicts of Interest: The authors declare no conflict of interest.

References

1. Chen, C.Y.; Hung, F.Y.; Lui, T.S.; Chen, L.H. Microstructures and Mechanical Properties of Austempering SUS440 Steel Thin Plates. *Metals* **2016**, *6*, 35–46. [CrossRef]
2. Salleh, S.H.; Omar, M.Z.; Syarif, J.; Abdullah, S. Carbide Formation during Precipitation Hardening of SS440C Steel. *Eur. J. Sci. Res.* **2009**, *34*, 83–91.
3. Yao, D.; Li, J.; Li, J.; Zhu, Q.T. Effect of Cold Rolling on Morphology of Carbides and Properties of 7Cr17MoV Stainless Steel. *Mater. Manuf. Proc.* **2014**, *30*, 111–115. [CrossRef]
4. Zhu, Q.T.; Li, J.; Shi, C.B.; Yu, W.T. Effect of Quenching Process on the Microstructure and Hardness of High-Carbon Martensitic Stainless Steel. *J. Mater. Eng. Perform.* **2015**, *24*, 4313–4321. [CrossRef]
5. Rajasekhar, A.; Madhusudhan, R.G.; Mohandas, T.; Murti, V.S.R. Influence of austenitizing temperature on microstructure and mechanical properties of AISI 431 martensitic stainless steel electron beam welds. *Mater. Design.* **2009**, *30*, 1612–1624. [CrossRef]
6. Salleh, S.H. Investigation of Microstructures and Properties of 440C Martensitic Stainless Steel. *Int. J. Mech. Mater. Eng.* **2009**, *4*, 123–126.
7. Zhu, Q.T.; Li, J.; Shi, C.B.; Yu, W.T. Effect of electroslag remelting on carbides in 8Cr13MoV martensitic stainless steel. *Int. J. Miner. Metall. Mater.* **2015**, *22*, 1149–1156. [CrossRef]
8. Lin, T.L.; Lin, C.C.; Tsai, T.H.; Lai, H.J. Microstructure and Mechanical Properties of 0.63C-12.7Cr Martensitic Stainless Steel during Various Tempering Treatments. *Mater. Manuf. Proc.* **2010**, *25*, 246–248. [CrossRef]
9. Zhou, B.; Shen, Y.; Chen, J.; Cui, Z.S. Breakdown Behavior of Eutectic Carbide in High Speed Steel during Hot Compression. *J. Iron Steel Res. Int.* **2011**, *18*, 41–48. [CrossRef]
10. Karagöz, S.; Fischmeister, H.F. Cutting performance and microstructure of high speed steels: Contributions of matrix strengthening and undissolved carbides. *Metall. Mater. Trans. A* **1998**, *29*, 205–216. [CrossRef]
11. Gahr, K.H.Z.; Scholz, W.G. Fracture-toughness of white cast irons. *J. Met.* **1980**, *32*, 38–44. [CrossRef]

12. Chumanov, V.I.; Chumanov, I.V. Control of the carbide structure of tool steel during electroslag remelting: Part I. *Russ. Metall.* **2011**, *7*, 515–521. [CrossRef]
13. Zhou, X.; Fang, F.; Li, G.; Jiang, J.Q. Morphology and Properties of M_2C Eutectic Carbides in AISI M2 Steel. *ISIJ Int.* **2010**, *50*, 1151–1157. [CrossRef]
14. Cho, K.S.; Sang, I.K.; Park, S.S.; Choi, W.S.; Moon, H.K.; Kwon, H. Effect of Ti Addition on Carbide Modification and the Microscopic Simulation of Impact Toughness in High-Carbon Cr-V Tool Steels. *Metall. Mater. Trans. A* **2016**, *47*, 26–32. [CrossRef]
15. Eiselstein, L.E.; Ruano, O.A.; Sheby, O.D. Structural characterization of rapidly solidified white cast iron powders. *J. Mater. Sci.* **1983**, *18*, 483–492. [CrossRef]
16. Molian, P.A.; Wood, W.E. Rapid solidification of laser-processed chromium steels. *Mater. Sci. Eng.* **1984**, *62*, 271–277. [CrossRef]
17. Laird, G., II. Microstructures of Ni-hard I, Ni-hard IV and high-Cr white cast irons. *AFS Trans.* **1991**, *99*, 339–357.
18. Wu, X.; Xing, J.; Fu, H.; Zhi, X. Effect of titanium on the morphology of eutectic M7C3 carbides in hypereutectic high chromium white iron. *Mater. Sci. Eng. A* **2007**, *457*, 180–185. [CrossRef]
19. Mirzaee, M.; Momeni, A.; Keshmiri, H.; Razavinejad, R. Effect of titanium and niobium on modifying the microstructure of cast K100 tool steel. *Metall. Mater. Trans. B* **2014**, *45*, 2304–2314. [CrossRef]
20. Maldonadoruiz, S.I.; Orozcogonzález, P.; Baltazarhernández, V.H.; Bedollajacuinde, A.; Hernándezrodríguez, M.A.L. Effect of V-Ti on the Microstructure and Abrasive Wear Behavior of 6CrC Cast Steel Mill Balls. *J. Miner. Mater. Charact. Eng.* **2014**, *2*, 383–391.
21. Bratberg, J.; Frisk, K. An experimental and theoretical analysis of the phase equilibria in the Fe-Cr-V-C system. *Metall. Mater. Trans. A* **2004**, *35*, 3649–3663. [CrossRef]
22. Wang, M; Mu, S.; Sun, F.; Wang, Y. Influence of rare earth elements on microstructure and mechanical properties of cast high-speed steel rolls. *J. Rare Earths* **2007**, *25*, 490–494. [CrossRef]
23. Fu, H.G.; Fu, D.M.; Zou, D.N.; Xing, J.D. Structures and Properties of High-Carbon High Speed Steel by RE-Mg-titanium Compound Modification. *J. Wuhan Univ. Technol.* **2004**, *19*, 48–51. [CrossRef]
24. Duan, J.T.; Jiang, Z.Q.; Fu, H.G. Effect of RE-Mg complex modifier on structure and performance of high speed steel roll. *J. Rare Earths* **2007**, *25*, 259–263.
25. Yu, S.C.; Zhu, Q.H.; Wu, S.Q.; Gong, Y.J.; Gong, Y.S.; Lian, M.S.; Ye, G.; Chang, Y.J. Microstructure of Steel 5Cr21Mn9Ni4N Alloyed by Rare Earth. *J. Iron Steel Res. Int.* **2006**, *13*, 40–44. [CrossRef]
26. Razavinejad, R.; Firoozi, S.; Mirbagheri, S.M.H. Effect of titanium addition on as cast structure and macrosegregation of high-carbon high-chromium steel. *Steel Res.* **2012**, *83*, 861–869. [CrossRef]
27. Bjärbo, A.; Hättestrand, M. Complex carbide growth, dissolution, and coarsening in a modified 12 pct chromium steel-an experimental and theoretical study. *Metall. Mater. Trans. A* **2001**, *32*, 19–27. [CrossRef]
28. Bramfitt, B.L. The effect of carbide and nitride additions on the heterogeneous nucleation behavior of liquid iron. *Metall. Mater. Trans. B* **1970**, *1*, 1987–1995. [CrossRef]
29. Bedolla-Jacuinde, A.; Correa, R.; Mejia, I.; Quezada, J.G.; Rainforth, W.M. The effect of titanium on the wear behavior of a 16%Cr white cast iron under pure sliding. *Wear* **2007**, *263*, 808–820. [CrossRef]

metals

MDPI

Article

Correlation between Zn-Rich Phase and Corrosion/Oxidation Behavior of Sn–8Zn–3Bi Alloy

Xin Zhang [1], Chong Li [1,*], Zhiming Gao [1], Yongchang Liu [1], Zongqing Ma [1], Liming Yu [1] and Huijun Li [2]

1   Tianjin Key Lab of Composite and Functional Materials, School of Materials Science and Engineering, Tianjin University, Tianjin 300350, China; zhang.xin.dream@163.com (X.Z.); gaozhiming@tju.edu.cn (Z.G.); licmtju@163.com (Y.L.); mzq0320@163.com (Z.M.); lmyu@tju.edu.cn (L.Y.)
2   School of Mechanical, Materials and Mechatronic Engineering, University of Wollongong, Wollongong, New South Wales 2522, Australia; huijun@uow.edu.au
*   Correspondence: lichongme@tju.edu.cn; Tel.: +86-130-2139-8676

Academic Editor: Soran Birosca
Received: 13 June 2016; Accepted: 20 July 2016; Published: 25 July 2016

Abstract: The microstructure of Sn–8Zn–3Bi alloy was refined by increasing the solidification rate and the correlation between Zn-rich phase and the corrosion/oxidation behavior of the alloy was investigated. The Zn-rich phase transforms from coarse flakes to fine needles dispersed in the β-Sn matrix with the increase of the cooling rate. The transformation of Zn-rich precipitates enhances the anticorrosive ability of Sn–8Zn–3Bi alloy in 3.5 wt.% NaCl solution. On the contrary, Sn–8Zn–3Bi alloy with a fine needle-like Zn-rich phase shows poor oxidation resistance under air atmosphere, due to the fast diffusion of Zn atoms in Sn matrix.

Keywords: Sn–Zn–Bi alloys; microstructure; Zn-rich phase; corrosion; oxidation

1. Introduction

Sn–Pb alloys have long been extensively used in microelectronic packaging industries due to their good wettability, excellent physical properties, low cost and low melting point [1]. However, in recent years, solder materials design has shifted focus to lead-free solder due to lead toxicity and its harmful effects on human health and the environment [2]. Accordingly, many lead-free Sn-based alloys with the addition of alloying elements have been investigated, such as Zn, Ag, Cu, Bi, In, Ni. Also, their microstructures, mechanical properties and solderability have been reported [3–10].

Among these lead-free solder alloys, the Sn–Zn based solder alloy has been found to be a promising candidate to replace the existing Sn–Pb eutectic solder, due to its relatively low melting temperature, superior mechanical properties at room temperature and relatively low cost [11–15]. Moreover, Bi is a surface active element and can reduce surface tension and eutectic temperature of the Sn–Zn system. So it has been added to the Sn–Zn alloy to form the ternary eutectic Sn–8Zn–3Bi with improved wettability and lower melting point [16,17].

To achieve high reliability in the long-term use process, solder materials are also required to exhibit high resistance to corrosion and oxidation conditions such as moisture, air pollutants and oceanic environments. In addition, relatively limited studies addressing their corrosion and oxidation behaviors are available [18–21]. So in the present work, the microstructure of Sn–8Zn–3Bi was manipulated by changing the solidification rate. We also concentrated on the relationship between the Zn-rich phase and the corrosion/oxidation behavior of the Sn–8Zn–3Bi alloy. These results provide crucial experimental input as a guide for fabricating Sn–Zn–Bi alloys in various harsh environments. It may be of reference value for researching the corrosion/oxidation behavior of other lead-free Sn-based alloys.

2. Experimental Procedures

The Sn–8Zn–3Bi alloy was prepared from the commercial pure Sn, Zn and Bi (Sn, Zn and Bi ingots with purity of 99.9%). Required quantities of the ingot were melted at 450 °C in a 25 kW medium frequency induction furnace under an inert argon atmosphere. The molten Sn–8Zn–3Bi alloy was homogenized at 500 °C, and then poured in a steel mold and porcelain crucible to prepare the chill cast ingot with rapid cooling rate (about 8 °C/S) and slow cooling rate (about 1 °C/S), respectively. Metallographic specimens were polished by a standard procedure. The microstructure characteristics of the specimens were observed using scanning electron microscope (Hitachi Model No. S4800).

Corrosion test was performed with samples in 3.5 wt.% NaCl solution, and its results were evaluated according to weight loss measurement. For the weight loss-measuring test, the samples (30 mm × 30 mm × 1 mm) were immersed in the NaCl solution for three, five, seven days. All electrochemical measurements were performed at room temperature (25 ± 2 °C) in a conventional three-electrode cell, a platinum plate as the counter electrode (CE) and a saturated calomel electrode (SCE) as the reference electrode (RE). The polarization curves were recorded in the anodic direction in the range of −1500 mV to 2000 mV at a scan rate of 0.5 mV/s. The oxidation property of the Sn–8Zn–3Bi solder alloy was conducted by a thermal gravimetric analyzer (TGA). The specimens were heated to 170 °C at a heating rate of 10 °C/min and hold for 2 h under air atmosphere.

Auger electron spectroscopy (AES) sputtering was used to analyze the distribution of Sn, Zn, Bi and O on the solder surface. The sputtering speed was 4 nm/m. Both the compositions and the chemical valence states of the elements in the corroded alloy were examined by using X-ray photoelectron spectroscopy (XPS) technique. And the calculation of the data was performed using XPS Peak 4.1 peak-fitting software.

3. Results and Discussion

Figure 1 shows the microstructures of slow cooling (SC) and rapid cooling (RC) Sn–8Zn–3Bi alloy. As shown in Figure 1a,c, it can be seen that coarse, long, flake-like Zn-rich phase uniformly distributes in the β-Sn matrix in the SC Sn–8Zn–3Bi alloy. Due to the large solid solubility of Bi in Sn, the addition of 3 wt.% Bi forms a solid solution of Bi in the Sn matrix, resulting in the invisibility of Bi in the microstructure. Notably, the rapid cooling rate obviously refines the solidification microstructure of Sn–8Zn–3Bi alloy (Figure 1b,d). The Zn-rich phase transforms to fine, short, needle-like precipitates dispersed in the β-Sn matrix in the rapid cooling Sn–8Zn–3Bi alloy.

Figure 1. Microstructures of Sn–8Zn–3Bi alloy: (**a**) slow cooling rate (SC); (**b**) rapid cooling rate (RC); (**c**) enlarged morphology of Zn-rich phase in (**a**); (**d**) enlarged morphology of Zn-rich phase in (**b**).

Figure 2 shows the weight loss of the samples as a function of the duration time in 3.5 wt.% NaCl solution. For the slow cooling and rapid cooling Sn–8Zn–3Bi alloy, the weight loss increased with the duration time. Moreover, it needs to be pointed out that compared to Sn–8Zn–3Bi (SC), the alloy with the rapid cooling rate suffers from comparatively less corrosion attack and shows better corrosion resistance. The open circuit potential (OCP) against time (Eocp-t) curves of both alloys measured in NaCl 3.5 wt.% solution are shown in Figure 3. As seen from Figure 3, the values of OCP (E_{OCP}) of Sn–8Zn–3Bi alloy gradually tend to be constant. From a comparison of curves, a positive shift in the E_{OCP} was clearly found when the cooling rate was increased, which reveals that RC Sn–8Zn–3Bi alloy provides better protection.

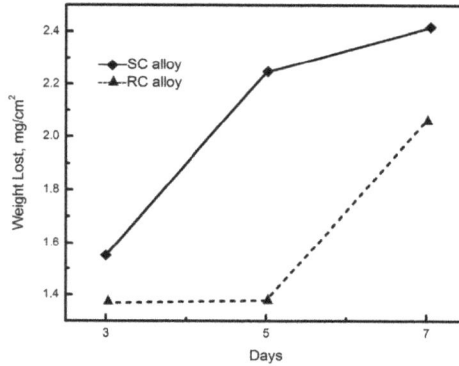

Figure 2. Weight loss of Sn–8Zn–3Bi alloy with two solidification rates as a function of duration time in 3.5% NaCl environment.

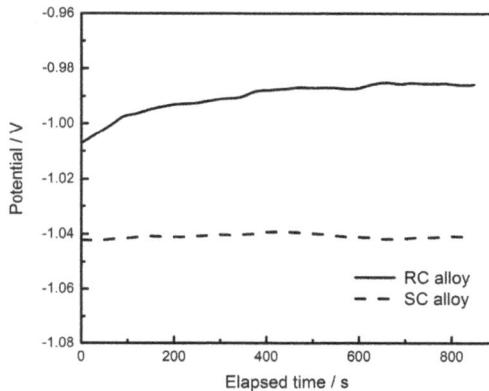

Figure 3. Open circuit potential curves for Sn–8Zn–3Bi solder alloy.

Moreover, polarization curves for the samples are given in Figure 4, and related values extrapolated from the polarization curves are summarized in Table 1. The corrosion potentials (E_{corr}) are −1.047 (SC) and −0.999 V_{SCE} (RC) for the Sn–8Zn–3Bi alloys, respectively. It indicates that the value of E_{corr} shifts towards the noble value with the increase of the cooling rate. The values of corrosion current density (I_{corr}) and polarization resistance (Rp) indicate that RC Sn–8Zn–3Bi alloy presents a lower corrosion rate and enhanced corrosion resistance compared to the SC alloy. Due to the same composition of the alloy, these results indicate that the better anticorrosive ability of the RC Sn–8Zn–3Bi alloy is strongly dependent on the transformation of the Zn-rich phase.

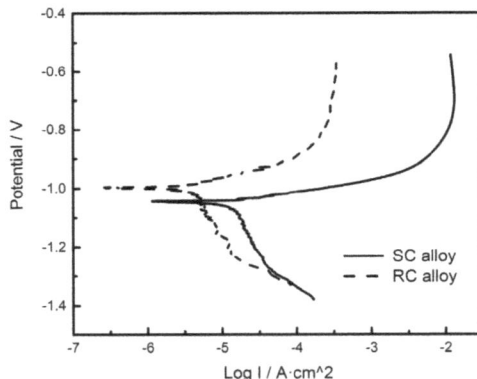

Figure 4. Potentiodynamic polarization curves of Sn–8Zn–3Bi solder alloy with two cooling rates in 3.5 wt.% NaCl solution.

Table 1. Corrosion parameters of the studied Sn–8Zn–3Bi solder in 3.5 wt.% NaCl solution at room temperature.

Sample	E_{corr} vs. SCE/ (V)	I_{corr} (A·cm^{-2})	Rp/(Ω·cm^2)	OCP/(V)
SC	−1.047	1.298×10^{-5}	1490.5	−1.042
RC	−0.999	3.907×10^{-6}	9832.9	−0.985

To better reveal the correlation between the Zn-rich phase and the corrosion behavior of Sn–8Zn–3Bi alloy, SEM micrographs are displayed in Figure 5. The pronounced degradation of the surface morphology with immersion time can be observed. At 5 h of immersion in NaCl 3.5 wt.% solution, the corrosion reactions occurred along the large plate-like Zn-rich precipitates in SC Sn–8Zn–3Bi (Figure 5a). Notably, the fine needle-like Zn-rich particles in RC alloy showed no obvious change (Figure 5b). When further increasing the duration time to 12 h in 3.5% NaCl solution, a concomitantly low level of surface attack was observed for RC Sn–8Zn–3Bi alloy (Figure 5d). In addition, many pits and grooves could be observed, but shorter and shallower in depth compared to those in SC alloy (Figure 5c).

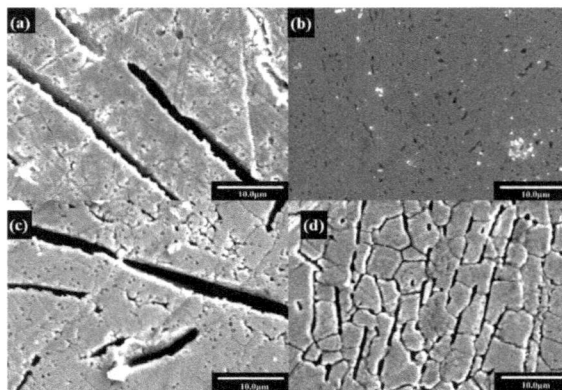

Figure 5. SEM images of Sn–8Zn–3Bi solder alloy after immersed in 3.5% NaCl solution for 5 h ((**a**): SC alloy; (**b**): RC alloy) and 12 h ((**c**): SC alloy; (**d**): RC alloy).

During the process of Sn–8Zn–3Bi immersed in 3.5% NaCl solution, the Cl⁻ can adsorb onto the passive film, followed by penetration through intergranular boundaries of the oxides [22]. Also, the Cl⁻ can migrate to the film/alloy interface to react with Zn to form chloride complexes [23]. Continuous growth of these complexes gives rise to internal stress at the interface due to volume expansion, and thus results in the breakdown of localized regions, followed by pit initiation at the initial Zn sites. Moreover, because of the large solid solubility of Bi in Sn, Bi forms a solid solution of Sn matrix in Sn–8Zn–3Bi alloy, which is expected to affect the difference of the corrosion potential between the Zn-rich phase and the Sn matrix to some extent. So, in the case of corrosion susceptibility, Zn still exhibits a more negative corrosion potential value than Sn, and preferentially dissolves due to galvanic coupling and thus promotes the pitting process inwards in the alloys. This is consistent with the observation in Figure 5. Figure 6 depicts the XPS peak-fitting analysis for Zn2p$_{3/2}$ obtained from SC and RC Sn–8Zn–3Bi in 3.5 wt.% NaCl solution, which confirms that the Zn element exists as Zn^{2+} and Zn° on the surface of corroded Sn–8Zn–3Bi alloy. These results indicate that the amount of Zn has been preferentially dissolved. The formation of pits and grooves (Figure 5) is attributed to the removal of active materials, particularly Zn.

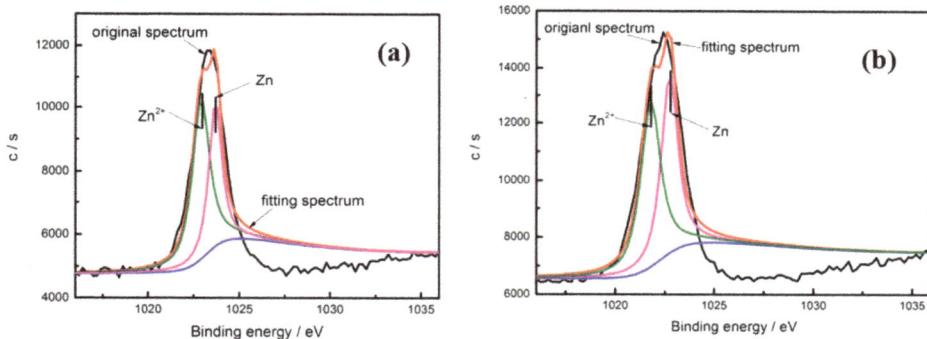

Figure 6. XPS spectrum for Zn2p$_{3/2}$ obtained from Sn–8Zn–3Bi solder alloy in 3.5 wt.% NaCl solution at room temperature: (**a**) SC alloy; (**b**) RC alloy.

In addition, Cl⁻ anions further migrate along the intergranular boundaries between Zn-rich precipitates and the Sn matrix, where the boundaries can provide an efficient diffusion path to the transport of Cl⁻ and promote the pitting process. Coarse, long, Zn-rich precipitates distribute in SC Sn–8Zn–3Bi, corresponding to larger intergranular boundaries with lower activation energy for the dissolution of Zn, compared to the fine, needle-like Zn-rich phase [22,24]. The corrosion oxide films around the boundaries show high activity and more defects may be formed in those regions due to the weakening of the bond between Zn-rich precipitates and the Sn matrix, which could promote the transport of Cl⁻ inwards in the alloy. Consequently, the pitting susceptibility of Sn–8Zn–3Bi increases with the increase of the size of the Zn-rich precipitates. In other words, the Sn–8Zn–3Bi alloy with fine Zn-rich precipitates shows superior protective behavior against the transport of Cl⁻ and pitting. Sn–8Zn–3Bi alloys with a fine Zn-rich phase are highly beneficial for the solder fabrication industries to achieve a superior anticorrosion behavior.

On the other hand, oxidation resistance also plays a significant role in the practical application of the lead-free solder. Figure 7 shows two samples of the time-dependence of the oxidized weight gain for Sn–8Zn–3Bi exposed to air atmosphere by TGA analysis. The weight gain of the alloy increases due to the oxidation of the solder. Interestingly, a more rapid increase is observed for the RC Sn–8Zn–3Bi alloy with a fine Zn-rich phase compared to the SC alloy.

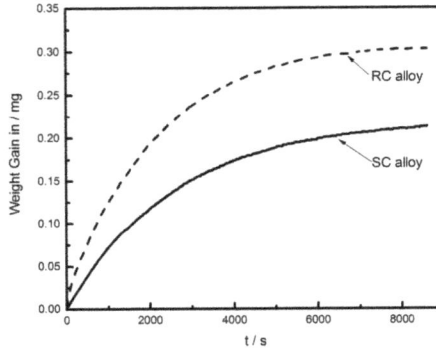

Figure 7. TGA curves of lead-free Sn–8Zn–3Bi solder alloy.

To understand the mechanism of the oxidation of the Sn–8Zn–3Bi solder, the oxidation surfaces of the samples were investigated by Auger analysis and scanning electron microscope. Figure 8 shows the results of the Auger analysis and Auger depth profiles of the two samples. In both specimens, only Zn and O elements were observed on the oxide surface without Bi and Sn. After etching a certain depth (44 nm), Bi and Sn were detected, as shown in Figure 8a,c. The result implies that the oxide ZnO prefers to form on the free surface of the samples.

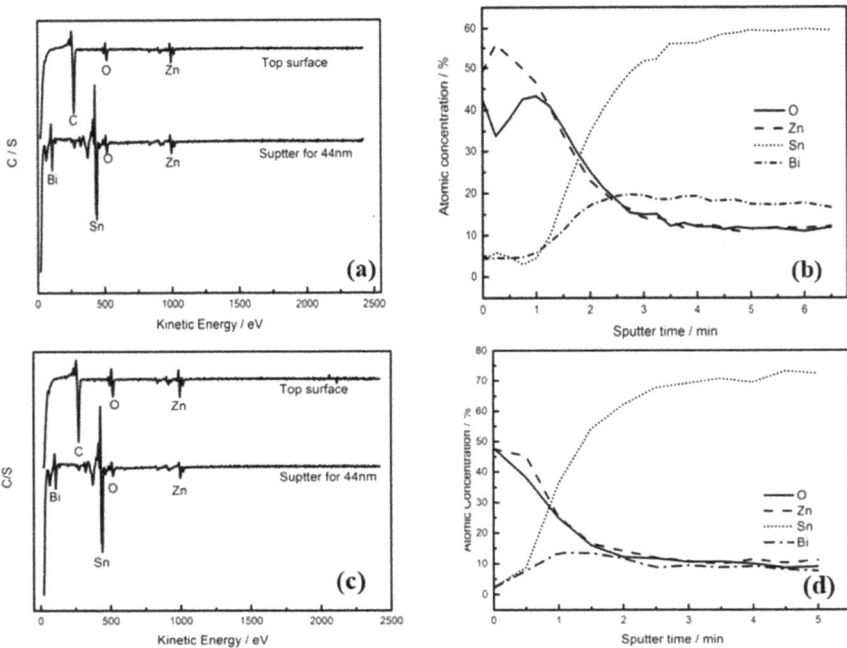

Figure 8. Auger surface analysis of RC alloy (**a**) and SC alloy (**c**); Auger depth profiles of RC alloy (**b**) and SC alloy (**d**) (from surface to etching about 20 nm).

The standard Gibbs free energies of the formation of ZnO and SnO_2 are -300 kJ/mol and -237 kJ/mol, respectively [25]. As the samples were exposed to air atmosphere at 170 °C, the

Zn on the free surface soon reacted with oxygen and formed ZnO on the free surface. The phase transformation from Zn to ZnO is accompanied by a volume expansion and the stress is induced by the volume expansion of ZnO. As a result, cracks occurred on the surface of the Zn-rich phase, as shown in Figure 9. Moreover, Zn massively diffused in the Sn matrix and exhibited strong segregation from the Sn grain boundaries to form oxides [26,27]. Fast diffusion of Zn or O dominated the oxidation process. As a result, an oxide ZnO layer was formed as a uniform layer, and Sn and Bi were not segregated on the oxide surface (Figure 8a,c). As shown in Figure 8b,d, the oxygen concentration tends to reduce gradually with the increase of the sputtering time, indicating that the oxidation propagation takes place from the free surface to the inside of the alloy. Bi can dissolve into the Sn matrix, forming a solid solution, which causes the Sn matrix to become liable to form cracks and promotes the diffusion of Zn [27]. Furthermore, a high distortion energy is expected in the Sn lattice due to the solid solution formation. The high distortion energy of the Sn matrix can also promote Zn diffusion [28].

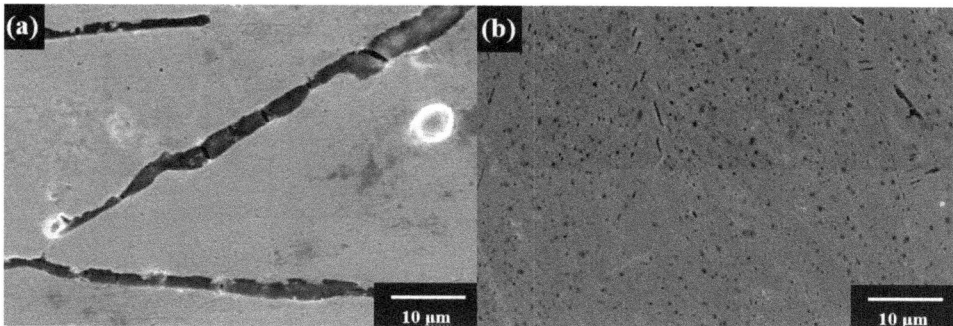

Figure 9. SEM images of Sn–8Zn–3Bi solder alloy after 170 °C 2 h exposure under air atmosphere: (**a**) SC alloy; (**b**) RC alloy.

Moreover, the oxidized layer thickness is also evaluated by the product of the sputtering time and sputtering rate. The RC Sn–8Zn–3Bi alloy has a thicker oxide layer (16.3 nm) than the SC alloy (10 nm). In the RC Sn–8Zn–3Bi alloy, the Zn-rich phase has a smaller size, higher density and more uniform distribution, in comparison to that in the SC alloy (Figure 9). Therefore, fast diffusion of Zn atoms in the Sn matrix with a short diffusion distance can be expected. Meanwhile, because the oxygen diffused along the interfaces between the Zn particles and the Sn matrix [26,27], the interdiffusion of O atoms is also accelerated, thus increasing the oxidation rate of the alloy. As a result, Sn–8Zn–3Bi alloy with a fine needle-like Zn-rich phase shows poor oxidation resistance under air atmosphere. However, there is no reference data for diffusion coefficients obtained at the corresponding exposure condition. The detail kinetics and behavior of the Sn–8Zn–3Bi alloy remain a topic to be researched and determined.

4. Conclusions

The increase of the cooling rate significantly refines the solidification microstructure of Sn–8Zn–3Bi alloy. The Zn-rich phase transforms from coarse, long flakes to fine, short needles distributed in the β-Sn matrix. In the case of corrosion susceptibility in 3.5% NaCl solution, Zn exhibits a more negative corrosion potential value and presents preferential dissolution. The transformation of the Zn-rich precipitates results in the Sn–8Zn–3Bi alloy exhibiting a superior protective performance. On the contrary, as the alloy is exposed to air atmosphere at 170 °C, diffusion of Zn or O dominates the oxidation process, and the Zn on the free surface reacts with oxygen and forms ZnO on the free surface. As a result, the refined Zn-rich phase weakens the oxidation resistance of the Sn–8Zn–3Bi alloy, due to the fast diffusion of Zn atoms in the Sn matrix.

Acknowledgments: This work was supported by a grant from National Science Fund for Distinguished Young Scholars of China (No. 51325401), the National Natural Science Foundation of China (No. 51404169) and the Natural Science Foundation of Tianjin (No. 15JCQNJC03200).

Author Contributions: Chong Li, Yongchang Liu and Huijun Li conceived and designed the experiments; Xin Zhang performed the experiments and wrote the manuscript; Zhiming Gao, Zongqing Ma and Liming Yu analyzed the data and revised the manuscript.

Conflicts of Interest: The authors declare no conflicts of interest.

References

1. Tu, K.N.; Zeng, K. Tin-lead (SnPb) solder reaction in flip chip technology. *Mater. Sci. Eng. R* **2001**, *34*, 1–58. [CrossRef]

2. Wood, E.P.; Nimmo, K.L. In search of new lead-free electronic solders. *J. Electron. Mater.* **1994**, *23*, 709–713. [CrossRef]

3. Guang, R.; Wilding, I.J.; Collins, M.N. Alloying influences on low melt temperature SnZn and SnBi solder alloys for electronic interconnections. *J. Alloy. Compd.* **2016**, *665*, 251–260.

4. Fawzy, A. Effect of Zn addition, strain rate and deformation temperature on the tensile properties of Sn–3.3 wt.% Ag solder alloy. *Mater. Charact.* **2007**, *58*, 323–331. [CrossRef]

5. Pereira, P.D.; Spinelli, J.E.; Garcia, A. Combined effects of Ag content and cooling rate on microstructure and mechanical behavior of Sn–Ag–Cu solders. *Mater. Des.* **2013**, *45*, 377–383. [CrossRef]

6. El-Daly, A.A.; Desoky, W.M.; Saad, A.F.; Mansor, N.A.; Lotfy, E.H.; Abd-Elmoniem, H.M.; Hashem, H. The effect of undercooling on the microstructure and tensile properties of hypoeutectic Sn–6.5Zn–*x*Cu Pb-free solders. *Mater. Des.* **2015**, *80*, 152–162. [CrossRef]

7. Osório, W.R.; Peixoto, L.C.; Garcia, L.R.; Noël, N.M.; Garcia, A. Microstructure and mechanical properties of Sn–Bi, Sn–Ag and Sn–Zn lead-free solder. *J. Alloy. Compd.* **2013**, *572*, 97–106. [CrossRef]

8. Liu, J.C.; Zhang, G.; Wang, Z.H.; Ma, J.S.; Suganuma, K. Thermal property, wettability and interfacial characterization of novel Sn–Zn–Bi–In alloys as low-temperature lead-free solders. *Mater. Des.* **2015**, *84*, 331–339. [CrossRef]

9. Spinelli, J.E.; Silva, B.L.; Garcia, A. Assessment of tertiary dendritic growth and its effects on mechanical properties of directionally solidified Sn-0.7Cu-*x*Ag solder alloys. *J. Electron. Mater.* **2014**, *43*, 1347–1361. [CrossRef]

10. Silva, B.L.; Cheung, N.; Garcia, A.; Spinelli, J.E. Sn–0.7 wt%Cu–(*x*Ni) alloys: Microstructure–mechanical properties correlations with solder/substrate interfacial heat transfer coefficient. *J. Alloy. Compd.* **2015**, *632*, 274–285. [CrossRef]

11. Garcia, L.R.; Osório, W.R.; Peixoto, L.C.; Garcia, A. Mechanical properties of Sn–Zn lead-free solder alloys based on the microstructure array. *Mater. Charact.* **2010**, *61*, 212–220. [CrossRef]

12. Lin, S.; Xue, S.B.; Xue, P.; Luo, D.X. Present status of Sn–Zn lead-free solders bearing alloying elements. *J. Mater. Sci. Mater. Electron.* **2015**, *26*, 4389–4411.

13. El-Daly, A.A.; Hammad, A.E.; Al-Ganainy, G.S.; Ibrahiem, A.A. Design of lead-free candidate alloys for low-temperature soldering applications based on the hypoeutectic Sn–6.5Zn alloy. *Mater. Des.* **2014**, *56*, 594–603. [CrossRef]

14. Garcia, L.R.; Peixoto, L.C.; Osório, W.R.; Garcia, A. Globular to needle Zn-rich phase transition during transient solidification of a eutectic Sn–9% Zn solder alloy. *Mater. Lett.* **2009**, *63*, 1314–1316. [CrossRef]

15. Garcia, L.R.; Osório, W.R.; Peixoto, L.C.; Garcia, A. Wetting behavior and mechanical properties of Sn-Zn and Sn-Pb solder alloys. *J. Electron. Mater.* **2009**, *38*, 2405–2414. [CrossRef]

16. Zhou, J.; Sun, Y.S.; Xue, F. Properties of low melting point Sn–Zn–Bi solders. *J. Alloy. Compd.* **2005**, *397*, 260–264. [CrossRef]

17. Song, J.M.; Wu, Z.M. Variable eutectic temperature caused by inhomogeneous solute distribution in Sn–Zn system. *Scr. Mater.* **2006**, *54*, 1479–1483. [CrossRef]

18. Marcus, P.; Maurice, V.; Strehblow, H.H. Localized corrosion (pitting): A model of passivity breakdown including the role of the oxide layer nanostructure. *Corros. Sci.* **2008**, *50*, 2698–2704. [CrossRef]

19. Liu, N.S.; Lin, K.L. Effect of Ga on the Oxidation properties of Sn–8.5Zn–0.5Ag–0.1Al–*x*Ga solders. *Oxid. Met.* **2012**, *78*, 285–294. [CrossRef]

20. Kim, S.H.; Hui, K.N.; Kim, Y.J.; Lim, T.S.; Yang, D.Y.; Kim, K.B.; Kim, Y.J.; Yang, S. Oxidation resistant effects of Ag$_2$S in Sn–Ag–Al solder: A mechanismfor higher electrical conductivity and less whisker growth. *Corros. Sci.* **2016**, *105*, 25–35. [CrossRef]
21. Nazeri, M.F.M.; Mohamad, A.A. Corrosion resistance of ternary Sn–9Zn–xIn solder joint in alkaline solution. *J. Alloy. Compd.* **2016**, *661*, 516–525. [CrossRef]
22. Liu, J.C.; Park, S.W.; Nagao, S.; Nogi, M.; Koga, H.; Ma, J.S.; Zhang, G.; Suganuma, K. The role of Zn precipitates and Cl$^-$ anions in pitting corrosion of Sn–Zn solder alloys. *Corros. Sci.* **2015**, *92*, 263–271. [CrossRef]
23. Burstein, G.; Liu, C.; Souto, R.; Vines, S. Origins of pitting corrosion. *Corros. Eng. Sci. Technol.* **2004**, *39*, 25–30. [CrossRef]
24. Liu, J.C.; Zhang, G.; Ma, J.S.; Suganuma, K. Ti addition to enhance corrosion resistance of Sn–Zn solder alloy by tailoring microstructure. *J. Alloy. Compd.* **2015**, *644*, 113–118. [CrossRef]
25. Barin, I. *Thermochemical Data of Pure Substances*; VCH: New York, NY, USA, 1995; Volume 2, p. 1549.
26. Lee, J.E.; Kim, K.S.; Inoue, M.; Jiang, J.X.; Suganuma, K. Effects of Ag and Cu addition on microstructural properties and oxidation resistance of Sn–Zn eutectic alloy. *J. Alloy. Compd.* **2008**, *454*, 310–320. [CrossRef]
27. Jiang, J.X.; Lee, J.E.; Kim, K.S.; Suganuma, K. Oxidation behavior of Sn–Zn solders under high-temperature and high-humidity conditions. *J. Alloy. Compd.* **2008**, *462*, 244–251. [CrossRef]
28. Mrowec, S. *Defects and Diffusion in Solids: An Introduction*; Elsevier Scientific: New York, NY, USA, 1980.

![metals logo] *metals*

MDPI

Article

Effect of Hydrogen and Strain-Induced Martensite on Mechanical Properties of AISI 304 Stainless Steel

Sang Hwan Bak, Muhammad Ali Abro and Dong Bok Lee *

School of Advanced Materials Science and Engineering, Sungkyunkwan University, Suwon 16419, Korea;
sanghwan.bak@hanmail.net (S.H.B.); muhammad_aliabro@inbox.com (M.A.A.)
* Correspondence: dlee@skku.ac.kr; Tel.: +82-31-290-7355

Academic Editor: Soran Birosca
Received: 12 May 2016; Accepted: 16 July 2016; Published: 20 July 2016

Abstract: Plastic deformation and strain-induced martensite (SIM, α') transformation in metastable austenitic AISI 304 stainless steel were investigated through room temperature tensile tests at strain rates ranging from 2×10^{-6} to 2×10^{-2}/s. The amount of SIM was measured on the fractured tensile specimens using a feritscope and magnetic force microscope. Elongation to fracture, tensile strength, hardness, and the amount of SIM increased with decreasing the strain rate. The strain-rate dependence of RT tensile properties was observed to be related to the amount of SIM. Specifically, SIM formed during tensile tests was beneficial in increasing the elongation to fracture, hardness, and tensile strength. Hydrogen suppressed the SIM formation, leading to hydrogen softening and localized brittle fracture.

Keywords: stainless steel; hydrogen embrittlement; strain-induced martensite; magnetic force microscopy

---

1. Introduction

Metastable austenitic stainless steel belongs to transformation-induced plasticity (TRIP) steel, which has high strength and good ductility owing to the formation of strain-induced martensite (SIM, α') [1–5]. This, however, leads to hydrogen embrittlement (HE), limiting the application of austenitic stainless steel such as AISI 301, 304, and 316 in hydrogen-containing environments since SIM acts as a diffusion path of hydrogen [6,7]. HE depends on many factors such as the composition, temperature, residual stress, microstructure, and surface condition of steels. To explain the effect of hydrogen on the mechanical properties of steel, mechanisms such as internal pressure [8–11], surface energy [12], decohesion [13], hydrogen-enhanced plasticity [14], and hydrogen-enhanced strain-induced vacancies [15–17] have been proposed. However, HE has not yet been fully clarified due to the following uncertainties: firstly, why SIM increases the ductility of austenitic stainless steel, and secondly, why hydrogen suppresses the SIM formation in H-charged austenitic stainless steel. In this study, room-temperature (RT) tensile tests were conducted on bare and H-charged AISI 304 at the strain rate of 2×10^{-6}/s–2×10^{-2}/s in order to clarify the first uncertainty. To clarify the second uncertainty, SIM was measured in situ for bare and H-charged AISI 304 during RT tensile tests, and the volume fraction and distribution of SIM were measured for AISI 304 deformed to the same strain. The aim of this study is to correlate the amount of SIM with mechanical properties, and to correlate the distribution of the SIM formed and the HE with the role of hydrogen in suppressing the SIM formation in AISI 304.

2. Materials and Methods

2.1. Materials and Hydrogen Charging

AISI 304 was solution-annealed at 1100 °C for 1 h and water quenched. Its chemical composition is listed in Table 1. Its microstructure was examined by optical microscopy. After etching with 8:1:1 vol. ratio of $H_2O:HF:HNO_3$ solution for 5 min, its grain size was determined using intercept measurement. Hydrogen was charged through electrolytic charging at 80 °C with the current density of 500 A/m^2 for 20 h in 1 N H_2SO_4 solution containing 0.25 g/L $NaAsO_2$. The amount of hydrogen charged to the AISI 304 was analyzed with a hydrogen analyzer for two different samples.

Table 1. Chemical composition of AISI 304 stainless steels (wt. %).

Fe	Cr	Ni	Mn	Si	Cu	Co	Mo	N	C	S	P
Bal.	18.17	8.03	1.04	0.46	0.18	0.12	0.11	0.047	0.043	0.04	0.001

2.2. Determination of the Amount of Strain-Induced Martensite (SIM) and Tensile Testing Procedure

RT tensile tests were conducted at strain rates varying from 2×10^{-6}/s to 2×10^{-2}/s on bare and H-charged AISI 304 immediately after hydrogen charging. The rod-shaped samples were 4 mm in diameter with a gauge length of 25.4 mm. Tensile tests were performed twice to get the average value. The amount of SIM was measured in situ for bare and H-charged samples during RT tensile testing using a feritscope (Fisher MP30, Aberdeenshire, Scotland). The values measured by the feritscope agreed well with those measured by the magnetic force microscope (MFM, Park systems, Suwon, Korea), when considering the dependence of strain rate on the martensite concentration. After calibrating the feritscope using the standard sample, the average amount of the formed SIM was measured either ex situ within 1 mm of the fracture surface five times, or in situ at the center of the gauge section during tensile testing. For the sample deformed to the same strain, the volume fraction and distribution of SIM on the surface were measured using the MFM by applying the distinct magnetic properties of nonmagnetic austenite (γ) and ferromagnetic α' [17–19]. Images of MFM and atomic force microscope (AFM, Park systems, Suwon, Korea) were obtained in air using a scanning probe microscope (Nanoscope IIIa) in the tapping/lift mode with a lift height of 100 nm.

3. Results and Discussion

3.1. Hydrogen Diffusion in AISI 304

Figure 1 shows an optical microstructure of the H-charged sample. The mean grain size of 67 ± 5 μm was finer than other reported sizes (80–150 μm) [20,21]. Hydrogen charging did not change the microstructure. The average amount of hydrogen immediately after H-charging was 51 ± 5 ppm, which was smaller than the 80–110 ppm reported under similar charging conditions [22,23]. This was due to the finer mean grain size, i.e., hydrogen discharging out of the surface increased with the increasing density of the grain boundaries [24]. It is also worth noting that the mean grain size of a few tens of microns enhanced the hydrogen diffusion along the grain boundaries by several orders of magnitude [25].

Hydrogen was generally confined to the surface during charging. The hydrogen diffusion coefficient, D, was 4.0×10^{-15} m^2/s at the hydrogen-charging temperature of 80 °C [26]. The penetration depth of hydrogen, d, can be roughly expressed by $d = \sqrt{2Dt}$, where t is the charging time (h). It was about 24 μm, implying that hydrogen was present only around the surface. However, brittle facture occurred not only around the surface but also inside the sample where hydrogen was free, as will be explained later.

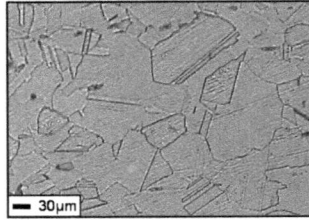

Figure 1. Optical microstructure of H-charged AISI 304 (etched).

3.2. Hydrogen Effect on Mechanical Property

The mechanical property strongly depended on the strain rate and hydrogen, as shown in Figure 2. H-free AISI 304 displayed a negative strain rate sensitivity in that the increment of the strain rate led to the decrement of the tensile strength, elongation to fracture, and hardness. Its yield stress, however, decreased at the slowest strain rate of 2×10^{-6}/s. The hydrogen effect varied, depending on the strain rate. The H-charging decreased the tensile strength, elongation to fracture, and hardness, but increased the yield stress to a small extent except at the slowest strain rate.

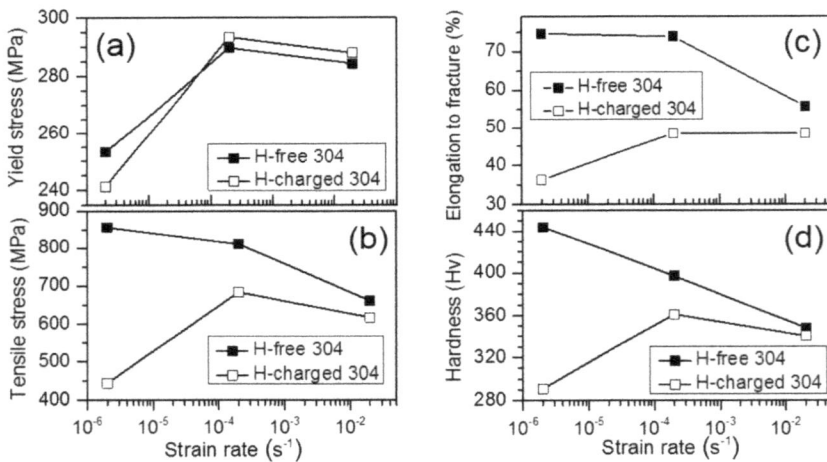

Figure 2. Mechanical properties of H-free and H-charged AISI 304 as a function of strain rate: (a) yield stress; (b) tensile strength; (c) elongation to fracture; (d) Vickers microhardness.

Figure 3 shows the fracture surfaces of H-free and H-charged AISI 304 after RT tensile tests. H-free samples displayed uniform ductile fracture with fine dimples at all strain rates (Figure 3a,b). In contrast, H-charged samples changed the fracture mode from ductile fracture with localized shear dimples at the strain rate of 2×10^{-2}/s (Figure 3c) to enhanced localized brittle fracture in a rectilinear shape parallel to each other at the strain rates of 2×10^{-4}/s and 2×10^{-6}/s (Figure 3d). The localized brittle fracture became more pronounced with decreasing the strain rate. Especially in Figure 3d, cracks developed around the boundary between the localized brittle fracture region and the ductile fracture region with dimples. Clearly, hydrogen induced localized brittle fracture in the H-charged AISI 304. Since SIM increased the elongation to fracture (Figure 2) and H-free samples displayed a uniform ductile fracture with fine dimples irrespective of the strain rates (Figure 3a,b), the γ-α transformation was proposed to suppress the formation of the localized brittle deformation. A similar mechanism was proposed to explain the enhanced elongation to fracture of the TRIP steel [5]. Since SIM delayed the formation of necking through the γ-α transformation during plastic deformation [3,5,27,28], the localized brittle fracture might be controlled by the γ-α transformation, which increased ductility.

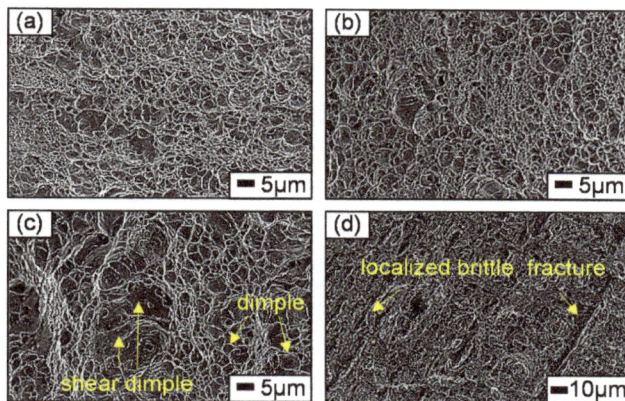

Figure 3. Fracture surfaces of AISI 304 after tensile test: (**a**) H-free (strain rate = 2 × 10^{-2}/s); (**b**) H-free (strain rate = 2 × 10^{-6}/s); (**c**) H-charged (strain rate = 2 × 10^{-2}/s); (**d**) H-charged (strain rate = 2 × 10^{-6}/s).

3.3. Hydrogen Effect on Formation of SIM

Figure 4 shows the amount of SIM that formed on the fracture surface at different strain rates for H-free and H-charged AISI 304. The amount of SIM that formed on the surface increased in the H-free sample, but decreased in the H-charged sample, with decreasing the strain rate. The H-free sample had a larger amount of SIM than the H-charged sample at the same strain rate. The amount of SIM was the smallest in the H-charged sample at the slowest strain rate of 2 × 10^{-6}/s. This suggested that hydrogen suppressed the SIM formation during tensile testing, which became more pronounced with decreasing the strain rate.

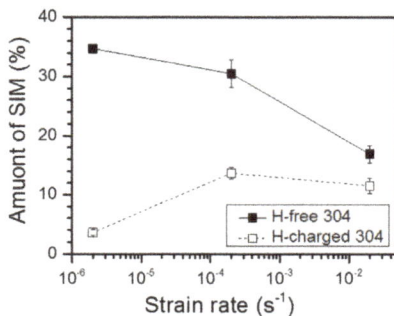

Figure 4. Amount of SIM as a function of the strain rate for H-free and H-charged AISI 304 after tensile tests.

To confirm the above suggestion, SIM was measured in situ for H-free and H-charged samples at the strain rate of 2 × 10^{-4}/s using the feritscope. Figure 5 indicates that the amount of SIM was decided by the plastic deformation and hydrogen. Hydrogen charging decreased the amount of SIM during the plastic deformation, displaying the largest decrement at the strain of 35%–45%.

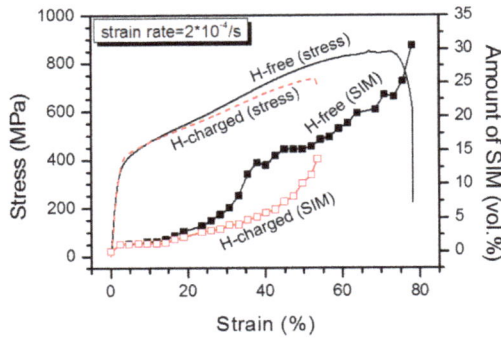

Figure 5. Stress (solid and dotted lines) and the amount of SIM formed (enclosed and open circles) as a function of strain for H-free and H-charged AISI 304. Strain rate = 2×10^{-4}/s.

In order to validate that hydrogen suppressed SIM formation at the same plastic strain, the volume fraction and distribution of SIM in H-free and H-charged AISI 304 were examined before and after tensile tests using MFM and the image analyzer, as shown in Figure 6. Dark and light phases corresponded to martensite (α') and austenite (γ), respectively, in the MFM images that were taken at 2.5 mm apart from the fracture surface. They were more distinct in the image analyzer images. The average amount of SIM was 1.5%, whereas SIM aligned along the tensile axis, tens to hundreds of micrometers apart, before tensile testing (Figure 6a). It increased with increasing the strain, regardless of the presence of hydrogen. In Figure 6b,c, SIM distributed non-uniformly in a ribbon shape several hundred micrometers in width. At 45% plastic strain, the H-free sample had more SIM than the H-charged one. Figures 5 and 6 suggested that hydrogen suppressed SIM formation during tensile testing. This was plausible considering the ductility loss and the beneficial effects of SIM on the mechanical property of the H-free samples. In addition, the localized brittle fracture became more pronounced with decreasing the strain rate in the H-charged sample (Figure 3). This was related to the decreased amount of SIM that formed during tensile deformation. As will be explained below, the decrement of the microhardness owing to the hydrogen softening in the H-charged sample provided further evidence for the suppression of SIM formation by hydrogen.

The relationship between the amount of SIM and the mechanical property is depicted in Figure 7. The elongation to fracture, tensile strength, and microhardness of H-free and H-charged AISI 304 positively depended on the amount of SIM that formed during RT tensile testing. SIM beneficially increased elongation to fracture, tensile strength, and hardness, with and without hydrogen. This was also consistent with the TRIP effect, which increased elongation to fracture and tensile strength through the deformation-induced martensite transformation [3–5]. Tensile properties of the H-free samples depended on the strain rate, as shown in Figures 2 and 7. This was also related to the amount of SIM formed during plastic deformation. However, the hardness of the H-free sample increased with increasing the amount of SIM (Figure 2d). This implied that SIM was harder than austenite [27]. The cause of the dependence of the strain rate on the amount of SIM formed in austenitic stainless steel, as shown in Figure 4, was still unresolved. The reason for the enhancement of elongation to fracture by the hard SIM was also unclarified. In Figure 7, H-charged samples showed more sensitive linear dependence than H-free samples, implying that the suppression of SIM formation by hydrogen had a decisive effect on the mechanical property. The dependence of mechanical properties on the amount of SIM (Figure 7) and the suppression of SIM formation by hydrogen (Figures 5 and 6) indicated that hydrogen decreased the amount of SIM formed during tensile testing, which led to the loss of elongation to fracture and tensile strength. Hence, the hydrogen embrittlement and softening were inevitable in the H-charged samples.

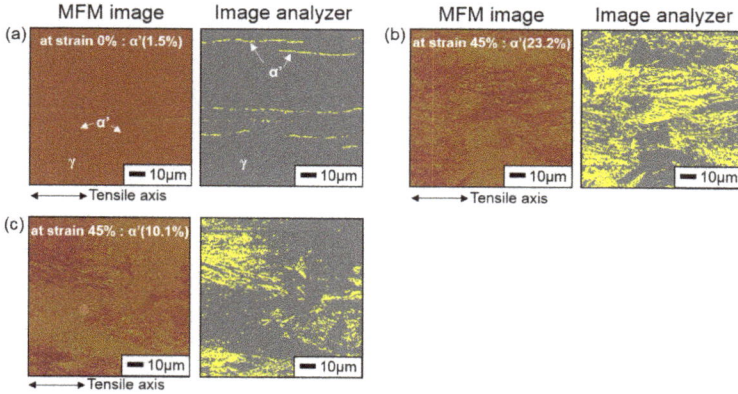

Figure 6. Distribution of austenite and martensite in AISI 304: (**a**) before tensile test (strain = 0%); (**b**) H-free (strain = 45%) at the strain rate of 2×10^{-4}/s; (**c**) H-charged (strain = 45%) at the strain rate of 2×10^{-4}/s.

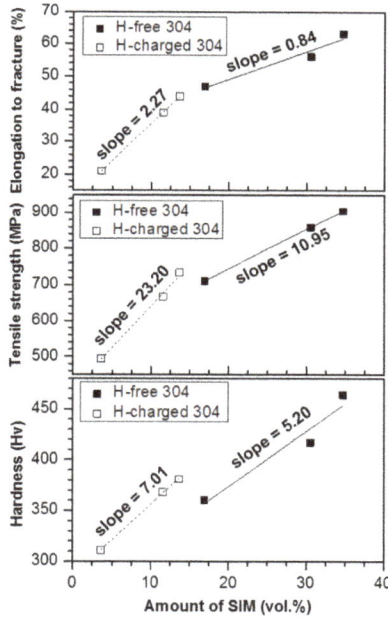

Figure 7. Linear dependence of mechanical properties on the amount SIM formed in H-free and H-charged AISI 304 during tensile testing: (**a**) elongation to fracture; (**b**) tensile strength; (**c**) microhardness.

4. Conclusions

The plastic deformation behavior and strain-induced martensite transformation were investigated through tensile testing AISI 304 metastable austenitic stainless steel at a slow strain rate at room temperature. The H-free AISI 304 showed a decreased yield stress and increased elongation to fracture and tensile strength with decreasing the strain rate. The strain-rate dependence of the tensile properties of H-free AISI 304 was closely related to the amount of SIM. In contrast, in H-charged samples, hydrogen decreased the yield stress slightly, but tensile strength and elongation to fracture significantly at all strain rates. However, the loss in tensile strength and elongation to fracture by hydrogen in H-charged samples showed strong strain-rate dependence. Considering the beneficial role of SIM in the tensile properties of H-free AISI 304, the ductility loss and hydrogen softening in the hydrogen-charged samples were related to the suppressed SIM formation by hydrogen. This suppression by hydrogen was considered as the main cause of hydrogen embrittlement and hydrogen softening in austenitic stainless steels.

Acknowledgments: This work was supported by the Human Resource Development Program (No. 20134030200360) of the Korea Institute of Energy Technology Evaluation and Planning (KETEP) grant funded by the Korea government Ministry of Trade, Industry and Energy.

Author Contributions: Sang Hwan Bak and Dong Bok Lee conceived and designed the experiments; Sang Hwan Bak performed the experiments and wrote the paper. Muhammad Ali Abro helped the discussion. Dong Bok Lee reviewed and contributed to the final manuscript.

Conflicts of Interest: The authors declare no conflict of interest.

References

1. Antolovich, S.D.; Fahr, D. An experimental investigation of the fracture characteristics of TRIP alloys. *Eng. Fract. Mech.* **1972**, *4*, 133–144. [CrossRef]
2. Perdahcıoğlu, E.S.; Geijselaers, H.J.M.; Groen, M. Influence of plastic strain on deformation-induced martensitic transformations. *Scr. Mater.* **2008**, *58*, 947–950. [CrossRef]
3. Hwang, J.; Son, I.; Yoo, J.; Zargaran, A.; Kim, N. Effect of reduction of area on microstructure and mechanical properties of twinning-induced plasticity steel during wire drawing. *Met. Mater. Int.* **2015**, *21*, 815–822. [CrossRef]
4. Oliver, E.C.; Withers, P.J.; Daymond, M.R.; Ueta, S.; Mori, T. Neutron-diffraction study of stress-induced martensitic transformation in TRIP steel. *Appl. Phys. A* **2002**, *74*, S1143–S1145. [CrossRef]
5. Tamura, I. Deformation-induced martensitic transformation and transformation-induced plasticity in steels. *Met. Sci.* **1982**, *16*, 245–253. [CrossRef]
6. Briant, C.L. Hydrogen assisted cracking of type 304 stainless steel. *Metall. Trans. A* **1979**, *10*, 181–189. [CrossRef]
7. Hanninen, H.; Hakkarainen, H. On the effects of α' martensite in hydrogen embrittlement of a cathodically charged AISI type 304 austenitic stainless steel. *Corrosion* **1980**, *36*, 47–51. [CrossRef]
8. Zapffe, C.A.; Sims, C.E. Hydrogen embrittlement internal stress and defects in steels. *Trans. Am. Inst. Min. Eng.* **1941**, *145*, 225–259.
9. Garofalo, F.; Chou, Y.T.; Ambegaokar, V. Effect of hydrogen on stability of micro cracks in iron and steel. *Acta Metall.* **1960**, *8*, 504–512. [CrossRef]
10. Bilby, B.A.; Hewitt, J. Hydrogen in steel—The stability of micro-cracks. *Acta Metall.* **1962**, *10*, 587–600. [CrossRef]
11. Escobar, D.P.; Miñambres, C.; Duprez, L.; Verbeken, K.; Verhaege, M. Internal and surface damage of multiphase steels and pure iron after electrochemical hydrogen charging. *Corros. Sci.* **2011**, *53*, 3166–3176. [CrossRef]
12. Simmons, G.W.; Pao, P.S.; Wei, R.P. Fracture mechanics and surface chemistry studies of subcritical crack growth in AISI 4340 steel. *Metall. Trans. A* **1978**, *9*, 1147–1158. [CrossRef]
13. Takano, N. First principles calculation of hydrogen embrittlement in iron. *Key Eng. Mater.* **2010**, *417–418*, 285–288. [CrossRef]
14. Birnbaum, H.K.; Sofronis, P. Hydrogen-enhanced localized plasticity—A mechanism for hydrogen-related fracture. *Mater. Sci. Eng. A* **1994**, *176*, 191–202. [CrossRef]
15. Takai, K.; Shoda, H.; Suzuki, H.; Nagumo, M. Lattice defects dominating hydrogen-related failure of metals. *Acta Mater.* **2008**, *56*, 5158–5167. [CrossRef]
16. Nagumo, M. Hydrogen related failure of steels—A new aspect. *Mater. Sci. Technol.* **2004**, *20*, 940–950. [CrossRef]

17. Zhang, L.; An, B.; Fukuyama, S.; Iijima, T.; Yokogawa, K. Characterization of hydrogen-induced crack initiation in metastable austenitic stainless steels during deformation. *J. Appl. Phys.* **2010**, *108*, 0635226. [CrossRef]

18. Miller, A.; Estrin, Y.; Hu, X.Z. Magnetic force microscopy of fatigue crack tip region in a 316L austenitic stainless steel. *Scr. Mater.* **2002**, *47*, 441–446. [CrossRef]

19. Sort, J.; Concustell, A.; Menéndez, E.; Suriñach, S.; Baró, M.D.; Farran, J.; Nogués, J. Selective generation of local ferromagnetism in austenitic stainless steel using nanoindentation. *Appl. Phys. Lett.* **2006**, *89*, 032509. [CrossRef]

20. Minkovitz, E.; Eliezer, D. Grain-size and heat-treatment effects in hydrogen-assisted cracking of austenitic stainless steels. *J. Mater. Sci.* **1982**, *17*, 3165–3172. [CrossRef]

21. Ralston, K.D.; Birbilis, N. Effect of grain size on corrosion: A review. *Corrosion* **2006**, *66*, 075005. [CrossRef]

22. Pan, C.; Chu, W.Y.; Li, Z.B.; Liang, D.T.; Su, Y.J.; Gao, K.W.; Qiao, L.J. Hydrogen embrittlement induced by atomic hydrogen and hydrogen-induced martensites in type 304L stainless steel. *Mater. Sci. Eng. A* **2003**, *351*, 293–298. [CrossRef]

23. Au, M. High temperature electrochemical charging of hydrogen and its application in hydrogen embrittlement research. *Mater. Sci. Eng. A* **2007**, *454–455*, 564–569. [CrossRef]

24. Yao, J.; Cahoon, J.R. Experimental studies of grain boundary diffusion of hydrogen in metals. *Acta. Metall. Mater.* **1991**, *39*, 119–126. [CrossRef]

25. Oudriss, A.; Creus, J.; Bouhattate, J.; Savall, C.; Peraudeau, B.; Feaugas, X. The diffusion and trapping of hydrogen along the grain boundaries in polycrystalline nickel. *Scr. Mater.* **2012**, *66*, 37–40. [CrossRef]

26. Austin, J.H.; Elleman, T.S.; Verghese, K. Surface effects on the diffusion of tritium in 304-stainless steel and zircaloy-2. *J. Nucl. Mater.* **1973**, *48*, 307–316. [CrossRef]

27. Mine, Y.; Horita, Z.; Murakami, Y. Effect of hydrogen on martensite formation in austenitic stainless steels in high-pressure torsion. *Acta Mater.* **2009**, *57*, 2993–3002. [CrossRef]

28. Spencer, K.; Embury, J.D.; Conlon, K.T.; Veron, M.; Brechet, Y. Strengthening via the formation of strain-induced martensite in stainless steels. *Mater. Sci. Eng. A* **2004**, *387*, 873–881. [CrossRef]

metals

MDPI

Article

Effect of Microstructure on Fracture Toughness and Fatigue Crack Growth Behavior of Ti17 Alloy

Rong Liang [1], Yingping Ji [1,2,*], Shijie Wang [3] and Shuzhen Liu [1]

[1]   College of Mechanical Engineering, Ningbo University of Technology, Ningbo 315211, China;
      mikesmile@163.com (R.L.); shuzhenl@163.com (S.L.)
[2]   School of Materials Science and Engineering, Beihang University, Beijing 100191, China
[3]   China National Heavy Machinery Research Institute Co., Ltd., Xi'an 710032, China; xzswangsj@163.com
[*]   Correspondence: yingping04@163.com; Tel.: +86-574-8235-1508

Academic Editor: Soran Birosca
Received: 18 April 2016; Accepted: 20 July 2016; Published: 12 August 2016

Abstract: Ti-5Al-2Sn-2Zr-4Mo-4Cr (Ti17) is used extensively in turbine engines, where fracture toughness and fatigue crack growth (FCG) resistance are important properties. However, most research on the alloy was mainly focused on deformation behavior and microstructural evolution, and there have been few studies to examine the effect of microstructure on the properties. Accordingly, the present work studied the influences of the microstructure types (bimodal and lamellar) on the mechanical properties of Ti17 alloy, including fracture toughness, FCG resistance and tensile property. In addition, the fracture modes associated with different microstructures were also analyzed via the observation of the fracture surface. The results found that the lamellar microstructure had a much higher fracture toughness and superior resistance to FCG. These results were discussed in terms of the tortuous crack path and the intrinsic microstructural contributions.

Keywords: titanium alloys; microstructure; fracture toughness; fatigue crack growth behavior

1. Introduction

With the continuing desire to make engines with a high thrust-to-weight advantage, titanium alloys are the metals of choice for the gas turbine engine [1], where fracture toughness and resistance against fatigue failure are important properties. The optimized mechanical properties of titanium alloys can be achieved via controlling the microstructure. For example, Nalla et al. found that Ti-6Al-4V with a coarse lamellar microstructure had superior toughness while FCG (fatigue crack growth) behavior in large cracks was compared with a finer bimodal microstructure [2]. Verdhan et al. also found that Ti-6Al-2Zr-1.5Mo-1.5V with lamellar and acicular microstructures had lower FCG rates than that with the bimodal microstructure [3]. In general, three microstructures can be obtained in ($\alpha + \beta$) and β titanium alloys: equiaxed, lamellar, and bimodal, which can be obtained by different thermomechanical processing [4]. The first is the result of a recrystallization process after minimal plastic deformation. The second one can be attained by an annealing process above the β-transus temperature (T_β) with subsequent cooling to the ($\alpha + \beta$) phase. The third is the result of the annealing of a sufficiently plastically deformed ($\alpha + \beta$) titanium alloy below the T_β [5].

As a replacement for the conventionally used $\alpha + \beta$ titanium alloys in high-strength airframe and jet engine structural parts, β titanium alloys have attracted a great deal of interest recently. Take Ti-5Al-2Sn-2Zr-4Mo-4Cr (Ti17), for example: due to its higher strength level as compared to the typical $\alpha + \beta$ Ti-6Al-4V alloy [6–8], the alloy has been favored as a jet engine compressor and has attracted more attention in recent times [6–16]. Luo et al. conducted isothermal compression tests on Ti17 alloy and observed the microstructural evolution [11]. Wang et al. studied the grain growth kinetics of Ti17 alloys in the β phase [12]. Teixeira et al. investigated the transformation kinetics and microstructural evolution of Ti17 titanium alloy during continuous cooling [13]. Tarín et al. assessed the ultimate tensile strength, yield strength and elongation of Ti17 alloy [14]. Xu et al. studied the effect of globularization behavior of the lamellar alpha on the tensile properties of Ti17 alloy and found that there are linear relationships between tensile properties and the globularization fraction [15]. Moshier et al. investigated the load history effects on the fatigue crack growth threshold of Ti17 alloy [16]. Shi et al. studied the effects of lamellar features on the fracture toughness of Ti17 titanium alloy and found that the microstructure with long and thick needle-like α platelets had higher fracture toughness and strength than that with short rod-like α platelets [9]. It can be found that the previous research on Ti17 alloy was mainly focused on deformation behavior and microstructural evolution. However, there have been few studies to examine the effects of microstructure on fracture toughness and FCG behavior.

Thus, in this paper, the influences of the microstructural types (bimodal and lamellar) on the mechanical properties of Ti17 alloy were investigated, including yield stress, ductility, and fracture toughness as well as FCG properties. To investigate the fracture mode, the fracture surfaces of specimens after tests were observed by scanning electron microscopy (SEM).

2. Materials and Methods

The material used in this study is Ti17 alloy, whose chemical composition is detailed in Table 1.

Table 1. Chemical composition of Ti17 (wt. %).

Element	Al	Sn	Zr	Mo	Cr	Fe	C	N	H	O	Ti
wt. %	5.01	1.98	2.02	4.15	4.33	0.30	0.05	0.05	0.0125	0.09	Balance

To obtain two different microstructures of interest, materials were subjected to α + β and β processing route respectively. In α + β process, Ti17 alloys as received condition were hot forged in the α + β phase field at 850 °C, followed by air cooling to room temperature. Subsequently, the materials were recrystallized in the α + β phase field at 870 °C for 1 h to adjust a low volume fraction of the primary α phase ($α_p$), followed by air cooling to room temperature. Finally, materials were aged 8 h at 580 °C to precipitate fine secondary α ($α_s$). In β process, as received materials were first α + β forged at 850 °C and then heated to 920 °C into the β phase field to coarsen the β grain structure for 1 h, followed by air cooling. At last, these materials were aged at 580 °C for 8 h to precipitate $α_s$. The microstructures of the α + β and β processed were observed by SEM on an Apollo 300 (Camscan, Cambridge, UK) and they are illustrated in Figure 1. It is obvious that the material of the α + β processed exhibits the bimodal microstructure, and that β processed yields the lamellar one.

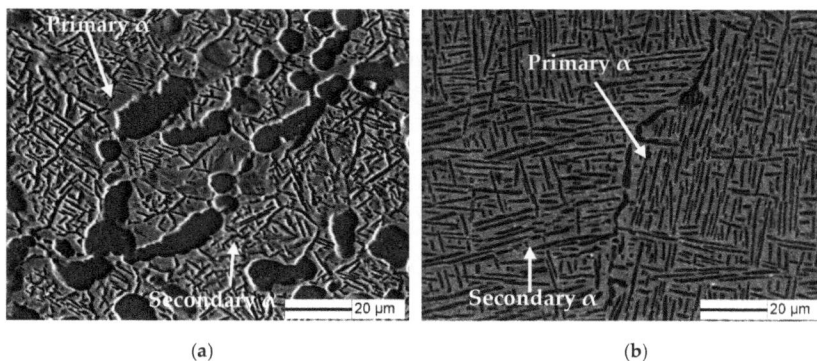

(a)   (b)

Figure 1. Microstructure of Ti17: (a) bimodal; (b) lamellar.

The quasi-static tensile tests were conducted at room temperature and in the laboratory air environment using a fully automated, closed-loop servo-hydraulic machine (INSTRON 8801). The specimens were made of a dog-bone shape (Figure 2a) and deformed to failure at a constant strain rate of $10^{-3} \cdot s^{-1}$. These tests yielded basic mechanical properties such as yield strength, tensile strength and ductility.

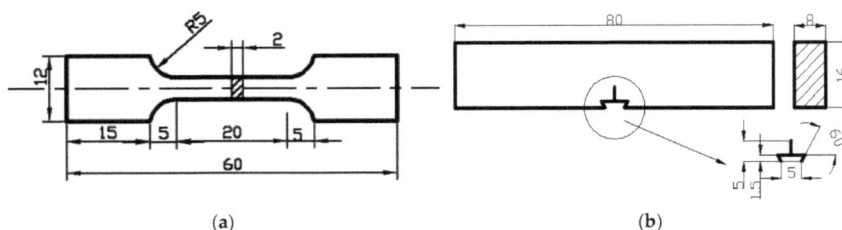

(a)   (b)

Figure 2. Specimens for (**a**) tensile; (**b**) fracture toughness and fatigue crack growth tests.

Fatigue and fracture toughness measurements were also carried out using the servo-hydraulic testing machine INSTRON 8801 and performed at room temperature. Mode I fracture toughness tests were performed using three-point (3P) bending specimens, with the geometry as described in Figure 2b. The specimens were loaded according to the standard three-point bending method gripped all along their width and fatigue precracks were initiated to a depth of 8 mm. Subsequently, monotonic loading was applied to the specimen until the crack propagated catastrophically to fracture. Further calculation details of fracture toughness were according to the BS7448-1:1991 standards. Fatigue crack growth experiments were conducted on the same specimens, in according with ASTM E647-93 standard. FCG experiments were carried out on with a frequency of 6 Hz under constants tress ratio $R = 0.1$ and the fatigue crack length (a) was monitored using compliance method.

For a given variant, three specimens were tested and the average values of the mechanical properties were determined. In addition to mechanical testing, the fracture surfaces of the failed specimens with different microstructures were examined by SEM to determine the macroscopic fracture mode and mechanisms.

3. Results and Discussion

3.1. Tensile Properties

The average values of the tensile properties of the tested specimens are summarized in Table 2, including yield strength (YS), ultimate tensile strength (US) and percentage elongation (A%). The values listed are an average of three tests. It can be found that the bimodal microstructure has a slightly increasing strength and higher plasticity than the lamellar one, indicative of a comparable tensile strength and a higher ductility of the former. The tensile ductility of α-β titanium alloys is mainly determined by two parameters, crack nucleation resistance and crack propagation resistance, and the former is the dominating parameter. Gungor et al. pointed out that colony boundaries and α at prior β grain boundaries served as void nucleation sites [17]. Thus, the increasing volume of α_p phases in the bimodal microstructure decreased the slip length and initiated more sites of void nucleation, leading to the higher ductility.

Table 2. Mechanical properties of Ti17 alloy with bimodal and lamellar structure.

Specimens	US/MPa	YS/MPa	A/%	CTOD/μm
Bimodal structure	1220	1165	13	12
Lamellar structure	1205	1150	9.8	68

Representative fractographs of the tensile specimens are shown in Figure 3, which reveal the specific role played by the microstructural effects on strength and ductility properties. As Figure 3a depicted, some randomly distributed ductile dimples in globular shapes cover the transgranular fracture regions, suggesting locally ductile failure occurred in the bimodal microstructure. In the case of the lamellar microstructure, the fracture surface was similar to the counterpart of the bimodal microstructure, i.e., some elongated dimples in the center of Figure 3b could also be noted. However, the dimples were smaller than in the former, corresponding to the lower value of percentage elongation. Assisting with the features observation of the fracture surfaces, it can be concluded that the ductile failure mechanism governed the tensile response of Ti17 alloy with both bimodal and lamellar microstructures.

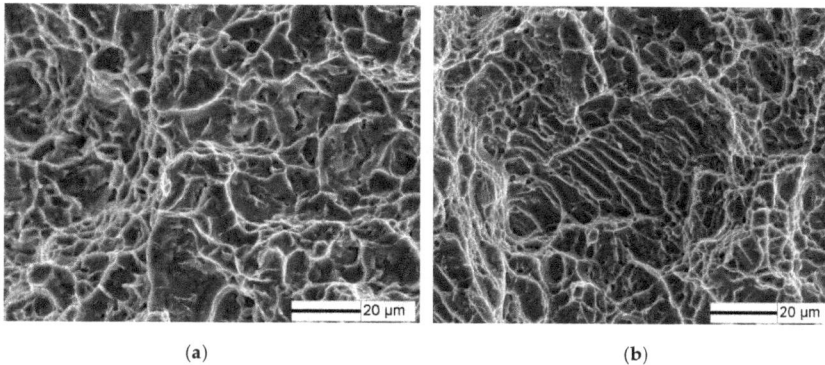

(a)   (b)

Figure 3. Tensile fracture surfaces of Ti17: (**a**) bimodal; (**b**) lamellar.

3.2. Fracture Toughness

A preliminary calculation revealed that, for the specimen size considered in these investigations, a valid plane strain fracture toughness, K_{IC}, could not be obtained. Hence, the crack tip opening displacement (CTOD), at the onset of crack initiation, was chosen as the critical fracture toughness of specimens. The average CTOD values were also enumerated in Table 2 and each value listed is an average of three tests. It can be observed that the lamellar microstructure exhibits higher fracture toughness as compared with the bimodal microstructure. Therefore, from the point of the damage tolerance design concept, titanium alloy with a lamellar microstructure is more significant than that with a bimodal microstructure, which is in agreement with the previous report [18].

Representative SEM micrographs depicting the fracture surfaces of bimodal and lamellar microstructures are shown in Figure 4. It is evident that fracture surfaces of both were covered with dense dimples. Crack propagated in ductile mode, which was also found in Reference [7]. However, some big differences between the two microstructures were observed. On one hand, the dimples on the fracture surfaces shown in Figure 4a are obviously bigger and deeper than those shown in Figure 4b. This indicates that the bimodal microstructure has higher plasticity than the lamellar microstructure, which is consistent with the tensile results shown in Table 2. On the other hand, the fracture surface of the lamellar microstructure is characterized by secondary cracks and a big fracture step, which increases the toughness.

Previous investigations found that there was a positive correlation between the crack path length and the fracture toughness [7,19,20]. Therefore, the crack path geometry has to be taken into account as a parameter responsible for the different fracture toughness. As depicted in Figure 5 (the nominal crack propagation direction is indicated by the arrow), bimodal and lamellar microstructures both fractured in tortuous paths during the fracture process, but the lamellar microstructure displays a more tortuous and deflected crack path than the former. Such increased crack-path tortuosity is a major contributor to the higher toughness of the lamellar microstructure. The tortuous and deflected crack path consumed much more energy than the flat one, leading to the higher fracture toughness. Cracks grew along a more tortuous and deflected crack path in lamellar microstructures, which over-compensated the lower ductility, resulting in the higher fracture toughness. The tortuous and deflected crack path was related with the more discontinuous α precipitates at the grain boundary. Therefore, it can be concluded that the long and thick α platelets in the lamellar microstructure can realize a good combination of fracture toughness and strength.

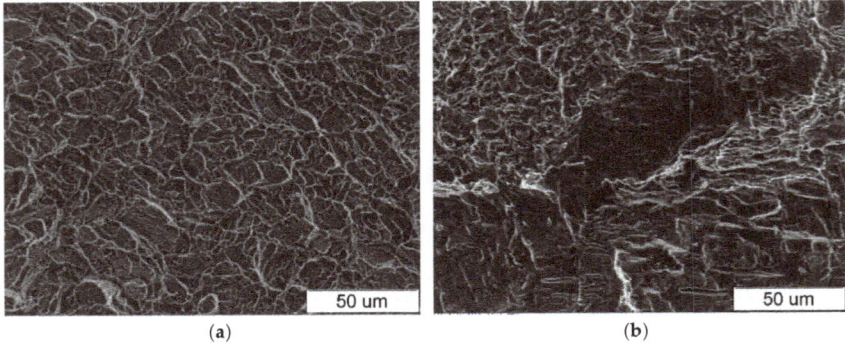

(a)   **(b)**

Figure 4. Fracture surfaces of specimens for fracture toughness test: (**a**) bimodal; (**b**) lamellar.

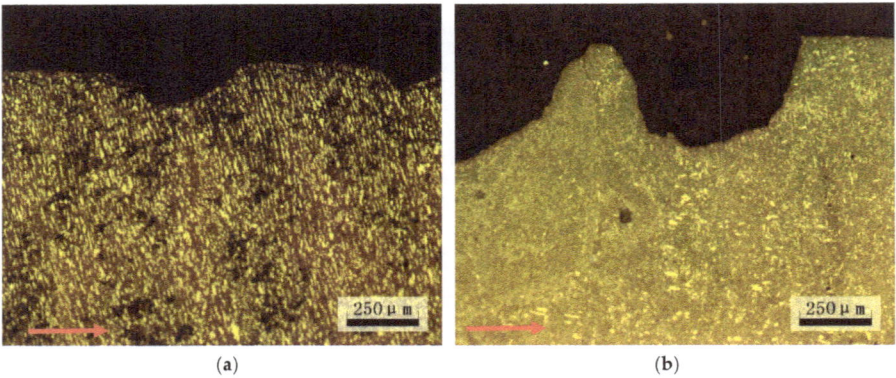

(a)   **(b)**

Figure 5. Metallographic profile of the crack path in Ti17: (**a**) bimodal; (**b**) lamellar.

3.3. Fatigue Crack Growth Behavior

The variation in FCG rates in bimodal and lamellar microstructures is presented in Figure 6. The fracture mechanics based on the Paris Power equation, given below, were used to analyze the experimental results.

$$da/dN = C(\Delta K)^m \qquad (1)$$

where da/dN is the FCG rate, ΔK is the stress intensity factor range, and the exponent "m" and coefficient "C" are the slope and the intercept of the line on the log-log plot respectively, and both are constant.

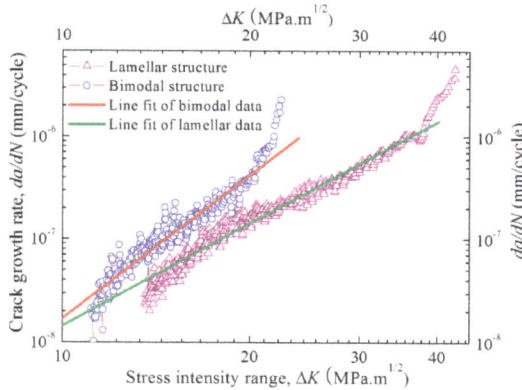

Figure 6. FCG behavior of bimodal and lamellar microstructures.

The Paris Power law relations (in units of mm/cycle, MPam$^{1/2}$) of bimodal and lamellar microstructures are shown as follows.

$$\frac{da}{dN} = 1.54 \times 10^{-8} \times (\Delta K)^{4.63} (\text{Bimodal structure}) \tag{2}$$

$$\frac{da}{dN} = 1.14 \times 10^{-8} \times (\Delta K)^{3.44} (\text{Lamellar structure}) \tag{3}$$

Although the differences are not large, in the lamellar structure, FCG rates (da/dN) were consistently lower and the threshold is higher than that in the bimodal microstructure. Thus, the lamellar structure clearly provides superior FCG resistance compared to the bimodal structure, which can primarily be traced to the increased tortuosity in the crack path in the lamellar structure. As shown in Figure 7, the crack path in the bimodal microstructure is straight, but it appears more deflected in the lamellar microstructure (the nominal crack propagation direction is indicated by the arrow).

Figure 7. A typical crack profile observed for the (**a**) bimodal and (**b**) lamellar microstructure.

The tortuous crack path in the lamellar structure resulted in a far rougher fracture surface, as shown by the SEM in Figure 8a,b for bimodal and lamellar structures, respectively. Typical fatigue striations and secondary cracks were observed on both fracture surfaces. However, the fracture surface of the lamellar microstructure is much rougher than that of the bimodal microstructure. In addition, there are more microscopic cracks detected in Figure 8b. Due to the tortuous nature of the cracks, the lamellar microstructure had a higher FCG resistance than the bimodal microstructure, which was also observed in the FCG behavior of a near-α Ti alloy [3].

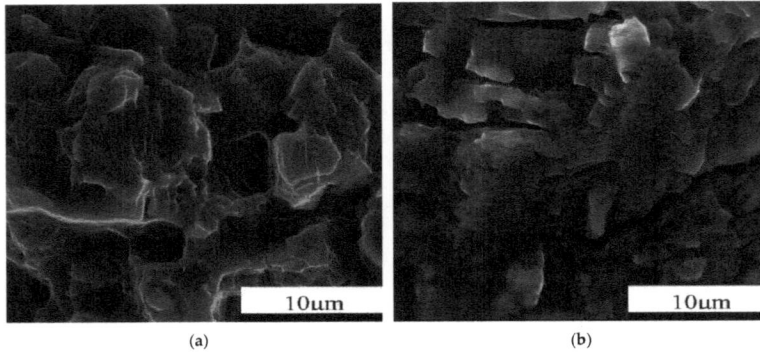

(a)   (b)

Figure 8. Fracture topography of Ti17 in Regime II of FCG: (**a**) bimodal; (**b**) lamellar.

The different damage modes are consistent with the microstructures. As Nalla et al. found, the crack paths in titanium alloys were strongly influenced by the orientation of neighboring colonies [2]. In the present study, the lamellar microstructure had big colonies of lath α and β, which acted as an effective slip barrier to prevent the transmission of slip to neighboring colonies. The cracks changed direction when they crossed the boundary between colonies. Thus, they caused crack branching and secondary crack creation. The occurrence of secondary cracking effectively reduced the local crack tip driving force due to the redistribution of stresses [21], resulting in the lower FCG rates of the lamellar microstructure. In the case of the bimodal microstructure, there were no obvious colonies and the fatigue crack grew in a straight way, resulting in a relatively higher FCG rate than the lamellar microstructure.

4. Conclusions

Based on the study of the mechanical properties of Ti17 alloys with bimodal and lamellar microstructure, the following conclusions can be made:

(1) The lamellar microstructure showed a superior fracture toughness over the bimodal structure, which was associated with the intrinsic microstructural contributions and tortuous crack path.

(2) The lamellar microstructure had a higher threshold stress intensity factor and a greater FCG resistance than the bimodal microstructure.

Acknowledgments: This work has been financially supported by "Fracture Mechanism of Dissimilar Titanium Alloy Welded Joints" Ningbo Natural Science Foundation program (NO.2015A610071) and the Zhejiang Natural Science Foundation program (NO.LQ13E050005). As part of these grants, we received funds for covering the costs to publish in open access.

Author Contributions: Rong Liang and Yingping Ji conceived and designed the experiments; Yingping Ji performed the experiments, analyzed the data and wrote the manuscript; Shijie Wang and Shunzhen Liu contributed to editing the manuscript.

Conflicts of Interest: The authors declare no conflict of interest.

Abbreviations

The following abbreviations are used in this manuscript:

α	Hexagonal phase in titanium alloys
β	Body-centered cubic phase in titanium alloys
T_β	Beta-transus temperature
a	Crack length
C, m	Parameters of the Paris-Erdogan equation
da/dN	Fatigue crack growth rate
ΔK	Cyclic stress intensity factor
N	Number of cycles

References

1. Elrod, C.W. Review of Titanium Application in Gas Turbine Engines. In Proceedings of the ASME Turbo Expo 2003, collocated with the 2003 International Joint Power Generation Conference, Atlanta, GA, USA, 16–19 June 2003; pp. 649–656.
2. Nalla, R.K.; Boyce, B.L.; Campbell, J.P.; Peters, J.O.; Ritchie, R.O. Influence of Microstructure on High-Cycle Fatigue of Ti-6Al-4V: Bimodal vs. Lamellar Structures. *Metall. Mater. Trans. A* **2002**, *33*, 899–917. [CrossRef]
3. Verdhan, N.; Bhende, D.D.; Kapoor, R.; Chakravartty, J.K. Effect of microstructure on the fatigue crack growth behaviour of a near-α Ti alloy. *Int. J. Fatigue* **2015**, *74*, 46–54. [CrossRef]
4. Lütjering, G. Influence of processing on microstructure and mechanical properties of (α + β) titanium alloys. *Mater. Sci. Eng. A* **1998**, *243*, 32–45. [CrossRef]
5. Krüger, L.; Grundmann, N.; Trubitz, P. Influence of Microstructure and Stress Ratio on Fatigue Crack Growth in a Ti-6-22-22-S alloy. *Mater. Today Proc.* **2015**, *2*, 205–211. [CrossRef]
6. Li, W.Y.; Ma, T.; Yang, S. Microstructure Evolution and Mechanical Properties of Linear Friction Welded Ti-5Al-2Sn-2Zr-4Mo-4Cr (Ti17) Titanium Alloy Joints. *Adv. Eng. Mater.* **2010**, *12*, 35–43. [CrossRef]
7. Shi, X.H.; Zeng, W.D.; Zhao, Q.Y. The effects of lamellar features on the fracture toughness of Ti17 titanium alloy. *Mater. Sci. Eng. A* **2015**, *636*, 543–550. [CrossRef]
8. Cadario, A.; Alfredsson, B. Fatigue growth of short cracks in Ti17: Experiments and simulations. *Eng. Fract. Mech.* **2007**, *74*, 2293–2310. [CrossRef]
9. García, A.M.M. *BLISK Fabrication by Linear Friction Welding*; INTECH Open Access Publisher: Rijeka, Croatia, 2011.
10. Kumar, B.V.R.R. A Review on Blisk Technology. *Int. J. Innov. Res. Sci. Eng. Technol.* **2013**, *2*, 1353–1358.
11. Luo, J.; Li, L.; Li, M.Q. The flow behavior and processing maps during the isothermal compression of Ti17 alloy. *Mater. Sci. Eng. A* **2014**, *606*, 165–174. [CrossRef]
12. Wang, T.; Guo, H.Z.; Tan, L.J.; Yao, Z.K.; Zhao, Y.; Liu, P.H. Beta grain growth behaviour of TG6 and Ti17 titanium alloys. *Mater. Sci. Eng. A* **2011**, *528*, 6375–6380. [CrossRef]
13. Teixeira, J.D.C.; Appolaire, B.; Aeby-Gautier, E.; Denis, S.; Cailletaud, G.; Späth, N. Transformation kinetics and microstructures of Ti17 titanium alloy during continuous cooling. *Mater. Sci. Eng. A* **2007**, *448*, 135–145. [CrossRef]
14. Tarín, P.; Fernández, A.L.; Simón, A.G.; Badía, J.M.; Piris, N.M. Transformations in the Ti-5Al-2Sn-2Zr-4Mo-4Cr (Ti-17) alloy and mechanical and microstructural characteristics. *Mater. Sci. Eng. A* **2006**, *438–440*, 364–368. [CrossRef]
15. Xu, J.W.; Zeng, W.D.; Zhao, Y.W.; Jia, Z.Q.; Sun, X. Effect of globularization behavior of the lamellar alpha on tensile properties of Ti-17 alloy. *J. Alloy. Compd.* **2016**, *673*, 86–92. [CrossRef]
16. Moshier, M.A.; Nicholas, T.; Hillberry, B.M. Load history effects on fatigue crack growth threshold for Ti-6Al-4V and Ti-17 titanium alloys. *Int. J. Fatigue* **2001**, *23*, 253–258. [CrossRef]
17. Gungor, M.N.; Ucok, I.; Kramer, L.S.; Dong, H.; Martin, N.R.; Tack, W.T. Microstructure and mechanical properties of highly deformed Ti-6Al-4V. *Mater. Sci. Eng. A* **2005**, *410–411*, 369–374. [CrossRef]
18. Chandravanshi, V.K.; Bhattacharjee, A.; Kamat, S.V.; Nandy, T.K. Influence of thermomechanical processing and heat treatment on microstructure, tensile properties and fracture toughness of Ti-1100-0.1B alloy. *J. Alloy. Compd.* **2014**, *589*, 336–345. [CrossRef]
19. Xue, Y.L.; Li, S.M.; Zhong, H.; Fu, H.Z. Characterization of fracture toughness and toughening mechanisms in Laves phase Cr_2Nb based alloys. *Mater. Sci. Eng. A* **2015**, *638*, 340–347. [CrossRef]

20. Shi, X.H.; Zeng, W.D.; Zhao, Q.Y. The effect of surface oxidation behavior on the fracture toughness of Ti-5Al-5Mo-5V-1Cr-1Fe titanium alloy. *J. Alloy. Compd.* **2015**, *647*, 740–749. [CrossRef]
21. Stephens, R.R.; Stephenst, R.I.; Veitt, A.L.; Albertson, T.P. Fatigue crack growth of Ti-62222 titanium alloy under constant amplitude and minitwist flight spectra at 25 °C and 175 °C. *Int. J. Fatigue* **1997**, *19*, 301–308. [CrossRef]

Article

Impurity Antimony-Induced Creep Property Deterioration and Its Suppression by Rare Earth Cerium for a 9Cr-1Mo Ferritic Heat-Resistant Steel

Yewei Xu and Shenhua Song *

Shenzhen Key Laboratory of Advanced Materials, Department of Materials Science and Engineering, Shenzhen Graduate School, Harbin Institute of Technology, Shenzhen 518055, China; xywgk123@sina.com
* Correspondence: shsong@hitsz.edu.cn; Tel.: +86-755-2603-3465; Fax: +86-755-2603-3504

Academic Editor: Soran Birosca
Received: 9 July 2016; Accepted: 9 August 2016; Published: 12 August 2016

Abstract: The high temperature creep properties of three groups of modified 9Cr-1Mo steel samples, undoped, doped with Sb, and doped with Sb and Ce, are evaluated under the applied stresses from 150 MPa to 210 MPa and at the temperatures from 873–923 K. The creep behavior follows the temperature-compensated power law as well as the Monkman-Grant relation. The creep activation energy for the Sb-doped steel (519 kJ/mol) is apparently lower than that for the undoped one (541 kJ/mol), but it is considerably higher for the Sb+Ce-doped steel (621 kJ/mol). Based on the obtained relations, both the creep lifetimes under 50 MPa, 80 MPa, and 100 MPa in the range 853–923 K and the 10^5 h creep rupture strengths at 853 K, 873 K, and 893 K are predicted. It is demonstrated that the creep properties of the Sb-doped steel are considerably deteriorated but those of the Sb+Ce-doped steel are significantly improved as compared with the undoped steel. Microstructural and microchemical characterizations indicate that the minor addition of Ce can stabilize the microstructure of the steel by segregating to grain boundaries and dislocations, thereby offsetting the deleterious effect of Sb by coarsening the microstructure and weakening the grain boundary.

Keywords: heat-resistant steels; creep properties; impurities; rare earths; grain boundaries; segregation

1. Introduction

In recent years, the problems of energy consumption and environmental pollution require advanced power plants with high efficient utilization of fossil energy, which prompts the development of materials for power plants with higher steam temperature and operating pressure. The modified 9Cr-1Mo ferritic heat-resistant steel, i.e., P91 steel, is extensively employed in the power generation and nuclear energy industries owing to its outstanding high temperature mechanical properties that enable higher operating temperatures and pressures to enhance the thermal efficiency [1–3]. Consequently, it is of great significance to explore the creep properties of this steel.

There have been a number of studies concerning the creep properties of the 9Cr steel, most of which focus on the evolution of martensitic microstructures including coarsening of precipitates, decreasing in dislocation density, and growing of martensite lath subgrains. Abe et al. [4] investigated the microstructural instability and its effect on the creep behavior of 9Cr-2W steel and observed that the recovery of dislocations, the agglomeration of carbides, and the growth of martensite lath subgrains occurred during creep with the aid of stress. Spigarelli et al. [5] also found that both large precipitates mainly present on the subgrain boundaries and fine ones mainly present in the subgrain interiors coarsened during creep. Other researchers further found that the creep rupture strength of the high Cr ferritic steel decreased due to the coarsening of precipitates during creep. After a long-term creep process, a new phase, such as Z-phase, formedat the cost of fine precipitates (MX-type carbides) [6,7]. However, there have been few reports concerning the effect of impurity elements or rare earths on the creep properties of P91 steel. Theoretical calculations indicate that the grain boundary cohesion can be increased by the grain boundary segregation of rare earth elements in iron [8,9], while it can be decreased by the grain boundary segregation of impurity elements, such as tin, phosphorus, and antimony in iron [10]. Our previous studies [11,12] regarding the effects of impurity tin and rare earth cerium on the creep properties of P91 heat-resistant steel demonstrated that tin can significantly deteriorate the creep properties of the steel as it can promote the nucleation of cavities or microcracks by grain or subgrain boundary segregation which weakens the boundary cohesion [11]. However, it is indicated that rare earth cerium can evidently improve the creep properties of the steel due to its segregation to grain or subgrain boundaries [12]. Until the present time, the combined effect of impurity elements and rare earths on the creep properties of P91 steel is little reported.

In the present work, we conducted high-temperature creep tests on the P91 steel samples of three groups: undoped, doped with impurity antimony, and doped with both impurity antimony and rare earth cerium. According to the creep rupture data which were extracted from the creep tests, parameters in the temperature-compensated power law [13] and Monkman-Grant relation [14] were determined. With the two relations above, the creep lifetimes of the steel were predicted at several combinations of stress and temperature within the valid range. After that, microstructural and microchemical observations were carried out so as to explore the effect of impurity antimony and the combined effect of impurity antimony and rare earth cerium on the creep properties of the steel.

2. Materials and Methods

Three heats of P91 steel: undoped, doped with impurity antimony, and doped with both impurity antimony and rare earth cerium, were prepared by vacuum induction melting with 50 kg ingots. The chemical compositions of the steels are listed in Table 1. The ingots were homogenized at 1523 K and then hot rolled in the range of 1273–1373 K into a plate 18 mm in thickness. After being austenitized at 1328 K for 1 h, the steels were air cooled to room temperature, and then tempered at 1033 K for 3 h, followed by water cooling to create a tempered martensite microstructure.

Table 1. Chemical compositions of undoped, Sb-doped, and Sb+Ce-doped steels (wt. %).

Elt.	C	Si	Mn	Sb	Ce	S	P	Cr	Mo	N	Nb	Ni	V	Fe
1	0.12	0.30	0.46	—	—	0.015	0.011	8.85	1.03	0.024	0.075	0.15	0.23	Bal.
2	0.12	0.31	0.47	0.048	—	0.013	0.012	8.95	0.99	0.022	0.071	0.12	0.22	Bal.
3	0.12	0.34	0.46	0.051	0.055	0.011	0.010	9.15	0.97	0.027	0.088	0.16	0.23	Bal.

Note: 1: undoped; 2: Sb-doped; 3: Sb+Ce-doped.

Creep tests were conducted with a high temperature creep testing machine. Since the P91 steel is usually used below 923 K in engineering [15] and the time to rupture is very long if the testing temperature is below 873 K, the creep tests were conducted between 873 K and 923 K. The circular creep specimens with a gage length of 50 mm and a gage diameter of 10 mm were divided into two groups. For one group, the specimens were creep tested at a constant engineering stress of 150 MPa at 873 K, 885 K, 898 K, 910 K and 923 K, respectively, so as to obtain the apparent activation energy (Q_c) for creep in the temperature-compensated power law. For the other group, the tests were conducted at a constant temperature of 873 K under the engineering stresses of 210 MPa, 195 MPa, 180 MPa, 165 MPa, and 150 MPa, respectively, so as to obtain the creep stress exponent (n) in the temperature-compensated power law. At each condition, two specimens were used except testing at a low temperature or stress where one specimen was only employed because the time to rupture was too long.

After creep rupture, to examine the evolution of microstructures and the compositional features at grain or subgrain boundaries, field emission gun scanning transmission electron microscopy equipped with energy dispersive X-ray spectroscopy (FEGSTEM-EDS) was employed. The instrument used in the present work was JEM-2100F FEGSTEM (JEOL, Tokyo, Japan) equipped with an Oxford INCA energy dispersive X-ray spectrometer (Oxford Instruments, Oxford, UK), operating under a voltage of 200 kV. The disc specimens for FEGSTEM microanalysis were cut from the specimens near the fracture tip as-tempered and creep-tested at 873 K under 150 MPa. More details on FEGSTEM grain boundary microanalysis can be seen in [16,17]. In addition, the morphology of the specimens near the fracture tip was observed using optical microscopy.

3. Results and Discussion

After creep tests, creep data, such as minimum creep rate ($\dot{\varepsilon}_m$) and rupture time (t_r), for all three groups of P91 steel specimens were extracted from the creep curves, which are listed in Table A1 (see Appendix A). Based on these data, ln(minimum creep rate) is plotted as a function of reciprocal temperature ($1/T$) and ln(stress), as shown in Figure 1a,b, respectively. Clearly, there is a linear relationship between ln(minimum creep rate) and reciprocal temperature with R-squared values of 0.977, 0.997, and 0.991 (see Figure 1a) or between ln(minimum creep rate) and ln(stress) with R-squared values of 0.976, 0.994, and 0.998 (see Figure 1b) for the undoped, Sb-doped, and Sb+Ce-doped steels, respectively. As can be seen, the fitted line for the undoped steel is located under those for the Sb-doped and Sb+Ce-doped steels, showing a better creep behavior. The fitted lines for the Sb-doped and Sb+Ce-doped steels intersect at a certain temperature, indicating that they have different slopes, i.e., different creep activation energy (Q_c) and stress exponent (n) values. Since the lines for the Sb+Ce-doped steel are steeper than those for the Sb-doped one, the former has higher Q_c and n values.

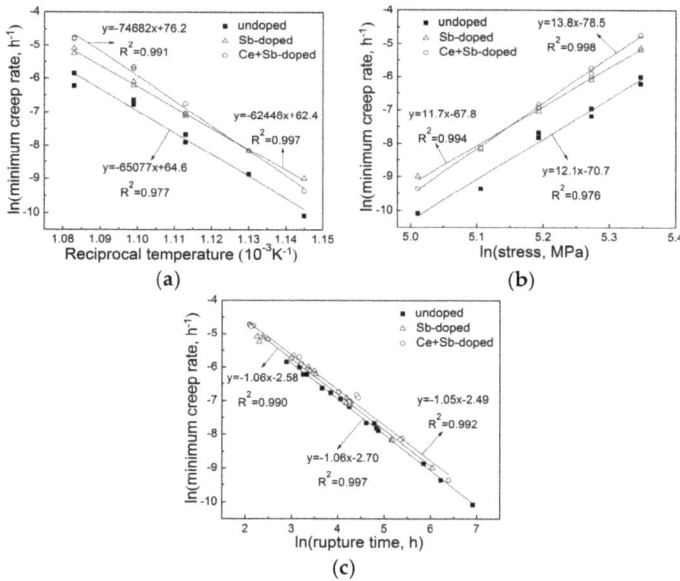

Figure 1. Variation of minimum creep rate with: (**a**) reciprocal temperature under a stress of 150 MPa; (**b**) stress at a temperature of 873 K; and (**c**) rupture time for the undoped, Sb-doped, and Sb+Ce-doped steels.

For the undoped steel, the linear fitting gives:

$$\ln\dot{\varepsilon}_m = -65077/T + 64.6 \qquad (\sigma = 150 \text{ MPa}) \qquad (1)$$

$$\ln\dot{\varepsilon}_m = 12.2\ln\sigma - 70.7 \qquad (T = 873 \text{ K}) \qquad (2)$$

Experimentally, it is frequently observed that, during the steady-state creep, the strain rate or the minimum creep rate ($\dot{\varepsilon}_m$) depends on the applied stress (σ) and exponentially on the absolute temperature (T) in the form of the temperature-compensated power law [13]:

$$\dot{\varepsilon}_m = A\sigma^n \exp\left(-Q_c/RT\right) \tag{3}$$

where A is a constant, n is the stress exponent for creep, R is the gas constant (8.314 J/mol K), and Q_c is the apparent activation energy for creep. These values are frequently used to estimate the creep properties of steel. The temperature-compensated power law can be rewritten as:

$$\ln\dot{\varepsilon}_m = \ln A + n\ln\sigma - Q_c/RT \tag{4}$$

Comparing Equation (4) with Equations (1) and (2), one can obtain $Q_c = 541$ (kJ/mol), $n = 12.1$ and:

$$\ln A + n\ln 150 = 64.6 \tag{5}$$

$$\ln A - Q_c/7258 = -70.7 \tag{6}$$

with Equations (5) and (6), the two values of A may be acquired as $A_1 = 61.2$ h^{-1} and $A_2 = 48.2$ h^{-1}. Evidently, these two values are roughly close to each other, implying that there is a fixed value of A in the temperature and stress ranges considered. Hence, the mean value of the two values may be taken as the measured one, i.e., $A=54.7$ h^{-1}. Similarly, for the Sb-doped steel and Sb+Ce-doped steel, one can obtain $Q_c = 519$ (kJ/mol), $n = 11.7$ and $A = 42.7$ (Sb-doped steel), and $Q_c = 621$ (kJ/mol), $n = 13.8$ and $A = 1152.3$ (Sb+Ce-doped steel). Thus, the temperature-compensated power law of creep for the steel may be expressed as:

$$\dot{\varepsilon}_m = 54.7\sigma^{12.1}\exp\left(-541050/RT\right) \qquad \text{(undoped steel)} \tag{7}$$

$$\dot{\varepsilon}_m = 42.7\sigma^{11.7}\exp\left(-519192/RT\right) \qquad \text{(Sb-doped steel)} \tag{8}$$

$$\dot{\varepsilon}_m = 1152.3\sigma^{13.8}\exp\left(-620906/RT\right) \qquad \text{(Sb+Ce-doped steel)} \tag{9}$$

where $\dot{\varepsilon}_m$ is in h^{-1}, σ is in MPa, and T is the absolute temperature.

As shown in Equations (7)–(9), the obtained values of stress exponent ($n = 12.1$, $n = 11.7$ and $n = 13.8$ for the undoped, Sb-doped, and Sb+Ce-doped steels, respectively) indicate that all steels deform by the dislocation creep mechanism [18,19]. The acquired values of apparent activation energy ($Q_c = 541$ kJ/mol, $Q_c = 519$ kJ/mol and $Q_c = 621$ kJ/mol for the undoped, Sb-doped, and Sb+Ce-doped steels, respectively) are apparently higher than the activation energy for self-diffusion in iron (close to 240 kJ/mol), which also indicates a dislocation creep mechanism. In addition, the value of Q_c for the undoped steel is apparently higher than that for the Sb-doped one while the values of A and n are close to each other, indicating that at the same temperature and engineering stress the minimum creep rate of the Sb-doped steel is much higher than that of the undoped one. As for the Sb+Ce-doped steel, although the values of A and n are higher than those of the other two steels, the value of Q_c is much higher. This means that, at relatively low temperature and engineering stress where the creep process is mainly dominated by the value of Q_c, the minimum creep rate of the Sb+Ce-doped steel is much lower than that of the other two steels. The changes of stress exponent n and apparent activation energy Q_c in the three steels indicate that the minor additions of Sb and Ce significantly affect the creep process of P91 steel.

As is well known [20], the selection of a material employed at high temperatures needs to take its creep properties into account in engineering design. The creep lifetime of the material under a certain engineering stress or the creep rupture strength of 10^5 h creep lifetime is usually used as a reference.

To estimate the above creep properties of the steel we need to further consider the Monkman-Grant relation, an empirical equation that links the minimum creep rate to the rupture time, which is given by:

$$\dot{\varepsilon}_m \times t_r^m = C \tag{10}$$

where $\dot{\varepsilon}_m$ is the minimum creep rate, t_r is the rupture time, and m and C are constants. According to the data in Table A1, ln(minimum creep rate, $\dot{\varepsilon}_m$) is plotted as a function of ln(rupture, t_r) for the three steels in Figure 1c, demonstrating a sound linear relationship with R-square values of 0.997, 0.990; and 0.992, respectively. The linear fitting gives:

$$\ln\dot{\varepsilon}_m = -1.06\ln t_r - 2.70 \qquad \text{(undoped steel)} \tag{11}$$

$$\ln\dot{\varepsilon}_m = -1.06\ln t_r - 2.58 \qquad \text{(Sb-doped steel)} \tag{12}$$

$$\ln\dot{\varepsilon}_m = -1.05\ln t_r - 2.49 \qquad \text{(Sb+Ce-doped steel)} \tag{13}$$

Equation (10) can be rewritten as:

$$\ln\dot{\varepsilon}_m = -m\ln t_r + \ln C. \tag{14}$$

with Equations (11)–(14), one can obtain $m = 1.06$ and $C = 0.067$ for the undoped steel, $m = 1.06$ and $C = 0.076$ for the Sb-doped steel, and $m = 1.05$ and $C = 0.083$ for the Sb+Ce-doped steel. As a consequence, there are:

$$\dot{\varepsilon}_m \times t_r^{1.06} = 0.067 \qquad \text{(undoped steel)} \tag{15}$$

$$\dot{\varepsilon}_m \times t_r^{1.06} = 0.076 \qquad \text{(Sb-undoped steel)} \tag{16}$$

$$\dot{\varepsilon}_m \times t_r^{1.05} = 0.083 \qquad \text{(Sb+Ce-undoped steel)} \tag{17}$$

where $\dot{\varepsilon}_m$ is in h^{-1} and t_r is in h.

As seen, the values of parameter m or C are very close to each other for the three steels, indicating that impurity Sb or rare earth Ce does not influence the Monkman-Grant relation of the material. Using Equations (7)–(9) along with Equations (15)–(17), one can estimate the creep lifetime at any temperature under any stress with the assumption that both the temperature-compensated power law and the Monkman-Grant equations are valid under the temperature and stress conditions considered. Owing to the fact that the P91 steel is usually employed at a temperature below 923 K and at a stress below 100 MPa in engineering [15], we estimate the 50 MPa, 80 MPa, and 100 MPa stress creep lifetimes at temperatures between 853 K and 923 K, and the creep rupture strengths of 10^5 h creep lifetime at 853 K, 873 K, and 893 K, respectively. The estimated results are shown in Figure 2. Obviously, the 100 MPa stress creep lifetime for the Sb-doped steel is much shorter (up to fifty percent shorter) than that for the undoped steel in the range 853–923 K, while the 100 MPa stress creep lifetime for the Sb+Ce-doped steel is close to that for the undoped steel in the range 893–923 K. However, in the range 853–893 K where the P91 steel is usually employed in engineering, the 100 MPa stress creep lifetime for the Sb+Ce-doped steel is apparently longer (1.1 to 1.8 times longer) than that for the undoped steel. Similarly, under 80 MPa and 50 MPa which are closer to the service stress of the P91 steel, the creep lifetime for the Sb-doped steel is much shorter (about fifty percent shorter) than that for the undoped-steel (see Figure 2b,c), indicating that the creep properties of the Sb-doped steel worsen significantly due to the effect of impurity Sb. As for the Sb+Ce-doped steel under the above condition, the creep lifetime is much longer (about 10 times longer) than that for the undoped steel (see Figure 2b,c), implying that the creep properties of the steel are significantly improved due to the effect of rare earth Ce. The estimates of the creep rupture strength of 10^5 h creep lifetime at 853 K, 873 K, and 893 K also show that the creep rupture strength for the Sb-doped steel is about 7% lower than that for the undoped steel, but it is about 3% higher for the Sb+Ce-doped steel (see Figure 2d).

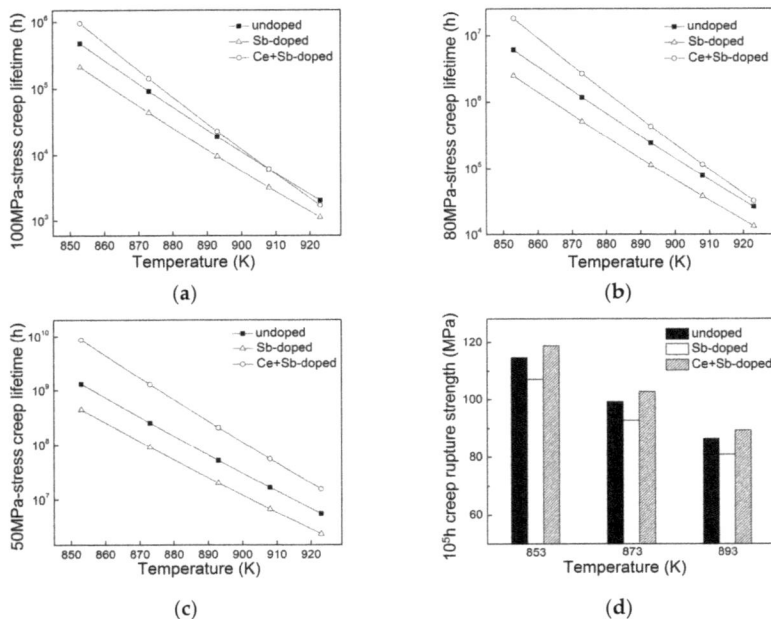

Figure 2. (**a**) 100 MPa-stess creep lifetime; (**b**) 80 MPa-stess creep lifetime; (**c**) 50 MPa-stess creep lifetime; and (**d**) 10^5 h creep rupture strength of the steels at 853 K, 873 K, and 893 K for the undoped, Sb-doped, and Sb+Ce-doped steels.

The evolution of the microstructure of P91 steel plays a crucial role in the creep processes of the material. Normally, during creep, microstructure coarsening, such as precipitate coarsening, dislocation density decrease, and subgrain disappearance, occurs, which deteriorates the creep properties of the steel. In order to explore the effect of impurity Sb and rare earth Ce on the microstructure evolution, TEM micrographs were taken from the as-tempered specimens, as well as the specimens crept at 873 K under 150 MPa for the three steels, which are shown in Figure 3. As seen, the as-tempered specimens (Figure 3a,c,e) exhibit a typical tempered martensite microstructure. The microstructure is with a high density of dislocations and precipitates shown as black dots, which are distributed along prior austenite grain boundaries or martensite lath boundaries. It is indicated by EDS microanalysis that there are two types of precipitates: coarser ones with sizes ranging from 100 nm to 200 nm, which mainly contain Cr, and finer ones with sizes less than 100 nm, which mainly contain Nb and V. Accordingly, the coarser precipitates, such as those marked by square in Figure 3, could be $M_{23}C_6$-type carbides and the finer ones, such as those marked by circle in Figure 3, could be MX-type carbides [6]. The precipitates in the undoped (Figure 3a) and Sb+Ce-doped steels (Figure 3e) are a little more than those in the Sb-doped steel (Figure 3c) and their dislocation density are somewhat higher. As shown in Figure 3b,d,f, the dislocation density is reduced apparently and the precipitates as well as the martensite laths are coarsened after the creep test. However, there are still some dislocations and subgrains remaining in the undoped and Sb+Ce-doped steels after the creep test, but there are almost no dislocations or subgrains remaining in the Sb-doped steel. This means that the degradation in microstructure is accelerated due to impurity Sb in the creep process, while rare earth Ce can stabilize the microstructure, thus significantly offsetting the detrimental effect of impurity Sb.

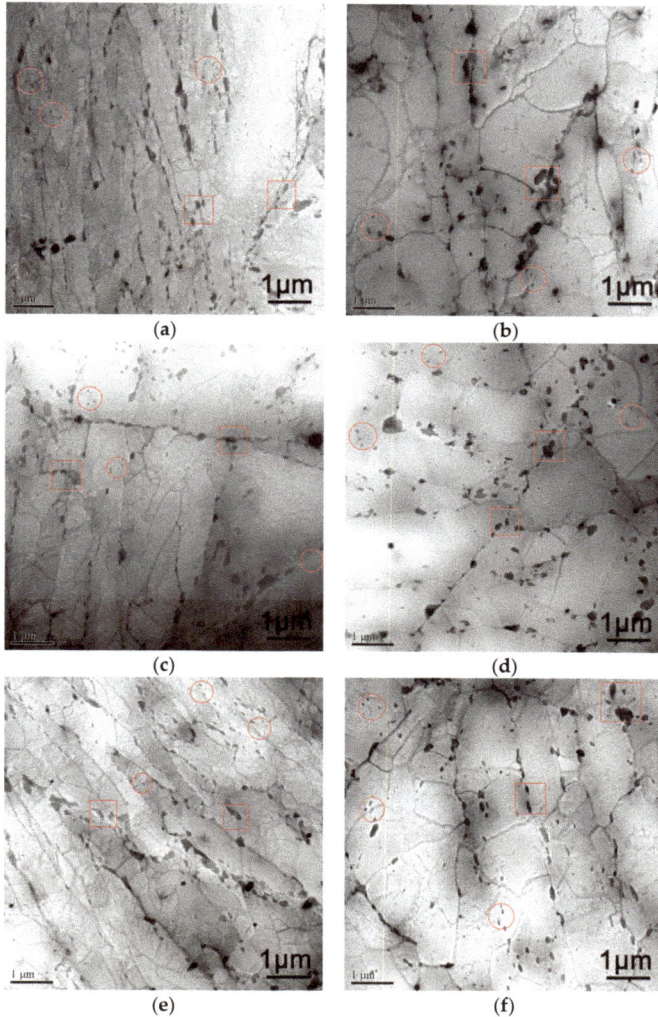

Figure 3. Transmission electron micrographs of the specimens as-tempered and creep-tested under 150 MPa at 873 K (the coarser precipitates, such as those marked by a square, are $M_{23}C_6$-type, while the finer ones, such as those marked by a circle are MX-type): (**a**) undoped steel as tempered; (**b**) undoped steel after creep test; (**c**) Sb-doped steel as tempered; (**d**) Sb-doped steel after creep test; (**e**) Sb+Ce-dopedsteel as tempered; and (**f**) Sb+Ce-doped steel after creep test.

Creep rupture is also related to the voids that nucleate, grow and cluster during creep. Study by Otto et al. [21] indicates that the small addition of impurity elements significantly deteriorates the creep properties in Cu-0.008 wt. % Bi and Cu-0.092 wt. % Sb, which is associated with the promotion of the formation of intergranular creep damage such as cavities and microcracks by Bi and Sb. Therefore, we observed the morphology of voids near the fracture tip for the three kinds of specimens tested at 873 K under 150 MPa, which are shown in Figure 4. As can be seen, the voids in the Sb-doped specimens (see Figure 4b) are much more than those in the undoped (see Figure 4a) and Sb+Ce-doped specimens (see Figure 4c) where the void morphology is similar to each other. Accordingly, Sb can promote the nucleation of voids while the nucleation of voids is restrained in the Sb+Ce-doped steel. It may be claimed that the promoted formation of voids in the Sb-doped steel may deteriorate its creep properties and thus accelerate creep rupture while the minor addition of Ce can suppress the effect of Sb, thereby improving the creep properties. In addition, as shown in Figure 4b, many voids in the Sb-doped steel have linked together and formed linear cracks parallel to the applied stress. With consideration of the deformation texture of the specimens near the fracture tip where the grains are elongated along the direction of the applied stress unless recrystallization takes place, it can be envisaged that the voids are initiated from the boundaries parallel to the applied stress. This is reasonable because during creep the vacancies migrate from the boundaries perpendicular to the applied stress to those parallel to the applied stress. It may be stated from the aforementioned results that the enhancement of void formation in the Sb-doped specimen may be related to the Sb boundary segregation, while the significant decrease of voids in the Sb+Ce-doped specimen may be related to the effect of Ce boundary segregation.

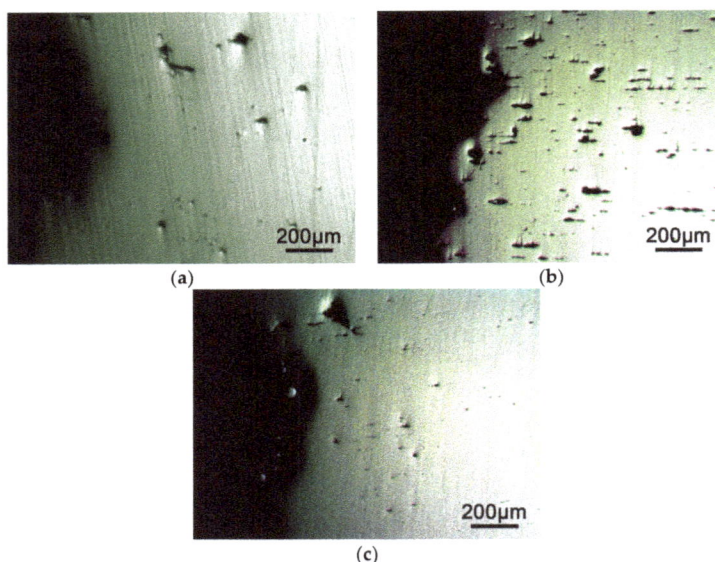

Figure 4. Morphology of voids near the fracture tip in the specimens tested under 150 MPa at 873 K: (a) undoped steel ruptured after 1010 h; (b) Sb-doped steel ruptured after 423 h; (c) Sb+Ce-doped steel ruptured after 601 h.

During creep, both impurity element tin and rare earth element cerium can segregate to the grain or subgrain boundaries [22,23]. The equilibrium segregation may occur during isothermal holding prior to creep loading and the non-equilibrium segregation may take place during creep, which is related to a vacancy-solute complex effect or boundary diffusion effect [24,25]. The segregation of impurity element can reduce considerably the grain boundary cohesion in steel, thus promoting the nucleation of cavities on the grain boundaries [10,21] and deteriorating the creep properties of the steel. Nevertheless, the rare earth element can increase the grain boundary cohesion in steel, thus improving the creep properties of the steel. To explore the segregation behavior in the Sb-doped and Sb+Ce-doped specimens, the concentrations of Sb and Ce at the boundaries were measured with the use of FEGSTEM after the specimens were tempered and creep-tested at 873 K under 150 MPa. Figure 5 represents an FEGSTEM image showing a segment of grain boundary analyzed, which is free from precipitates. Moreover, the boundary is quite parallel to the incident electron beam, which is suitable for FEGSTEM boundary microanalysis. The boundary concentrations of Sb and Ce are given in Figure 6 with its matrix concentration plotted for comparison. Obviously, both impurity Sb and rare earth Ce segregate apparently to the boundaries in the specimens as tempered or creep-tested. The boundary levels of both Sb and Ce are over 20 times higher than their matrix levels. The average boundary concentration of Sb in the Sb-doped specimen increases by approximately 40% after creep-tested (from 0.47 at. % to 0.64 at. %, see Figure 6a,c), while the average boundary concentration of Sb in the Sb+Ce-doped specimen as tempered (0.53 at. %) is almost the same as that in the Sb+Ce-doped specimen creep-tested (0.58 at. %). For clarity, the average boundary concentrations of Sb in the Sb-doped and Sb+Ce-doped specimens as tempered and creep-tested are represented in Figure 7.

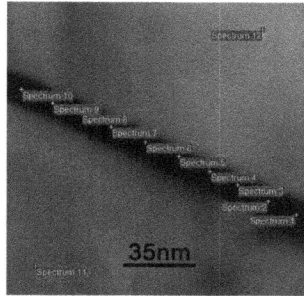

Figure 5. FEGSTEM image showing a segment of grain boundary analyzed.

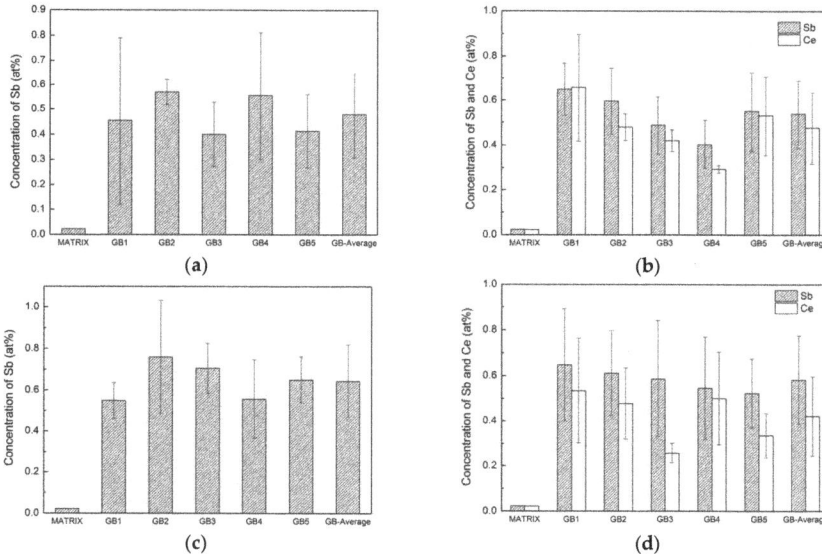

Figure 6. (**a**) Grain boundary concentration of Sb in the Sb-doped specimen as tempered; (**b**) grain boundary concentrations of Sb and Ce in the Sb+Ce-doped specimen as tempered; (**c**) grain boundary concentration of Sb in the Sb-doped specimen creep-tested under 150 MPa at 873 K (the specimen ruptured after 423 h); and (**d**) grain boundary concentrations of Sb and Ce in the Sb+Ce-doped specimen creep-tested under 150 MPa at 873 K (the specimen ruptured after 601 h), with the matrix concentration plotted for comparison (the error bar represents the standard deviation).

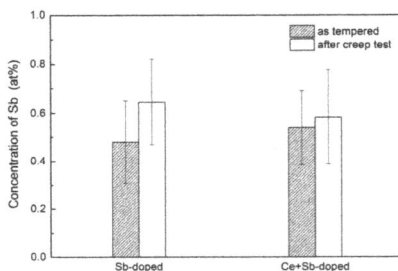

Figure 7. Grain boundary concentrations of Sb in the Sb-doped and Sb+Ce-doped specimens as tempered and creep tested (the Sb-doped specimen ruptured after 423 h and the Sb+Ce-doped specimen ruptured after 601 h. The error bar represents the standard deviation).

As discussed above, the boundary cohesion can be considerably increased by the boundary segregation of Ce, which can compensate the detrimental effect of Sb and thereby restrain the initiation of cavities and microcracks. In other words, rare earth Ce can strengthen the boundary in steel. Moreover, rare earth Ce could also segregate to dislocations, which may restrict the slip and climb of the dislocations, thus stabilizing the dislocations and leading to a higher apparent activation energy for creep (621 kJ/mol). With these mechanisms, rare earth Ce can stabilize the microstructure of the steel during creep, which may be the reason why there are more dislocations and subgrains remaining in the Sb+Ce-doped specimen than in the Sb-doped specimen after creep test, and the microstructure in the Sb+Ce-doped specimen are not apparently coarsened as compared with that in the Sb-doped specimen. In addition, as shown in Figure 7, the rare earth Ce in the Sb+Ce-doped steel can reduce the boundary segregation of Sb, thus suppressing the effect of Sb to a certain degree. Consequently, the creep properties of the Sb+Ce-doped steel are apparently better than those of the Sb-doped steel.

4. Conclusions

The creep properties of three groups of P91 ferritic heat-resistant steel specimens, undoped, doped with antimony, and doped with both antimony and cerium, are examined under the applied stresses from 150–210 MPa and at the temperatures from 873–923 K. The creep behaviors agree with the temperature-compensated power law and Monkman-Grant relation. For the temperature-compensated power law equation, the values of apparent activation energy and stress exponent for creep are about 541 kJ/mol and 12.1 for the undoped steel, 519 kJ/mol, and 11.7 for the Sb-doped steel, and 621 kJ/mol and 13.8 for the Sb+Ce-doped steel, respectively. For the Monkman–Grant equation, the values of constants m and C are around 1.06 and 0.067 for the undoped steel, 1.06 and 0.076 for the Sb-doped steel, and 1.05 and 0.083 for the Sb+Ce-doped steel, respectively. In terms of these relationships, the creep lifetimes for the three steels are predicted under several conditions. It is demonstrated that minor Sb can significantly deteriorate the creep properties of the steel, but they can, evidently, be improved by the minor addition of Ce. Rare earth Ce can both strengthen the grain boundary and stabilize the microstructure of the steel by segregating to grain boundaries and dislocations, which leads to a high apparent activation energy for creep (621 kJ/mol). This beneficial effect can offset the detrimental effect of Sb that enhances the coarsening of microstructure and deteriorates the grain boundary by segregation.

Acknowledgments: This work was supported by the National Natural Science Foundation of China (Grant No. 51071060).

Author Contributions: Yewei Xu contributed to experimental work and result analysis. Shenhua Song contributed to research design and result analysis.

Conflicts of Interest: The authors declare no conflict of interest.

Appendix A

After creep tests, the values of minimum creep rate and creep rupture time were obtained from the creep curves. These data are listed in Table A1.

Table A1. Minimum creep rate and rupture time for the undoped, Sb-doped, and Sb+Ce-doped specimens tested at various engineering stresses and temperatures.

Temp. (K)	Stress (MPa)	Creep Rate (h^{-1})			Rupture Time (h)		
		Undoped	Sb-Doped	Sb+Ce-Doped	Undoped	Sb-Doped	Sb+Ce-Doped
923	150 (test 1)	1.98×10^{-3}	5.27×10^{-3}	8.37×10^{-3}	26	10	8.4
923	150 (test 2)	2.90×10^{-3}	6.13×10^{-3}	8.59×10^{-3}	18	9.5	8.8
910	150 (test 1)	1.31×10^{-3}	2.24×10^{-3}	3.52×10^{-3}	39	30	21
910	150 (test 2)	1.15×10^{-3}	2.01×10^{-3}	3.33×10^{-3}	47	34	24
898	150 (test 1)	3.74×10^{-4}	8.13×10^{-4}	1.18×10^{-3}	131	71	56
898	150 (test 2)	4.70×10^{-4}	8.53×10^{-4}	8.77×10^{-4}	101	66	70
885	150 (test 1)	1.42×10^{-4}	2.86×10^{-4}	2.87×10^{-4}	347	175	210
873	150 (test 1)	4.20×10^{-5}	1.25×10^{-4}	8.60×10^{-5}	1010	423	601
873	165 (test 1)	8.60×10^{-5}	2.82×10^{-4}	2.97×10^{-4}	507	177	220
873	180 (test 1)	4.02×10^{-4}	8.80×10^{-4}	1.08×10^{-3}	126	72	82
873	180 (test 2)	4.64×10^{-4}	9.87×10^{-4}	9.94×10^{-4}	120	66	85
873	195 (test 1)	9.58×10^{-4}	2.51×10^{-3}	2.74×10^{-3}	58	29	25
873	195 (test 2)	7.62×10^{-4}	2.22×10^{-3}	3.21×10^{-3}	70	33	20
873	210 (test 1)	2.47×10^{-3}	6.01×10^{-3}	8.77×10^{-3}	24	11	8.2
873	210 (test 2)	1.99×10^{-3}	5.67×10^{-3}	8.96×10^{-3}	28	12	8.1

References

1. Mannan, S.L.; Chetal, S.C.; Raj, B.; Bhoje, S.B. Selection of materials for prototype fast breeder reactor. *Trans. Indian Inst. Met.* **2003**, *56*, 155–178.
2. Klueh, R.L.; Ehrlich, K.; Abe, F.J. Ferritic/Martensitic steels: Promises and problems. *Nucl. Mater.* **1992**, *191*, 116–124.
3. Klueh, R.L. Elevated temperature ferritic and martensitic steels and their application to future nuclear reactors. *Int. Mater. Rev.* **2005**, *50*, 287–310. [CrossRef]
4. Abe, F.; Nakazawa, S.; Araki, H.; Node, T. The role of microstructural instability on creep-behavior of a martensitic 9Cr-2W steel. *Mater. Trans. A* **1992**, *23*, 469–477. [CrossRef]
5. Spigarelli, S.; Cerri, E.; Bianchi, P.; Evangelista, E. Interpretation of creep behaviour of a 9Cr-Mo-Nb-V-N (T91) steel using threshold stress concept. *Mater. Sci. Technol.* **1999**, *15*, 1433–1440. [CrossRef]
6. Maruyama, K.; Sawada, K.; Koike, J. Strengthening mechanisms of creep resistant tempered martensitic steel. *ISIJ Int.* **2001**, *41*, 641–653. [CrossRef]
7. Hald, J. Microstructure and long-term creep properties of 9%–12% Cr steels. *Int. J. Press. Vessel. Pip.* **2008**, *85*, 30–37. [CrossRef]
8. Liu, G.L.; Zhang, G.Y.; Li, R.D.; Chin, J. Electronic theoretical study of the interaction between rare earth elements and impurities at grain boundaries in steel. *J. Rare earths* **2003**, *21*, 372–374.
9. Zhang, D.B.; Wu, C.J.; Yang, R. Interaction between cerium and phosphorus segregating in the grain boundaries in α-Fe studied by computer modelling and Auger electron spectroscopy. *Mater. Sci. Eng. A* **1991**, *131*, 93–97. [CrossRef]
10. Seah, M.P. Adsorption-induced interface decohesion. *Acta Metall.* **1980**, *28*, 955–962. [CrossRef]
11. Song, S.H.; Xu, Y.W.; Yang, H.F. Effect of impurity tin on the creep properties of a P91 heat-resistant steel. *Metall. Mater. Trans. A* **2014**, *45*, 4361–4370. [CrossRef]
12. Xu, Y.W.; Song, S.H.; Wang, J.W. Effect of rare earth cerium on the creep properties of modified 9Cr-1Mo heat-resistant steel. *Mater. Lett.* **2015**, *161*, 616–619. [CrossRef]
13. Brown, A.M.; Ashby, M.F. On the power-law creep equation. *Scr. Mater.* **1980**, *14*, 1297–1302. [CrossRef]
14. Monkman, F.C.; Grant, N.J. An empirical relationship between rupture life and minimum creep rate in creep-rupture tests. *Proc. Am. Soc. Test. Mater.* **1956**, *56*, 593–620.

15. Ennis, P.J.; Czyrska-Filemonowicz, A. Recent advances in creep-resistant steels for power plant applications. *Sadhana* **2003**, *28*, 709–730. [CrossRef]
16. Song, S.H.; Weng, L.Q. An FEGSTEM study of grain boundary segregation of phosphorus during quenching in a 2.25Cr-1Mo steel. *J. Mater. Sci. Technol.* **2005**, *21*, 445–450.
17. Chen, X.M.; Song, S.H.; Weng, L.Q.; Liu, S.J. Solute grain boundary segregation during high temperature plastic deformation in a Cr-Mo low alloy steel. *Mater. Sci. Eng. A* **2011**, *528*, 7663–7668. [CrossRef]
18. Kassner, M.E.; Perez-Prado, M.T. Five-power-law creep in single phase metals and alloys. *Prog. Mater. Sci.* **2000**, *45*, 1–102. [CrossRef]
19. Latha, S.; Mathew, M.D.; Parameswaran, P.; BhanuSankaraRao, K.; Mannan, S.L. Thermal creep properties of alloy D9 stainless steel and 316 stainless steel fuel clad tubes. *Int. J. Press. Vessel. Pip.* **2008**, *85*, 866–870. [CrossRef]
20. Wang, Y.; Mayer, K.H.; Scholz, A.; Berger, C.; Chilukuru, H.; Durst, K.; Blum, W. Development of new 11% Cr heat resistant ferritic steels with enhanced creep resistance for steam power plants with operating steam temperatures up to 650 °C. *Mater. Sci. Eng. A* **2009**, *510*, 180–184. [CrossRef]
21. Otto, F.; Viswanathan, G.B.; Payton, E.J.; Frenzel, J.; Eggeler, G. On the effect of grain boundary segregation on creep and creep rupture. *Acta Mater.* **2012**, *60*, 2982–2998. [CrossRef]
22. Matsuoka, H.; Osawa, K.; Ono, M.; Ohmura, M. Influence of Cu and Sn on hot ductility of steels with various C content. *ISIJ Int.* **1997**, *37*, 255–262. [CrossRef]
23. Song, S.H.; Yuan, Z.X.; Jia, J.; Guo, A.M.; Shen, D.D. The role of tin in the hot ductility deterioration of a low-carbon steel. *Metall. Meter. Trans. A* **2003**, *34*, 1611–1616. [CrossRef]
24. Song, S.H.; Wu, J.; Yuan, Z.X.; Weng, L.Q.; Xi, T.H. Non-equilibrium grain boundary segregation of phosphorus under a high applied tensile stress in a 2.25Cr1Mo steel. *Mater. Sci. Eng. A* **2008**, *486*, 675–679. [CrossRef]
25. Song, S.H.; Wu, J.; Weng, L.Q.; Liu, S.J. Phosphorus grain boundary segregation under different applied tensile stress levels in a Cr-Mo low alloy steel. *Mater. Lett.* **2010**, *64*, 849–851. [CrossRef]

metals

MDPI

Article

Dynamic Recrystallization and Hot Workability of 316LN Stainless Steel

Chaoyang Sun [1,*], Yu Xiang [1], Qingjun Zhou [2], Denis J. Politis [3], Zhihui Sun [1,*]
and Mengqi Wang [1]

1   School of Mechanical Engineering, University of Science and Technology Beijing, Beijing 100083, China;
    rane_glacier@sina.com (Y.X.); wmq19920302@163.com (M.W.)
2   Capital Aerospace Machinery Company, Beijing 100076, China; zhouqingjunxxx@163.com
3   Department of Mechanical Engineering, Imperial College London, London SW7 2AZ, UK;
    denis.politis06@imperial.ac.uk
*   Correspondence: suncy@ustb.edu.cn (C.S.); sunzhihui@ustb.edu.cn (Z.S.); Tel.: +86-10-62334197 (C.S. & Z.S.);
    Fax: +86-10-62329145 (C.S. & Z.S.)

Academic Editor: Soran Birosca
Received: 13 May 2016; Accepted: 27 June 2016; Published: 5 July 2016

Abstract: To identify the optimal deformation parameters for 316LN austenitic stainless steel, it is necessary to study the macroscopic deformation and the microstructural evolution behavior simultaneously in order to ascertain the relationship between the two. Isothermal uniaxial compression tests of 316LN were conducted over the temperature range of 950–1150 °C and for the strain rate range of 0.001–10 s^{-1} using a Gleeble-1500 thermal-mechanical simulator. The microstructural evolution during deformation processes was investigated by studying the constitutive law and dynamic recrystallization behaviors. Dynamic recrystallization volume fraction was introduced to reveal the power dissipation during the microstructural evolution. Processing maps were developed based on the effects of various temperatures, strain rates, and strains, which suggests that power dissipation efficiency increases gradually with increasing temperature and decreasing stain rate. Optimum regimes for the hot deformation of 316LN stainless steel were revealed on conventional hot processing maps and verified effectively through the examination of the microstructure. In addition, the regimes for defects of the product were also interpreted on the conventional hot processing maps. The developed power dissipation efficiency maps allow optimized processing routes to be selected, thus enabling industry producers to effectively control forming variables to enhance practical production process efficiency.

Keywords: 316LN; hot deformation; constitutive analysis; dynamic recrystallization; hot processing map

1. Introduction

Stainless steels, and in particular 316LN austenitic stainless steel, have been widely used as the piping material for nuclear power plants (NPPs) due to their high resistance to corrosion and oxidation, whilst retaining high strength and excellent ductility over a wide range of temperatures [1]. As of present, the main methods for manufacturing the pipes for the AP1000 nuclear power plants are the use of forging and extrusion technologies to take advantage of the extraordinarily high ductility of the material.

The well-developed microstructures and desired properties are usually formed by the careful control of the strain, strain rate, and temperature [2–5]. Over the past decade, numerous scholars have attempted to investigate the microstructure of alloys through experimental and theoretical modeling work. A lot of researchers have placed attention on using different measures to analyze flow stress and dynamic recrystallization. Cram [6] extended and successfully applied a physically based description for nucleation in static recrystallization to nucleation during dynamic recrystallization (DRX), and comparisons between experiment and model calculations showed good agreement over a wide range of deformation temperature, initial grain size, and applied strain rate. Beltran [7] developed and validated a model describing recrystallization in metallic materials capable of handling DRX, post-dynamic recrystallization (PDRX), and grain growth against experimental test cases for multi-pass hot deformation of 304L austenitic stainless steel, predicting microstructural evolution due to different recrystallization regimes through a modified Kocks-Mecking equation. Madej [8] compared the mean and full field dynamic recrystallization models and concluded that full field approaches additionally extend predictive capabilities of DRX models by incorporating microstructure evolution in an explicit manner. Many studies have already focused on the investigation of microstructural evolution during the hot deformation in order to obtain eminent mechanical properties for 316LN products. Pan [9] established the constitutive equation for the material, revealing that the best condition for forming through the forging process is to maintain the temperature below 1150 °C and the strain rate below 0.01 s^{-1}. However, the study neglected the effects of deformation processes on the processability of the material. Zhang [10] investigated the high temperature behavior of 316LN and simulated the microstructural evolution during deformation without an explanation of the microstructural evolution in locating the optimal regions for large plastic deformation processes. Bai [11] also constructed the constitutive equation for 316LN in a relatively narrow strain range, but did not explain the means in which microstructural evolution influences the macroscopic deformation. Zhang [12] studied the dynamic and post deformation recrystallization of 316LN stainless steel without clearly identifying the proper processing conditions for the material.

On the contrary to the research above, some researchers only studied the hot workability of the material. Processing maps, depicted by Prasad [13,14] firstly in 1984, can be employed to understand deformation mechanisms and determine the optimal deformation parameters for practical applications. Guo [15] constructed the power dissipation map of 316LN stainless steel without a clear description on its applications. He [16], Guo [17], and Liu [18] investigated the hot workability of 316LN and constructed the hot processing map, but the characteristics of different regions on the conventional processing map should be further detailed. Sun [19] also constructed the hot processing map with metallographic analysis.

In contrast to the large amount of research dedicated to separate investigations of microstructure or the macroscopic deformation of 316LN, the work conducted in this paper has attached importance to the application of the processing map and its characterization of the microstructure. In the interest of determining the optimal regimes for hot deformation of 316LN, it is important to discuss the deformation behavior of the material under different hot deformation conditions through the combined investigation on the macroscopic deformation and the microstructural evolution. Based on the experimental data, the kinetics of dynamic recrystallization for 316LN was illustrated by the use of the Avrami equation. Hot processing maps were established to locate the optimal regimes for the hot deformation of 316LN stainless steel, which were also verified effectively through the exploration of microstructure. To assist industry operators for practical production processes, the updated diagrams of the power dissipation efficiency and instability coefficient against the strain and the temperature were constructed, on which the optimized routes to obtain qualified products were presented.

2. Materials and Methods

The chemical composition of the 316LN specimen material used in the study is listed in Table 1. The present experiments were conducted on a Gleeble-1500 thermal-mechanical simulator (Dynamic Systems Inc., Poestenkill, NY, USA). To ensure the uniformity of the material, homogenizing annealing was employed before the preparation of the specimens. Carbon foils were placed on the top and the bottom of each specimen to reduce the friction between die and specimen during deformation. Cylindrical specimens (12 mm in height and 8 mm in diameter) were heated to 1200 °C at the rate of 10 °C/s and held for 300 s to obtain a homogenized microstructure, and then cooled to various testing temperatures at the rate of 10 °C/s and held for 30 s ahead of the isothermal compression at constant strain rate. The tests were then performed at temperatures between 950 °C and 1150 °C with intervals of 50 °C using uniaxial compression at the strain rates of 0.001 s^{-1}, 0.01 s^{-1}, 0.1 s^{-1}, 1 s^{-1}, and 10 s^{-1} up to the true (logarithmic) strain of 0.916 (the maximum compression ratio of all specimens was 60%), followed by water quenching. For microstructural investigations, the compressed specimens were then sectioned parallel to the deformation axis, mounted and then polished in aqua regia (HNO$_3$ 20 mL, HCl 60 mL).

Table 1. Chemical composition of 316LN austenitic stainless steel.

Component	C	Cr	Ni	Mo	Mn	P	S	Si	N	Fe
wt. %	0.017	17.03	12.71	2.53	1.29	0.020	0.001	0.34	0.12	Balance

3. Results and Discussion

3.1. Constitutive Analysis

The flow curves of experimental 316LN stainless steel at the temperature range of 950–1150 °C and strain rates of 0.001–10 s^{-1} are shown in Figure 1. As expected, the flow stress is dependent on both the deformation temperature and strain rate. Meanwhile, the flow stress increased with increase of the strain rate and decrease of the temperature. Furthermore, all the flow stress curves increased significantly at the initial stage of hot deformation, attributed to the work hardening, and then followed by a relatively steady state which indicates the occurrence of DRX [20,21].

The constitutive equation is an important mathematical model to predict and analyze the relationship among temperature, strain rate, and flow stress during hot deformation. Among all the equations, the Arrhenius-type equation proposed by Sellars [22] is the most widely used. Additionally, the well-known Zener-Hollomon parameter is also widely applied to describe the effects of temperature, strain rate, and apparent activation energy under different deformation conditions.

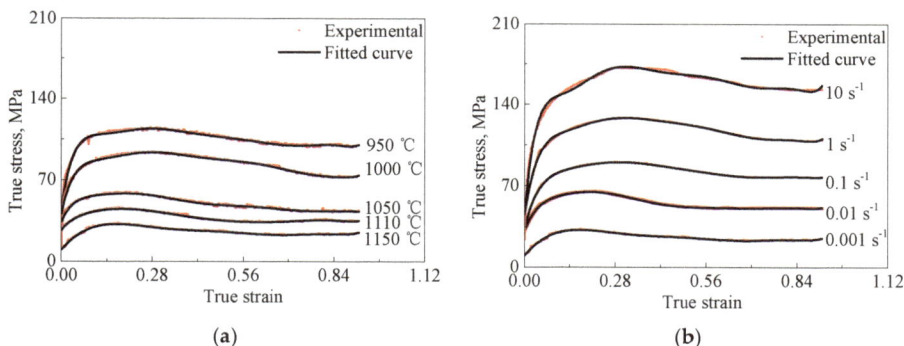

Figure 1. Flow curves for the 316LN stainless steel at: (a) $\dot{\varepsilon}$ = 0.001 s^{-1}; (b) T = 1150 °C.

The Zener-Hollomon parameter can be described as follows:

$$Z = \dot{\varepsilon}\exp\left[\frac{Q_{act}}{R(T + 273.15)}\right] = \begin{cases} A_1\sigma_p^{n_1} & \alpha\sigma < 0.8 \\ A_2\exp\left(\beta\sigma_p\right) & \alpha\sigma > 1.2 \\ A_3\left[\sinh\left(\alpha\sigma_p\right)\right]^n & \text{for all } \sigma \end{cases} \tag{1}$$

$$\ln\left(\dot{\varepsilon}\right) + \frac{Q_{act}}{R}\frac{1}{(T+273.15)} = \begin{cases} n_1\ln\left(\sigma_p\right) + \ln\left(A_1\right) & \alpha\sigma < 0.8 \\ \beta\sigma_p + \ln\left(A_2\right) & \alpha\sigma > 1.2 \\ n\ln\left[\sinh\left(\alpha\sigma_p\right)\right] + \ln\left(A_3\right) & for\ all\ \sigma \end{cases} \qquad (2)$$

where Q_{act} is determined as the activation energy of deformation which is the threshold of the dynamic recrystallization; σ_p is the first peak stress of each curve; $\dot{\varepsilon}$ is the strain rate; R is the universal gas constant (8.314 J·mol^{-1}·°C^{-1}); T is the temperature; A_1, A_2, A_3, n_1, n, β, and α ($\approx\beta/n_1$) are material constants. Equation (2) is a transformation of Equation (1). So the variables can be defined as:

$$\frac{\partial\ln\left(\dot{\varepsilon}\right)}{\partial\sigma_p} = \beta, \quad \frac{\partial\ln\left(\dot{\varepsilon}\right)}{\partial\ln\left(\sigma_p\right)} = n_1, \quad \frac{\partial\ln\left(\dot{\varepsilon}\right)}{\partial\ln\left[\sinh\left(\alpha\sigma_p\right)\right]} = n, \quad \frac{\partial\left\{\ln\left[\sinh\left(\alpha\sigma_p\right)\right]\right\}}{\partial\left(\frac{1}{(T+273.15)}\right)} = \frac{Q_{act}}{nR} \qquad (3)$$

Hence, the parameters required for describing the material behavior are plotted in Figure 2 and the Zener-Hollomon parameter of the 316LN stainless steel can be depicted as:

$$Z = \dot{\varepsilon}\exp\left[\frac{442,089.02}{R(T+273.15)}\right] \qquad (4)$$

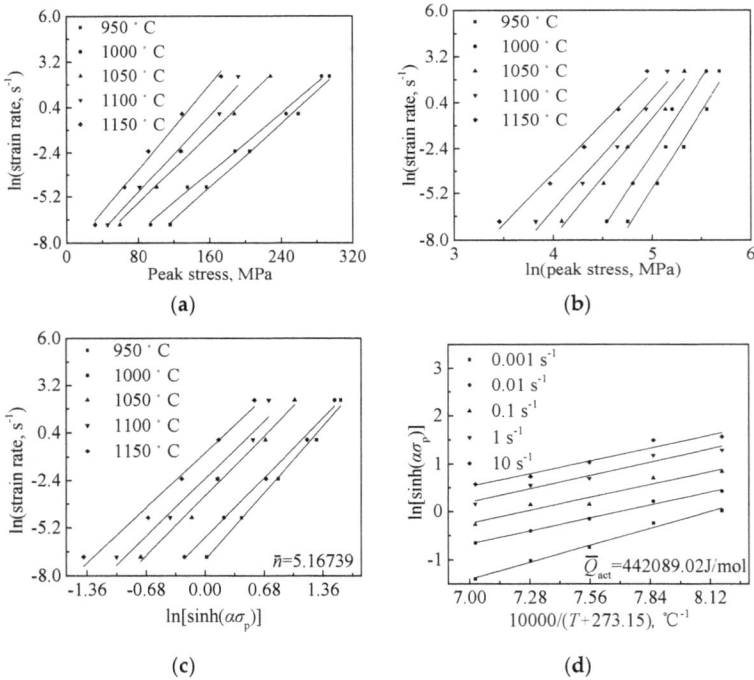

Figure 2. Relationships between: (a) peak stress and ln(strain rate); (b) ln(peak stress)and ln(strain rate); (c) ln[sinh($\alpha\sigma_p$)] and ln(strain rate); (d) 10,000/(T + 273.15) and ln[sinh($\alpha\sigma_p$)].

Therefore, the constitutive equation for hot deformation behavior of the present 316LN stainless steel is given by:

$$\dot{\varepsilon} = 9.612\times10^{15}[\sinh(0.00771\sigma_p)]^{5.167}\exp[-\frac{442,089.02}{R(T+273.15)}] \qquad (5)$$

$$\sigma_p = \frac{1}{0.00771} \ln \left\{ \left(\frac{Z}{9.612 \times 10^{15}} \right)^{1/5.167} + \left[\left(\frac{Z}{9.612 \times 10^{15}} \right)^{2/5.167} + 1 \right]^{1/2} \right\} \tag{6}$$

3.2. Microstructure Evolution

Dynamic recovery (DRV) is the softening mechanism during single-stage hot deformation in which the grains are elongated yet the metal seems to exhibit higher ductility, lower stress and lower product strength than in cold working. At other times, equiaxed grains are observed at the end of the process, and if formed during deformation, that would be dynamic recrystallization [20]. Dynamic recrystallization, which is one of the main softening mechanisms at high temperatures, takes place after a critical strain ε_c has been attained. It is the core characteristic of the softening mechanism of the materials with low and medium stacking fault energy (SFE). When the softening process is governed by dynamic recrystallization, the flow stress passes through a peak σ_p and drops to a steady state regime, as shown in Figure 3. Generally, the critical strain ε_c symbolizing the start of DRX can be obtained either by a direct microstructure observation or through an analysis of the flow stress curve. However, microstructure observation is a more complicated and time-consuming method compared to the flow stress curve analysis, as it requires a large number of samples for examination. The flow stress curve analysis method, firstly proposed by Kocks and Mecking [23], and then further developed by McQueen and Ryan [24], can be used to emphasize the point where DRX occurs on the flow stress curve.

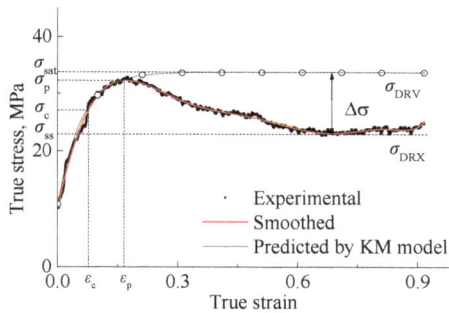

Figure 3. Schematic description of the flow behavior of 316LN at high temperature $T = 1150\ °C/\dot{\varepsilon} = 0.001\ \text{s}^{-1}$ (σ_{sat} is the saturation stress; σ_{ss} is the steady state stress of the experimental data; σ_c is the critical stress; σ_p is the peak stress; ε_c is the critical strain; ε_p is the peak strain; $\Delta\sigma = \sigma_{sat} - \sigma_{ss}$).

When the dislocation density ρ of the material reaches the dynamic balance, the flow stress tends towards the saturation stress σ_{sat}. During the entire process of deformation, work hardening occurs as a result of the refinement of the microstructure caused by the increased strain, and thus the flow stress can be described by:

$$\sigma\left(\dot{\varepsilon}, T\right) = \alpha_1 G b \sqrt{\rho\left(\dot{\varepsilon}, T\right)} \tag{7}$$

where σ is the flow stress, ε is the true strain, α_1 is dislocation interaction constant relating to dislocation spacing and stress component, G is the shear modulus, and b is the Burgers vector. Based on the Kocks-Mecking model, the apparent hardening depends on the dislocation density which is the result of interplay between storage and annihilation of dislocations as follows:

$$\frac{d\rho}{d\varepsilon} = k_1 \sqrt{\rho} - k_2 \rho \tag{8}$$

where k_1 and k_2 are constants related to microscopic parameters. Consequently, the Kocks-Mecking model assumes that hardening is caused by the increase of the average value of dislocation density, which is directly proportional to the square root of the dislocation density. Therefore, when combined with Equation (7), this can be written as:

$$\sigma = \alpha_1 G b \sqrt{\rho} = \alpha_1 G b \left(\frac{k_1}{k_2} - C_1 e^{-\frac{k_2 \varepsilon}{2}} \right) = \alpha_1 G b \left(\frac{k_1}{k_2} + \left(\sqrt{\rho_0} - \frac{k_1}{k_2} \right) e^{-\frac{k_2 \varepsilon}{2}} \right) \tag{9}$$

and:

$$\alpha_1 G b = \sigma_{\text{sat}} / (k_1 / k_2) \tag{10}$$

$$\sqrt{\rho_0} = \frac{\sigma_0}{\sigma_{\text{sat}}} \left(\frac{k_1}{k_2} \right) \tag{11}$$

where σ_0 is the initial real stress, C_1 is the integration constant and ρ_0 is the initial dislocation density. The equation can also be expressed as:

$$\sigma = \sigma_{\text{sat}} + (\sigma_0 - \sigma_{\text{sat}}) e^{-\frac{k_2 \varepsilon}{2}} \tag{12}$$

The research of Poliak and Jonas [25] states that the second derivative of the strain hardening rate $\theta = d\sigma/d\varepsilon$ against flow stress σ vanishes when DRX begins:

$$\kappa_\sigma = \frac{\partial}{\partial \sigma} \left(-\frac{\partial \theta}{\partial \sigma} \right) \bigg|_{\sigma = \sigma_c} = 0 \tag{13}$$

$$\kappa_\varepsilon = \frac{\partial}{\partial \sigma} \left(-\frac{\theta'}{\theta} \right) = \frac{(\theta')^2 - \theta'' \theta}{\theta^3} \tag{14}$$

where σ_c is the stress at critical strain ε_c. Figure 4a shows the relationship between ε_c and ε_p, and it is similar to the result studied by Ji [26]. Figure 4b shows the method to identify the saturation stress σ_{sat} and the steady flow stress σ_{ss} due to the dynamic recrystallization: the intersection point where a tangent line at the critical strain point cuts the x-axis ($\theta = 0$) is the σ_{sat}; the intersection point of the lower value where the σ-θ curve cuts the x-axis ($\theta = 0$) is the σ_{ss} [27,28].

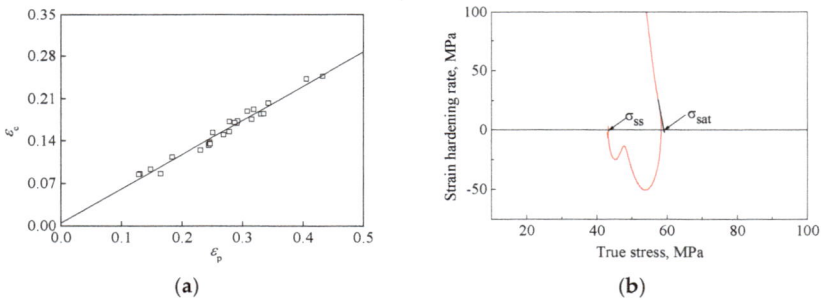

Figure 4. (a) Relationship between ε_c and ε_p; (b) the true stress-strain hardening rate curve at $T = 1050\,^\circ\text{C} / \dot{\varepsilon} = 0.001\,\text{s}^{-1}$.

In flow stress models that include DRX, the flow stress is typically modeled as a piecewise function consisting of a strain hardening model, which is valid up to the critical point, and a rule of mixture, which is active when recrystallized and unrecrystallized grains coexist. It is worth mentioning that, for the computation of the flow stress, the Taylor factor has to be considered, but its influence is neglected here for the sake of compact notation [20,25]. If only the first recrystallization cycle is taken into account, the flow stress can be written as follows:

$$\sigma(\varepsilon) = \begin{cases} h(\varepsilon) & \varepsilon < \varepsilon_c \\ h(\varepsilon)(1 - X(\varepsilon)) + r(\varepsilon) X(\varepsilon) & \varepsilon \geqslant \varepsilon_c \end{cases} \tag{15}$$

where $h(\varepsilon)$ describes the evolution of flow stress due to strain hardening as well as softening by DRV in the absence of DRX, $r(\varepsilon)$ describes the flow stress of recrystallized grains as a function of strain (in the first recrystallization cycle only), and $X(\varepsilon)$ is the dynamically recrystallized volume fraction. For simplicity, it is assumed that both temperature and strain rate are constant, so that the flow stress is a function of strain only. Thus Equation (14) can also be expressed as:

$$\kappa_\varepsilon = \begin{cases} \dfrac{h''(\varepsilon)^2 - h'''(\varepsilon) h'(\varepsilon)}{h'(\varepsilon)^3} & \varepsilon < \varepsilon_c \\ \dfrac{(h''(\varepsilon) - \Delta\sigma X''(\varepsilon))^2 - (h'''(\varepsilon) - \Delta\sigma X'''(\varepsilon))(h'(\varepsilon) - \Delta\sigma X'(\varepsilon))}{(h'(\varepsilon) - \Delta\sigma X'(\varepsilon))^3} & \varepsilon \geqslant \varepsilon_c \end{cases} \tag{16}$$

where $\Delta\sigma = \sigma_{sat} - \sigma_{ss}$. Therefore, the conditions in Equation (16) are only fulfilled if $X(\varepsilon_c) = X'(\varepsilon_c) = X''(\varepsilon_c) = X'''(\varepsilon_c) = 0$, so as to guarantee that κ_ε is continuous which is defined as:

$$\lim_{\varepsilon \to \varepsilon_c^-} \kappa_\varepsilon = \lim_{\varepsilon \to \varepsilon_c^+} \kappa_\varepsilon, \ \lim_{\varepsilon \to \varepsilon_c^-} \sigma = \lim_{\varepsilon \to \varepsilon_c^+} \sigma \tag{17}$$

Many models for DRX are the modified forms of the original JMAK equation $X(\varepsilon) = 1 - \exp(-B't^q)$, due to Kolmogorov, Johnson, and Mehl and Avrami [29] who studied the kinetics of phase transformations. For modified JMAK kinetics to analyze the dynamic recrystallization, the recrystallized volume fraction is given as a function of strain via:

$$X(\varepsilon) = 1 - \exp\left(-\alpha_2 \left(\frac{\varepsilon - \varepsilon_c}{\varepsilon_p}\right)^q\right) \tag{18}$$

$$\ln(-\ln(1 - X(\varepsilon))) = \ln(\alpha_2) + q\ln\left(\frac{\varepsilon - \varepsilon_c}{\varepsilon_p}\right) \tag{19}$$

where ε_p is the first peak strain, q is the constant corresponding to nucleation mode, and α_2 is the term associated with the nucleation and growth rates. These parameters can be used to represent the flow behavior of the material under consideration, although some rate equations must also be provided, as will be shown later in the paper. The reader must bear in mind that the latter approach neglects any possible hardening taking place concurrently with the deformation during the dynamic recrystallization regime. This approximation is valid when the deformation process is mainly governed by softening due to dynamic recrystallization, as is generally accepted once the peak stress has been attained. The experimental values of the Avrami exponents reported later also support the validity of this approach. Obviously, at the critical strain, $X(\varepsilon_c) = 0$.

The recrystallization fraction can also be described as the softening fraction of the deformation, which is given by:

$$X(\varepsilon) = \frac{\sigma_{DRV} - \sigma_{DRX}}{\sigma_{sat} - \sigma_{ss}} \tag{20}$$

where σ_{DRX} is the real stress derived from isothermal compression experimental data, σ_{DRV} is the predicted stress without the recrystallization when $\varepsilon > \varepsilon_c$. By plotting $\ln(-\ln(1 - X(\varepsilon)))$ against $\ln((\varepsilon - \varepsilon_c)/\varepsilon_p)$, the material constant α_2 and q can be calculated by the slope and the intercept of the linear regression line, as shown in Figure 5. The recrystallization fraction of the 316LN can be given by:

$$X(\varepsilon) = 1 - \exp\left(-0.4899 \left(\frac{\varepsilon - \varepsilon_c}{\varepsilon_p}\right)^{1.170}\right) \tag{21}$$

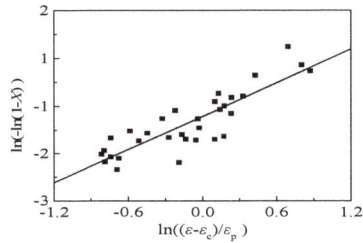

Figure 5. Dependence of $\ln(-\ln(1-X(\varepsilon)))$ with $\ln(\varepsilon-\varepsilon_c)/\varepsilon_p$.

Figure 6 displays some observed microstructures (taken by using DM4000M) at different strains during the isothermal compression experiments at 1150 °C and strain rate of 0.01 s^{-1}. Figure 6a displays the initial microstructure of the material, in which the average size of the initial grains is measured to be about 75 μm. Figure 6b displays microstructure after deformation to a strain of 0.1, and because of the orientation difference between adjacent grains, smaller grains emerge at the grain boundaries. The mixed partially recrystallized structure observed in Figure 6c was attained after being compressed to a strain of 0.4, and the occurrence of dynamic recrystallization (finer equiaxed grains) during deformation can be clearly recognized, showing that the average size of the newly formed grains is measured to be about 45 μm. But there are still some unexpected grains (grain diameters above 45 μm at spot A). Figure 6d displays recrystallized grains obtained after being compressed to a strain of 0.9, in which grains are 75% (average size is measured to be about 20 μm) smaller than that of the initial microstructure state, corresponding to steady state of stresses.

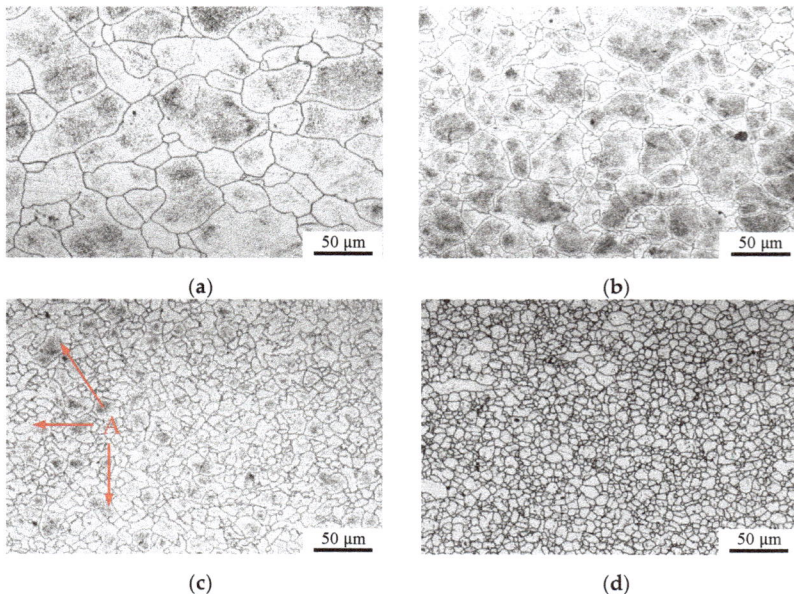

Figure 6. Microstructures of 316LN stainless steel at T = 1150 °C and $\dot{\varepsilon}$ = 0.01 s^{-1}: (**a**) initial microstructure; (**b**) ε = 0.1; (**c**) ε = 0.4; (**d**) ε = 0.9.

3.3. Hot Deformation Behavior of 316LN Stainless Steel

Processing maps are constructed based on the dynamic materials model (DMM) in which the materials are considered as a dissipater of power [30,31]. The dynamic power dissipation at a certain deformation condition, i.e., at a constant temperature and strain rate, is partitioned by strain rate sensitivity (SRS) *m*. It is clear that the majority of dissipated power is transformed into heat resulting in a temperature rise, and only a small fraction of energy causes microstructure evolution. The total power dissipation (*P*) consisting of two complementary functions *G* and *J* is given by:

$$P = \dot{\varepsilon} = G + J = \int_0^{\dot{\varepsilon}} \sigma d\dot{\varepsilon} + \int_0^{\sigma} \dot{\varepsilon} d\sigma \tag{22}$$

The quantity *G* which is given under the true stress-true strain rate curve, is designated as dissipater power content and its complementary component *J* as the dissipater power co-content. *J* in the model is assumed to be related to the microstructural changes occurring along with the deformation, as opposed to *G* which is related to continuum effects. As a rule of thumb, most of the dissipation is attributed to the temperature rise (*G* content), while a small amount is attributed to microstructural changes (*J* content). According to the dynamic constitutive equation for a certain condition of strain, temperature, and initial microstructure, the energy partitioning between *G* and *J* is determined by the strain rate sensitivity parameter (*m*) as follows:

$$m = \frac{\partial \ln \sigma}{\partial \ln \dot{\varepsilon}} = \frac{\dot{\varepsilon} \partial \sigma_{DRX}}{\sigma_{DRX} \partial \dot{\varepsilon}} = \frac{\dot{\varepsilon}}{\sigma_{DRX}} \frac{\partial \sigma_{DRX}}{\partial X_{DRX}} \frac{\partial X_{DRX}}{\partial \dot{\varepsilon}} = \frac{\dot{\varepsilon}}{\sigma_{DRX}} (\sigma_{ss} - \sigma_{sat}) \frac{dX_{DRX}}{\dot{\varepsilon}} \tag{23}$$

In Equation (23), dynamic recrystallization volume fraction is introduced to reveal the power dissipation during the microstructural evolution which is indicated by the strain rate sensitivity value. The SRS value distribution map at strain of 0.4 is constructed depending on various temperatures and strain rates in Figure 7, revealing that the higher SRS values correspond to the regions in which the temperatures are higher than a threshold and the strain rates are relatively lower, which also shows that the dynamic recrystallization volume fraction is relatively higher in these regions. Due to the large strain and the lower strain rate during the deformation, the dislocations in the alloy have enough time to slip and climb, inducing sufficient annihilation and rearrangement of the dislocations, for the recrystallized grains to undergo nucleation growth. On the other hand, as a result of the large amount of heat generated in higher strain rate deformation processes, the microstructure is not uniform and internal flow is disordered [3,5,13,32,33].

The variation of the dimensionless parameter η, called the efficiency of power dissipation, constitutes a processing map in terms of strain ε, strain rate $\dot{\varepsilon}$ and temperature *T*, given by:

$$\eta = \frac{J}{J_{max}} = \frac{2m}{m+1} \tag{24}$$

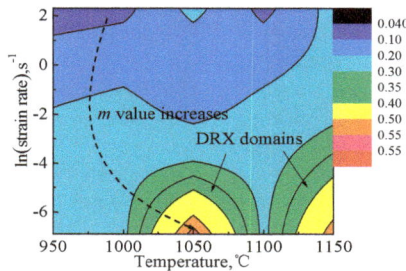

Figure 7. SRS values at the strain of 0.4 for 316LN stainless steel.

Furthermore, the principles of irreversible thermodynamics as applied to continuum mechanics of large plastic flow are explored to define a criterion for the onset of flow instability given by the equation for the instability parameter ξ which was first proposed by Ziegler [34] based on the DMM theory:

$$\xi\left(\dot{\varepsilon}\right) = \left. \frac{\partial \ln\left[m/\left(m+1\right)\right]}{\partial \ln\left(\dot{\varepsilon}\right)} + m \right|_{T} \leqslant 0 \tag{25}$$

If Q_{act} in the Equation (1) is replaced by the deformation energy Q and the σ_p is replaced by the real stress σ, the strain rate sensitivity m can be described as:

$$m = \frac{\partial \ln\left(\sigma\right)}{\partial \ln\left(\dot{\varepsilon}\right)} = \begin{cases} 1/n_1 & \alpha\sigma < 0.8 \\ \left[\ln\left(\dot{\varepsilon}\right) + \frac{Q}{R(T+273.15)} - \ln\left(A_2\right)\right]^{-1} & \alpha\sigma > 1.2 \end{cases} \tag{26}$$

It is obvious that Q and A_2 are the functions of strain ε, and it follows that m, η and ξ are also the functions against strain ε. Figure 8 demonstrates the hot processing map of 316LN stainless steel at $\varepsilon = 0.4$ and $\varepsilon = 0.9$. The number against each contour represents the power dissipation efficiency in the processing map, and the shaded area represents the flow instability (unstable) regime.

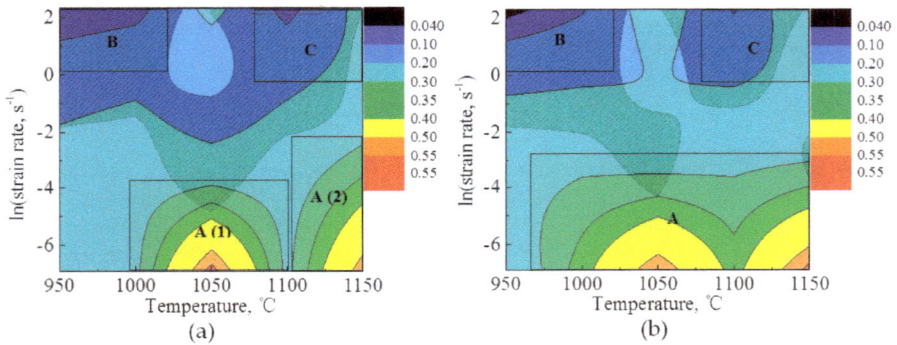

Figure 8. Hot processing map of 316LN stainless steel at: (**a**) $\varepsilon = 0.4$; (**b**) $\varepsilon = 0.9$ (The legend scale represents the power dissipation efficiency).

The map at a strain of 0.4 exhibits four domains in the following temperature ranges and strain rate ranges: (A(1)) 1010–1085 °C and 0.001–0.01 s^{-1}, with a peak efficiency of about 57% occurring at 1050 °C/0.001 s^{-1}; (A(2)) 1110–1150 °C and 0.001–0.03 s^{-1}, with a peak efficiency of about 53% occurring at 1150 °C/0.001 s^{-1}; (B) 950–1010 °C and 3.25–10 s^{-1}, with a peak efficiency of about 10%; as well as (C) 1100–1150 °C and 1–10 s^{-1}, with a peak efficiency of about 31%.

The microstructures at $\varepsilon = 0.4$ obtained on a specimen deformed at 1150 °C/0.01 s^{-1} (Figure 6c) corresponds to the domain A(2) in Figure 8a, representing dynamically recrystallized microstructure with typical wavy grain boundaries and showing fine grains (average grain diameter is about 45 μm), while there are still some unexpected grains (grain diameters above 45 μm at spot A) due to the insufficient deformation. Domains A(1) and A(2) in Figure 8a transform into one domain in Figure 8b owing to the abundant dynamic recrystallization in the temperature range of 990–1150 °C and the strain rate range of 0.001–0.01 s^{-1} with the peak efficiency of 0.57 occurring at 1150 °C/0.001 s^{-1} (domain A in Figure 8b). The domain B in Figure 8a is the regime assumed to be inappropriate for hot deformation because its power dissipation efficiency is very low and its instability parameter $\xi < 0$. Along with the increase of strain, the power dissipation efficiency becomes smaller (shown as the domain B in Figure 8b), which means it may cause tearing, surface cracking, inter-crystalline cracks, or inhomogeneous microstructure. The microstructure of the specimen deformed at the temperature range of 1100–1150 °C and strain rate range of 1–10 s^{-1} (shown as the domain C in Figure 8a,b) corresponds to the flow instability regime exhibiting flow localization. In the localized regions, the material has undergone static recrystallization during cooling resulting in fine grains along the localized bands [3,5].

The conventional hot processing map sheds light on the relationships between the processability of the material and the strain rate. But in the practical hot deformation of the material, the strain rate does not change significantly during the process, especially in tube extrusion and bar extrusion processes. Therefore, it is necessary to construct an updated hot processing map according to Equation (26) which reveals the relationship of the variation of the power dissipation efficiency with the increase of the strain (shown in Figure 9). In Figure 9a, there are some shaded areas which are inappropriate for processing, so it is practical to avoid these regions to control the microstructure evolution through reining in the temperature during the process. In Figure 9b, the instability regime becomes even bigger, and the eligible region for obtaining the products with fine grain microstructures becomes smaller.

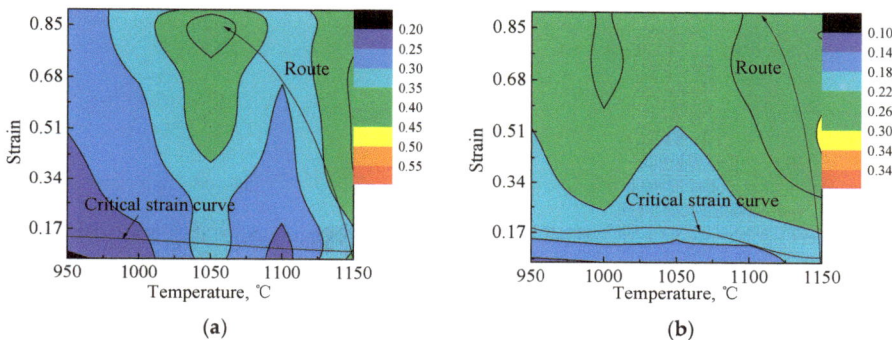

Figure 9. Updated hot processing map of 316LN stainless steel at (**a**) $\dot{\varepsilon} = 0.01$ s^{-1}; and (**b**) $\dot{\varepsilon} = 0.1$ s^{-1} (The legend scale represents the power dissipation efficiency. Critical strain curve is believed to be the start of the dynamic recrystallization).

4. Conclusions

The hot deformation behavior and dynamic recrystallization of 316LN stainless steel were studied by conducting isothermal compression tests over a wide range of temperatures and strain rates. The apparent activation energy was quantified, and a constitutive equation for hot deformation behavior was elaborated in the paper. The DRX of 316LN stainless steel was investigated and the parameters of JMAK equation were quantified. The following conclusions are drawn:

(1) Processing maps were produced that exhibit two domains for $\varepsilon = 0.4$ within the temperature and strain rate ranges: (a) 1010–1085 °C/0.001–0.01 s^{-1} and (b) 1110–1150 °C/0.001–0.03 s^{-1}, in which dynamic recrystallization is apt to occur. The dynamic recrystallization domain at $\varepsilon = 0.9$ in the temperature range of 990–1150 °C and the strain rate range of 0.001–0.01 s^{-1} with the peak efficiency of 0.57 occurring at 1150 °C/0.001 s^{-1} is illustrated on the processing map.

(2) The updated processing maps have demonstrated that careful process design has to be followed to enable successful deformation process and microstructural control, and the instability regions can be avoided along with the decrease of the temperature during deformation.

Acknowledgments: The work presented in this paper has been supported by National Science and Technology Major Projects "High-end CNC Machine Tools and Basic Manufacturing Equipment" (No. 2014ZX04014-51), National Natural Science Foundation of China (No. 51105029, 51575039) and NSAF (No. U1330121).

Author Contributions: Chaoyang Sun and Yu Xiang designed and conducted the experiments; Chaoyang Sun, Mengqi Wang, Zhihui Sun, Yu Xiang, and Qingjun Zhou analyzed the data; Chaoyang Sun and Yu Xiang contributed to writing and editing of the manuscript; Denis J. Politis polished the English.

Conflicts of Interest: The authors declare no conflict of interest.

References

1. Wang, S.; Yang, B.; Zhang, M.; Wu, H.; Peng, J.; Gao, Y. Numerical simulation and experimental verification of microstructure evolution in large forged pipe used for AP1000 nuclear power plants. *Ann. Nucl. Energ.* **2016**, *87*, 176–185. [CrossRef]
2. Suresh, K.; Rao, K.P.; Prasad, Y.V.R.K.; Hort, N.; Kainer, K.U. Study of hot forging behavior of as-cast Mg-3Al-1Zn-2Ca alloy towards optimization of its hot workability. *Mater. Des.* **2014**, *57*, 697–704. [CrossRef]
3. Prasad, Y.V.R.K.; Rao, K.P. Processing maps and rate controlling mechanisms of hot deformation of electrolytic tough pitch copper in the temperature range 300–950 °C. *Mater. Sci. Eng. A* **2005**, *391*, 141–150. [CrossRef]
4. Rao, K.P.; Prasad, Y.V.R.K.; Suresh, K. Materials modeling and simulation of isothermal forging of rolled AZ31B magnesium alloy: Anisotropy of flow. *Mater. Des.* **2011**, *32*, 2545–2553. [CrossRef]
5. Prasad, Y.V.R.K.; Rao, K.P. Materials modeling and finite element simulation of isothermal forging of electrolytic copper. *Mater. Des.* **2011**, *32*, 1851–1858. [CrossRef]
6. Cram, D.G.; Zurob, H.S.; Brechet, Y.J.M.; Hutchinson, C.R. Modelling discontinuous dynamic recrystallization using a physically based model for nucleation. *Acta Metall.* **2009**, *57*, 5218–5228. [CrossRef]
7. Beltran, O.; Huang, K.; Logé, R.E. A mean field model of dynamic and post-dynamic recrystallization predicting kinetics, grain size and flow stress. *Comp. Mater. Sci.* **2015**, *102*, 293–303. [CrossRef]
8. Madej, L.; Sitko, M.; Pietrzyk, M. Perceptive comparison of mean and full field dynamic recrystallization models. *Arch. Civ. Mech. Eng.* **2016**, *16*, 569–589. [CrossRef]
9. Pan, P.L.; Zhong, Y.X. Research on deformation property of 316LN nuclear main pipe steel at elevated temperature. *China Mech. Eng.* **2012**, *11*, 1354–1359.
10. Zhang, P.P.; Sui, D.S. Modeling of flow stress and dynamic recrystallization for 316LN steel during hot deformation. *J. Taiyuan Univ. Sci. Technol.* **2014**, *1*, 44–51.
11. Bai, Y.Q.; Chen, M.M. Hot deformation and dynamic recrystallization on behaviors of 316LN. *J. Taiyuan Univ. Sci. Technol.* **2009**, *5*, 424–427.
12. Zhang, R.H.; Wang, Z.H.; Shi, Z.P.; Wang, B.; Fu, W.T. Dynamic and post deformation recrystallization of nuclear-grade 316LN stainless steel. *Strength. Mater.* **2015**, *47*, 94–99. [CrossRef]
13. Prasad, Y.V.R.K.; Rao, K.P.; Gupta, M. Hot workability and deformation mechanisms in Mg/nano-Al$_2$O$_3$ composite. *Compos. Sci. Technol.* **2009**, *69*, 1070–1076. [CrossRef]
14. Prasad, Y.V.R.K.; Rao, K.P. Effect of homogenization on the hot deformation behavior of cast AZ31 magnesium alloy. *Mater. Des.* **2009**, *30*, 3723–3730. [CrossRef]
15. Guo, B.; Ji, H.; Liu, X.; Gao, L.; Dong, R.; Jin, M. Research on Flow Stress During Hot Deformation Process and Processing Map for 316LN Austenitic Stainless Steel. *J. Mater. Eng. Perform.* **2011**, *21*, 1455–1461. [CrossRef]
16. He, A.; Yang, X.; Xie, G.; Wang, X. Processing map and character of hot working of 316LN pipe during hot working process. *J. Iron Steel Res.* **2015**, *27*, 34–37.
17. Guo, M.W.; Wang, Z.H.; Zhou, Z.A.; Sun, S.H.; Fu, W.T. Effect of Nitrogen Content on Hot Deformation Behavior and Grain Growth in Nuclear Grade 316LN Stainless Steel. *Adv. Mater. Sci. Eng.* **2015**. [CrossRef]
18. Liu, X.G.; Ji, H.P.; Guo, H.; Jin, M.; Guo, B.F.; Gao, L. Study on hot deformation behavior of 316LN austenitic stainless steel based on hot processing map. *Mater. Sci. Technol.* **2013**, *29*, 24–29. [CrossRef]
19. Sun, C.Y.; Li, Y.M.; Xiang, Y.; Yang, J. Hot deformation behavior and hot processing maps of 316LN stainless steel. *Rare Met. Mater. Eng.* **2016**, *45*, 688–695.
20. Sakai, T.; Belyakov, A.; Kaibyshev, R.; Miura, H.; Jonas, J.J. Dynamic and post-dynamic recrystallization under hot, cold and severe plastic deformation conditions. *Prog. Mater. Sci.* **2014**, *60*, 130–207. [CrossRef]
21. Yang, L.C.; Pan, Y.T.; Chen, I.G.; Lin, D.Y. Constitutive relationship modeling and characterization of flow behavior under hot working for Fe-Cr-Ni-W-Cu-Co super-austenitic stainless steel. *Metals* **2015**, *5*, 1717–1731. [CrossRef]
22. Sellars, C.M.; McTegart, W.J. On the mechanism of hot deformation. *Acta Metall.* **1966**, *14*, 1136–1138. [CrossRef]
23. Mecking, H.; Kocks, U.F. Kinetics of flow and strain-hardening. *Acta Metall.* **1981**, *29*, 1865–1875. [CrossRef]
24. Ryan, N.D.; McQueen, H.J. Flow stress, dynamic restoration, strain hardening and ductility in hot working of 316 steel. *J. Mater. Process. Technol.* **1990**, *21*, 177–199. [CrossRef]

25. Poliak, E.I.; Jonas, J.J. A one-parameter approach to determining the critical conditions for the initiation of dynamic recrystallization. *Acta Metall.* **1996**, *44*, 127–136. [CrossRef]

26. Ji, H.P. Microstructure Prediction of 316LN Stainless Steel for Dynamic Recrystallization based on Cellular Automata Method. Ph.D. Thesis, Yanshan University, Qinhuangdao, China, December 2013.

27. Liu, X.; Zhang, L.; Qi, R.; Chen, L.; Jin, M.; Guo, B. Prediction of Critical conditions for dynamic recrystallization in 316LN austenitic steel. *J. Iron. Steel. Res. Int.* **2016**, *23*, 238–243. [CrossRef]

28. Prasad, Y.V.R.K.; Rao, K.P. *Hot Working Guide: A Compendium of Processing Maps*, 2nd ed.; ASM International: Cleveland, OH, USA, 2015; p. 261.

29. Avrami, M. Kinetics of phase change. I general theory. *J. Chem. Phys.* **1939**, *7*, 1103. [CrossRef]

30. Venugopal, S.; Mannan, S.L.; Prasad, Y.V.R.K. Optimization of cold and warm workability in stainless steel type AISI 316L using instability maps. *J. Nucl. Mater.* **1995**, *227*, 1–10. [CrossRef]

31. Venugopal, S.; Mannan, S.L.; Prasad, Y.V.R.K. Processing map for mechanical working of stainless steel type AISI 316 L. *Scr. Metall. Mater.* **1993**, *28*, 715–720. [CrossRef]

32. Venugopal, S.; Sivaprasad, P.V.; Prasad, Y.V.R.K. Validation of processing maps for 304L stainless steel using hot forging, rolling and extrusion. *J. Mater. Process. Technol.* **1995**, *59*, 343–350. [CrossRef]

33. Rao, K.P.; Prasad, Y.V.R.K.; Dzwonczyk, J.; Hort, N.; Kainer, K.U. Hot Deformation mechanisms in AZ31 magnesium alloy extruded at different temperatures: Impact of texture. *Metals* **2012**, *2*, 292–312. [CrossRef]

34. Ziegler, H. Some extremum principles in irreversible thermodynamics, with application to continuum mechanics. *Swiss Fed. Inst. Technol.* **1962**.

metals   **MDPI**

Article

Experimental Verification of Statistically Optimized Parameters for Low-Pressure Cold Spray Coating of Titanium

Damilola Isaac Adebiyi [1,2,*], Abimbola Patricia Popoola [1,*] and Ionel Botef [3]

[1] Department of Chemical, Metallurgical and Materials Engineering, Tshwane University of Technology, P.M.B. X680, Pretoria 0001, South Africa

[2] Department of Metallurgical Engineering, Vaal University of Technology, Vanderbijlpark 1911, South Africa

[3] School of Mechanical, Industrial and Aeronautical Engineering, University of the Witwatersrand, Johannesburg 2000, South Africa; ionel.botef@wits.ac.za

* Correspondence: adebiyidi@vut.ac.za (D.I.A.); PopoolaAPI@tut.ac.za (A.P.P.);
Tel.: +27-(0)16-950-7629 (D.I.A.); +27-(0)12-382-3513 (A.P.P.)

Academic Editor: Soran Birosca
Received: 19 April 2016; Accepted: 24 May 2016; Published: 2 June 2016

Abstract: The cold spray coating process involves many process parameters which make the process very complex, and highly dependent and sensitive to small changes in these parameters. This results in a small operational window of the parameters. Consequently, mathematical optimization of the process parameters is key, not only to achieving deposition but also improving the coating quality. This study focuses on the mathematical identification and experimental justification of the optimum process parameters for cold spray coating of titanium alloy with silicon carbide (SiC). The continuity, momentum and the energy equations governing the flow through the low-pressure cold spray nozzle were solved by introducing a constitutive equation to close the system. This was used to calculate the critical velocity for the deposition of SiC. In order to determine the input temperature that yields the calculated velocity, the distribution of velocity, temperature, and pressure in the cold spray nozzle were analyzed, and the exit values were predicted using the meshing tool of Solidworks. Coatings fabricated using the optimized parameters and some non-optimized parameters are compared. The coating of the CFD-optimized parameters yielded lower porosity and higher hardness.

Keywords: process parameters; cold spray; optimization; critical velocity; SiC; microstructure; properties

---

1. Introduction

Ti-6Al-4V is an alloy of titanium that is characterized by excellent bulk mechanical and chemical properties such as very good strength-to-weight ratio (due to low density), high specific strength, low elastic modulus, superior resistance to both corrosion and erosion in many environments, excellent high temperature resistance, and biocompatibility. Hence, the alloy is a favorite material for many applications in the automobile, aerospace, and aeronautical industries. It would have found more versatile applications in these industries were it not for its poor surface properties such as hardness and wear resistance which are due to the high friction coefficient, low hardenability, and the tendency to gall and smear [1]. These poor surface properties are traced to the low resistance to plastic shearing, low work hardening, and the low protection offered by its surface oxide [2]. The aforementioned setback of titanium alloy has necessitated various research works, including coating with powder, to enhance the surface properties and performance without altering the bulk composition/chemistry thereby retaining the excellent bulk properties [3].

Silicon carbide (SiC), due to its extreme hardness and high wear resistance [4], is one of such powders for coating titanium alloy in order to confer higher hardness and wear resistance on it.

However, SiC decomposes before melting [5] making high temperature coating processes unsuitable for its deposition. A promising option to high temperature deposition of SiC that will prevent its thermal decomposition is the cold spray process. Cold spray is a material deposition process in which relatively small particles (ranging from approximately 5 to 100 µm in diameter) in solid state are accelerated to a critical high velocity (typically 300–1400 m/s), and are subsequently plastically deformed to form a coating on the substrate [6,7]. In cold spray coating, attachment of powder to the substrate otherwise known as bonding is achieved by the kinetic energy of the powder particles rather than the thermal energy as the case in thermal spray processes [8]. Therefore, the fabrication of the coating of temperature sensitive and nano-structured material is possible by cold spray because no significant change in the microstructure of feedstock is involved in the process [9]. Bonding takes place when the velocity of the powder particles exceeds a certain value called the critical velocity (CV). Hence, the CV is defined as the velocity the particle must attain before deposition can take place after impacting the substrate [10]. The CV is a major parameter in cold spray; it is related to the mechanical and thermal properties of the spray materials and their particle sizes, and it determines which of particle deposition or substrate erosion will occur upon the impact of spray particles [11,12].

Typically, the CV is the velocity at which the transition from erosion of the substrate to deposition of the particle takes place [11]. Below the critical velocity, plastic deformation is too low to cause bonding, above the critical velocity, hydrodynamic penetration leads to strong erosion. Therefore, the optimum conditions for deposition lie between these two characteristic velocities [13]. According to Assadi [14], the value of CV is determined by the temperature, thermo-mechanical properties of the sprayed material [15] and the characteristics of the substrate [13,16,17]. Li *et al.* [18] measured the critical velocity for three metal alloys at different oxidizing conditions. The results showed that besides materials properties, the critical velocity was significantly influenced by the oxidation condition of the particles. The authors concluded that the material's properties influence the critical velocity more remarkably at low oxygen content than at high oxygen content. Yokoyama *et al.* [19] carried out the analysis of a metal particle impacting onto a metal substrate by using a dynamic finite element code (ABAQUS), and numerically studied the effects of substrate and particle temperature on the critical velocity. It was found that critical velocity decreases with higher stiffness of the substrate, higher particle temperature, and greater particle size. A decrease in gas temperature and pressure also affects coatings density and structure [20]. Li *et al.* [21] carried out both experimental and theoretical estimates of the critical velocity of copper (Cu) particles during cold spray coating. The authors concluded that the measured critical velocity was independent of the size and velocity distribution of the particle but depends on its temperature. Hence, in order to predict quality of the coating in cold spray, a good understanding of the particle velocity is vital.

Cold spray coating is carried out in a de Lava nozzle of which two variants are currently commercially available, *viz.*: the Low Pressure Cold Spray (LPCS) system and the High Pressure Cold Spray (HPCS) system. Inside the nozzle, the flow of high-velocity gas carrying very fine solid particles leads to a two-phase flow of gas-solid suspension. Such flow is usually extremely complex and difficult to understand due to the complexities arising mainly from particle-mean stream interactions, particle-flow turbulence interactions, inter-particle collisions, particle-wall interactions, and particle-shock wave interactions [22]. Dmitrienko and Uvarova [23] reported that mathematical modeling of heat and mass transfer in systems of dispersed particles, nano-particles, and nano-fluids, as in the cold spray nozzle, is one of the actual problems at present.

According to Vutova and Donchev [24], computational modeling and optimization are important for better process studying and understanding, and for quality improvement. Although few literatures are available on mathematical study and optimization of cold spray process parameters [18,19,21], as far as the authors of this work are aware, Computational Fluid Dynamics (CFD) analysis of input temperature to determine the critical velocity, which is the object of this investigation, has not been reported in the literature. The investigation consists of the following parts: (1) statistical computation of the critical velocity for cold spray coating of titanium alloy with SiC. This was done by using a constitutive equation to solve the continuity, momentum, and energy equations governing the flow of fluid inside the de Lava nozzle used in the LPCS system; (2) Computational fluid dynamics (CFD) analysis of the flow through the nozzle to determine the input temperature that will yield the computed velocity at the exit of the nozzle; and (3) verification of the mathematical model by comparing the properties of coatings deposited using the CFD-optimized parameters and some non-optimized parameters.

2. Mathematical Modeling

The low Pressure Cold Spray (LPCS) system is shown in Figure 1. The basic principle of the cold spray involves a high-velocity flow of the particles made possible by the high pressure and high velocity of the carrier gas. The high pressure jet is preheated to compensate for the adiabatic cooling due to expansion. The powder particles are transported by the energy of the preheated, high-pressure, high-velocity supersonic gas jet.

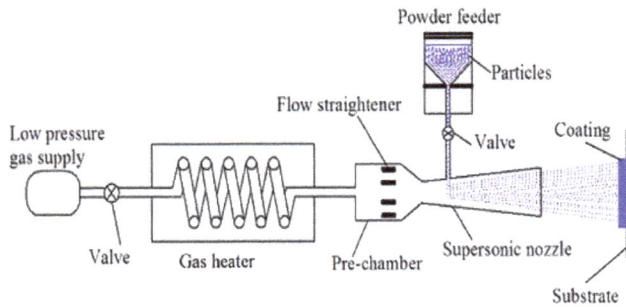

Figure 1. Schematic diagram of the low-pressure cold spray process [25].

2.1. Modeling Assumptions

In formulating this model, it is assumed that:

(1) Ideal gas law is obeyed by the carrier gas.
(2) The gas flow is one-dimensional, frictionless, and adiabatic.
(3) Steady-state conditions exist.
(4) Gas expansion is uniform; no shocks or discontinuities.
(5) Particles effect on gas conditions is negligible.
(6) Inter-particle collision is negligible.
(7) Particles effect on space charge is negligible.

2.2. Model Equations

According to Papyrin [26], Lee *et al.* [27] and Janzhong *et al.* [28], the flow through the nozzle in the LPGDS process is governed by the continuity, momentum and the energy equations. A constitutive equation is however required in order to close the system.

2.2.1. The Continuity Equation

A continuity equation is an equation that describes the transport of a conserved quantity. According to the continuity equation (otherwise known as the law of conservation of mass), the rate at which mass enters a system is equal to the rate at which mass leaves the system in any steady state process, *i.e.*, the total time rate of change of mass in a fixed region is zero. Therefore, the mass, energy and momentum of the powder in the gas stream are conserved. The continuity equation for the LPCS can be written as:

$$\frac{\partial \rho}{\partial t} + \frac{\partial (\rho v_i)}{\partial x_i} = 0 \tag{1}$$

where ρ is the density of the gas and v is the velocity.

2.2.2. The Momentum Equation

According to Chuayjan *et al.* [29], one of the governing equations describing the motion of a particle in the LPCS system is the principle of linear momentum. The principle states that the total momentum of a system of colliding objects remains constant provided no resultant external force acts on the system. In other words, when no external forces are acting, the time rate of change of the momentum is equal to the net force acting on the particle. Taking internal stress and the gravitational acceleration into account, the application of principle of conservation of translational momentum for the LPCS can be written as:

$$\frac{\partial \rho v_i}{\partial t} + \frac{\partial (\rho v_j v_i)}{\partial x_j} = \frac{\partial \tau_{ij}}{\partial x_j} - \frac{\partial p}{\partial x_i} + \rho g \tag{2}$$

τ is the internal stress and g is the gravitational acceleration.

2.2.3. The Energy Equation

The kinetic energy of the particles on impact is important for plastic deformation of the particles and formation of splats, which bond together to produce coatings. The energy equation is given in Equation (3):

$$\frac{\partial (\rho E)}{\partial t} + \frac{\partial (\rho v_j E)}{\partial x_j} = \frac{\partial}{\partial x_j} \left(k \frac{\partial T}{\partial x_j} \right) + \frac{\partial}{\partial x_j} (\tau_{ij} v_i) \tag{3}$$

2.2.4. The Constitutive Equation

Constitutive equations relate thermo-mechanical parameters, *i.e.*, strain (ε), strain rate ($\dot{\varepsilon}$), temperature (T), and flow stress (σ). Although the conserved quantitative (mass, momentum, and energy) are the basic quantities describing the flow through the LPCS system, in order to close the system, a constitutive equation is required for stress and flow/flux, otherwise stress and flow must be added to the list of variable. Equations (1)–(3) are solved in conjunction with an appropriate equation of state and the constitutive equation. The stress equation for Newtonian fluid is given by Equation (4):

$$\tau_{ij} = \mu \left[\left(\frac{\partial v_i}{\partial x_j} + \frac{\partial v_j}{\partial x_i} \right) - \frac{2 \partial v_k}{3 \partial x_k} \right] \tag{4}$$

where $i, j = (x, y, z)$.

The total stress tensor σ_{ij} in the fluid is given by the sum of internal stresses due to the fluid pressure p and the stress due to viscous forces as shown in Equation (5):

$$\sigma_{ij} = -p\delta_{ij} + \tau_{ij} \tag{5}$$

where δ_{ij} is the Kronecker delta, defined such that $\delta_{ij} = 1$ if $i = j$, otherwise $\delta_{ij} = 0$.

2.2.5. The Discrete Equations

The discrete approximation to the momentum, energy, and continuity equations can be written in a form shown in Equations (6) and (7):

$$Mu + A(u)u + GP = Ku + f(t) \tag{6}$$

$$Du = g(t) \tag{7}$$

M—Mass matrix (for equidistant discretization of the unit matrix), *A*—advection matrix, *G*—gradient matrix, *K*—diffusion matrix, *D*—divergence matrix, $f(t)$ and $g(t)$ represents the effects of the boundary conditions on velocity.

Using the notations and equalities in Equation (8), and If *G* is denoted with *C*, Equation (9) is obtained:

$$b \equiv Ku + f(t) - A(u)u, D \equiv G^T \tag{8}$$

$$\begin{bmatrix} M & C \\ C^T & 0 \end{bmatrix} \begin{bmatrix} P \\ u \end{bmatrix} = \begin{bmatrix} b \\ g \end{bmatrix} \tag{9}$$

Consequently, the discrete system for the constitutive equation is obtained as shown in Equation (10) whereas that of the momentum, energy, and continuity equations are represented as shown in Equation (11) [30]:

$$\begin{bmatrix} N & 0 & 0 \\ 0 & N & 0 \\ 0 & 0 & N \end{bmatrix} \begin{bmatrix} \tau_{11} \\ \tau_{22} \\ \tau_{12} \end{bmatrix} - \begin{bmatrix} 2L_1 & 0 & 0 \\ 0 & 2L_2 & 0 \\ 0 & 0 & 0 \end{bmatrix} \begin{bmatrix} V_i \\ V_j \\ P \end{bmatrix} + \begin{bmatrix} D_1 & 0 & 0 \\ 0 & D_2 & 0 \\ 0 & 0 & 0 \end{bmatrix} \begin{bmatrix} V_i \\ V_j \\ P \end{bmatrix} = \begin{bmatrix} 0 \\ 0 \\ 0 \end{bmatrix} \tag{10}$$

$$\begin{bmatrix} c_i u_i & 0 & 0 \\ 0 & c_i u_i & 0 \\ 0 & 0 & 0 \end{bmatrix} \begin{bmatrix} V_i \\ V_j \\ P \end{bmatrix} + \begin{bmatrix} 2K_{11} + 2K_{22} & K_{21} & Q_1 \\ K_{22} & 2K_{11} + 2K_{22} & Q_2 \\ Q_1^T & Q_2^T & 0 \end{bmatrix} \begin{bmatrix} V_i \\ V_j \\ P \end{bmatrix} = \begin{bmatrix} F_1 \\ F_2 \\ 0 \end{bmatrix} \tag{11}$$

Equation (11) is solved by using Continuum mechanics.

The finite element equation for the flow process can be written by defining the coefficient matrix as given in Equations (12)–(19) [30].

$$\begin{bmatrix} C_i U_i(V_i) \\ C_i V_i(V_j) \\ 0 \end{bmatrix} + \begin{bmatrix} (2K_{11} + K_{22})V_i - K_{21}(V_j) + Q_1 P \\ -K_{22}(V_i) + K_{11} + 2K_{22}(V_j) - Q_2 P \\ Q_1^T(V_i) - Q_2^T(V_j) + 0 \end{bmatrix} = \begin{bmatrix} F_1 \\ F_2 \\ 0 \end{bmatrix} \tag{12}$$

$$C_i U_i(V_i) - C_i U_i(V_j) + \begin{bmatrix} (2K_{ij} + K_{ij})V_i - K_{ij}(V_j) + Q_i P \\ -K_{ij}(V_i) + K_{ij} + 2K_{ij}(V_j) - Q_i P \\ Q_i^T(V_i) - Q_i^T(V_j) + 0 \end{bmatrix} = \begin{bmatrix} F_1 \\ F_2 \\ 0 \end{bmatrix} \tag{13}$$

$$C_i U_i(V_i - V_j) + \begin{bmatrix} 3K_{ij}V_i - K_{ij}(V_j) + Q_i P \\ -K_{ij}(V_i) + K_{ij} + 2K_{ij}(V_j) - Q_i P \\ Q_i^T(V_i) - Q_i^T(V_j) + 0 \end{bmatrix} = \begin{bmatrix} F_1 \\ F_2 \\ 0 \end{bmatrix} \tag{14}$$

Taken $(V_i - V_j) = U$, then Equation (15) is obtained:

$$C_i U_i + \begin{bmatrix} + \\ - \\ + \end{bmatrix} \begin{bmatrix} 2K_{ij}(U) + Q_i P \\ K_{ij}(1 + U) - Q_i P \\ Q_i^T(U) \end{bmatrix} = \begin{bmatrix} F_1 \\ F_2 \\ 0 \end{bmatrix} \tag{15}$$

$$C_i U_i + (2K_{ij}U - K_{ij}U + Q_i^T(U)) = F_i \tag{16}$$

$$C_i U_i + (K_{ij} U + Q_i^T (U)) = F_i \qquad (17)$$

$$C_i U_i + (K_{ij} U + Q_i^T (U)) = F_i \qquad (18)$$

$$F_i = CU + K_{ij} U + Q_i^T U \qquad (19)$$

Thus, the finite element Equations (10) and (11) become Equations (20) and (21):

$$C_{(u)} + KU = F \qquad (20)$$

$$N_\tau - LU + DU = 0 \qquad (21)$$

Equation (20) is the typical form of the Newtonian equation whereas Equation (21) represents the extra stress finite element analogue of the constitutive equation described in Section 2.2.4. These equations are solved by substituting the boundary conditions in tab:metals-06-00135-t001. The exit velocity was calculated from Equation (20).

Table 1. Thermo mechanical properties of the carrier gas (air) and the silicon carbide (SiC) powder.

Carrier Gas (Air)		SiC Powder	
Density	1.205 kg/m^3	Density	3160 kg/m^3
Specific heat capacity, Cp	1.005 J/kg· K	Heat capacity, C	675 J· kg^{-1}· K^{-1}
Thermal conductivity, h	0.0257 W/m· K	Thermal conductivity	490 W· m^{-1}· K^{-1}
Kinematic viscosity, v	15.11 ×10^{-6} m^2/s	Elastic modulus, E	570 MPa
Expansion coefficient	3.43 × 10^{-3} 1/K	Poisson ratio, μ	0.17
Prandtl's number, P_r	0.713		

2.3. Thermo-Physical Properties and Boundary Conditions

The boundary conditions are determined by the properties of the carrier gas and the feedstock powder. The carrier gas is air and the powder used in the calculation is SiC ceramic powder. The properties of the air and SiC are given in   tab:metals-06-00135-t001. The feedstock powder is at room temperature at inlet and its mass flow rate at room temperature is 10 g/min. At the outlet, the pressure of the nozzle, $u = v = w = 0$.

According to Lupoi [31], the supersonic nozzle is a critical component for the cold spray coating process. Lee *et al.* [32] reported that nozzle optimization is crucial to minimizing shock loss. Figure 2 shows the geometry of the nozzle whose parameters are in shown in tab:metals-06-00135-t002.

Table 2. Geometrical parameters of the nozzle.

Section of the Nozzle	(mm)
Throat diameter	4
Nozzle entrance diameter	8
Nozzle exit diameter	6
Length of converging section	50
Length of diverging section	125

Successful bonding of feedstock powder particle to the substrate demands that the particle must exceed a critical velocity, which is a function of the thermo-mechanical properties of the powder and the substrate [7–13]. Basically, the velocity of the particle is always less than that of the carrier gas. Therefore, the velocity of the carrier gas must be substantially above the critical velocity for deposition of high quality coating [18]. Li *et al.* [33] reported that particle temperature affects critical velocity.

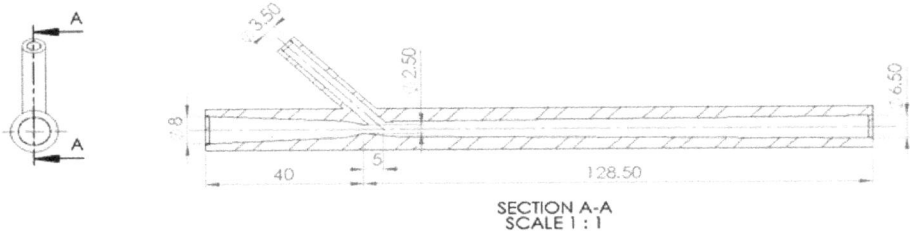

Figure 2. Geometry of the nozzle.

2.4. Experimental Details

The feedstock powder used for this investigation consists of 95 wt. % SiC of 53 μm particle size which was mechanically blended with 5 wt. % Al (Centerline SST-Al5001, CenterLine Limited, Windsor, ON, Canada) in a planetary ball mill. The addition of Al is necessary to act as a binder because SiC is a ceramic powder with limited plastic deformability. In order to simplify the mathematical model, the thermo-mechanical properties of Al are ignored. The weight percentage of Al is negligible as compared to SiC. Ti-6Al-4V with nominal composition: 6.10 wt. % Al, 4.01 wt. % V, 0.15 wt. % Fe, 0.007 wt. % C, 0.12 wt. % O, 0.005 wt. % N, Ti, balance and sectioned to $35 \times 35 \times 5$ mm^3 was used as the substrate. The substrates were grit blasted with 100–300 μm alumina grit (Centerline SST-G0002, CenterLine Limited, Windsor, ON, Canada) prior to coating deposition. This is necessary to facilitate adhesion of the feedstock powder to the substrate. The cold spray coating experiment was performed with the Centreline SST Series P low pressure cold spray machine. The carrier gas was air generated by a 10 bar-capacity CompAir external air compressor. Centreline SST Series P is equipped with an automated gun moving device which was set at a transverse velocity of 3 ms^{-1}. Thus, it takes approximately 18 s to travel the 35 mm long substrate. Five passes were made on the substrate to form an approximate total thickness of 3 mm. A somewhat parabolic coating shape was observed. This suggests that the density of particle at the center of the nozzle is greater than that at the periphery. The powder was delivered axially into the pre-chamber of the nozzle in the supersonic gas stream by an integrated, dual-hopper, non-pressurized vibratory powder feeder at a feed rate set at 30%. Coating deposition was performed at the system's maximum allowable operational pressure of 0.99 MPa. This is because much higher pressure is usually required for the deposition of carbides during the cold spray coating [34]. There was no direct measurement of velocity at the exit of the nozzle but several values of temperature were simulated using the Solidworks Flow Simulation CFD package to determine their corresponding velocities at the exit of the nozzle.

To verify the accuracy of the set of simulated parameters, coatings were deposited using three temperatures and their properties were compared. Temperatures were selected based on the simulation results and preliminary investigations [35]. The temperatures are: (1) the CFD-optimized temperature (*i.e.*, the temperature whose output velocity agrees with the calculated velocity), T_{opt} = 773 K; (2) T_1 = 723 K (50 K below T_{opt}); and (3) T_2 = 823 K (50 K above T_{opt}).

After the coating process, the samples were sectioned, mounted, and polished semi-automatically to mirror finish following standard metallographic procedure [36]. X-ray diffraction analysis was performed on the coatings using the Philips P1710 Panalytical diffractometer (PANalytical, The Analytical X-ray Company, Almelo, The Netherlands) with Cu target Kα radiation to identify the constituent phases. The samples were scanned at intervals of 2θ and a step size of 0.02. The phases present were identified using X' Pert High score plus software (Version 2.2.2, PANalytical B.V., Almelo, The Netherlands). The structural data of the identified phases were taken from the ICSD database. The hardness measurements were performed on the prepared cross-section of the coating with a Vickers hardness tester (Future Tech FM-800, Version 1.15, Future Tech. Corp., Tokyo, Japan) according to ASTM E384 standard [37] to ensure consistent result. A load of 100 g was allowed to dwell for 15 s.

A total of ten indentations were made on each coating sample and the average is reported as the surface hardness of the sample. In order to avoid strain hardening effects and possible cracking of the reinforced matrix, indentations were spaced at a distance of at least four times the diagonal of the previous indent as per ASTM C1327 standard [38].

ImageJ analysis software (Version 1.48k, 2013, National Institute of Mental Health, Bethesda, MD, USA) was used to estimate the percentage porosity of coatings by calculating the area fraction of the pores. In order to cater for porosity at varied scale (micrometric, submicron, and nanometer pores) in the coatings, two different magnifications (1000× and 10,000×) of the SEM image were used. The work of Konečná *et al.* [39] agrees with this practice. The use of two different magnifications ensures that features like column gaps and big cracks will be captured by the lower magnification whereas features like sub-micrometric and nanometeric pores will be captured by the higher magnification [40]. Five micrographs were taken for each magnification. This is because a large number of micrograph at higher magnification may lead to a more reliable assessment of porosity. Although this method of image analysis may be cumbersome and time consuming, Ganvir *et al.* [40] reported it will provide an estimate of all types of pores such as connected, non-connected, vertical and branching cracks in the coating which may not be otherwise possible.

3. Results and Discussion

3.1. Exit Velocity

Figure 3 shows the CFD analysis of gas flow in the nozzle for the three temperatures investigated. The distribution and exit values of velocity, temperature, and pressure are also shown in the Figure. The value of the velocity obtained from the mathematical calculation is in the range of that obtained from the CFD when the temperature is 500 °C (773 K). This is about 826 m/s as shown in Figure 3B. At 450 °C (723 K), the exit velocity is about 787 m/s (Figure 3A), whereas at 550 °C (823 K), the exit velocity is about 866 m/s (Figure 3C). According to the simulation result, Increase in input gas temperature led to increase in exit velocity. This is probably because, as shown in Equation (22), at a given mass flow rate, the velocity of the air increases as temperature increases [41]. This also confirms the work of Huang and Fukanuma [37], who reported that that higher gas temperature benefits impact velocity:

$$V_{rms} = \sqrt{\frac{3RT}{M}} \tag{22}$$

where, V_{rms} is the velocity, R is the universal gas constant, T is temperature, and M is mass.

Thus, a certain velocity range for efficient bonding and good coating properties called window of deposition has been defined. The window of deposition usually has both the minimum and maximum values. The minimum value is the critical velocity whereas the maximum value is called the erosion velocity. If the velocities of the particles are less than the critical velocity, there will be insufficient plastic deformation and no bonding. If the velocities of the particles are higher than critical velocity, there will be strong erosion as a result of hydrodynamic penetration. The optimum velocity for efficient deposition and good quality bonding lies between these two velocities [18]. The optimum velocity obtained from the statistical calculation is in good agreement with the value obtained from the CFD at temperature of 500 °C (773 K).

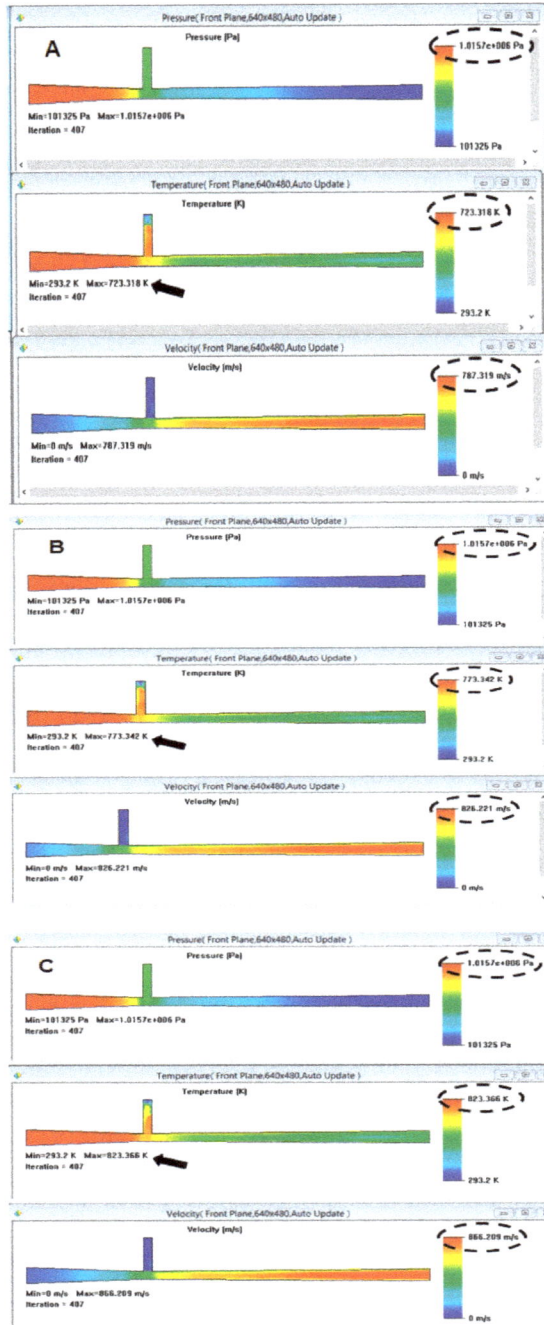

Figure 3. Computational fluid dynamics (CFD) predictions of the exit velocity for the three temperatures: (**A**) at $T_1 = 723$ K, (**B**) at $T_{opt} = 773$ K, and (**C**) at $T_2 = 823$ K.

3.2. Microstructure and Porosity of the Coatings

The SEMs of the cross-section of the coatings are shown in Figure 4 whereas Figure 5 shows the XRD in relation to the feedstock powder.

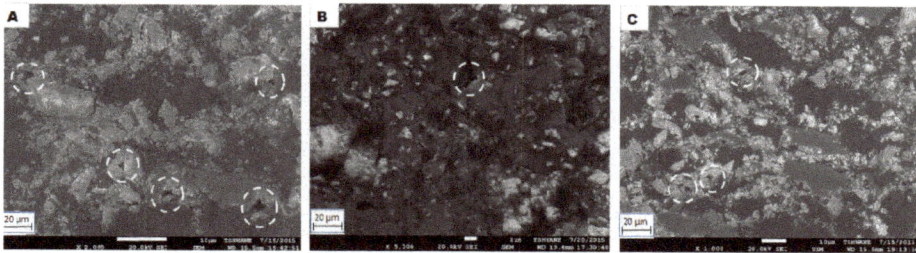

Figure 4. Scanning electron micrograph of the coatings: (**A**) 450 °C; (**B**) 500 °C; (**C**) 550 °C.

The micrographs of all the coatings are similar as they show partially homogenous distribution of SiC, and are generally characterized by fully dense structures and uneven surfaces. Although, no inter-connected pores are observed at any of the temperature investigated, a difference in porosity level is observed. The coating deposited using the CFD-optimized parameters yielded less porosity of 4.32% as compared to 5.05% and 4.97% respectively obtained at T_1 and T_2 (lower and higher than T_{opt}). The lower porosity obtained from the optimized velocity is probably due to the hammering effect of the SiC particles on the coating which subsequently leads to a more dense structure. Higher velocity during cold spray coating usually leads to lower porosity.

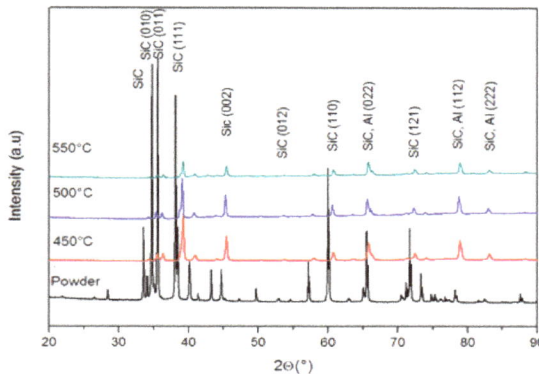

Figure 5. XRD of the coatings in comparison with the feedstock powder.

Huang and Fukanuma [42] observed that higher gas temperature benefits impact velocity and particle temperature, which leads to increase in the velocity of in-flight particles. Khalid [43] reported that impact energy is increased when the working gas temperature is increased. This increase in both the velocity of the in-flight particles and impact energy are expected to produce a decrease in volume fraction of porosity in the coatings. Thus the coating deposited at T_2 is expected to yield less porosity than that deposited at T_{opt}. Contrarily, it is observed that at temperature above the optimized value, there is increase in percentage porosity of the coating. This is probably due to fracturing of the particles of SiC powder upon impact at higher velocity which will lead to increase in porosity. SiC is a ceramic powder; it is rigid and brittle, and higher impact velocity has the tendency to fracture it [44]. Moreover, higher deposition temperature (and velocity) could have caused sufficient heat in the jetting region which could have aided thermal softening of the few Al particles in the feedstock powder. The increase in porosity at higher deposition temperature (and velocity) could also be because, at T_2, Al particles would have probably experienced thermal softening. Thermal softening causes reduced consolidation and increase porosity [42].

As shown in the XRD in Figure 5, there are no differences observed in the diffraction pattern of the coatings at the three temperatures investigated. This shows that the investigated temperature range has no effect on the phases formed in the coatings. Moreover, the diffraction patterns of the feed stock powder are identical to those of the coatings, although there is a noticeable peak shift between them. This peak shift could be traced to the high impact of the deposition process which would have caused micro-straining, amorphization and grain refinement of the powders particles.

3.3. Hardness of the Coatings

Although there is a general improvement in the surface hardness of the coatings, the sample deposited using the CFD-optimized parameter has the highest average hardness of 652 ± 12.7 $HV_{0.3}$ as compared to the 599 ± 14.8 $HV_{0.3}$, and 634 ± 13.5 $HV_{0.3}$ respectively obtained at T_1 and T_2. The general increase in hardness could be traced to the presence of hard SiC (Vickers hardness of 2350 [45]) particles in the feedstock powder, strain hardening effect of the plastically deformed Al particles during cold spray deposition, and accumulated strain caused by particle deformation during high impact blending and supersonic deposition processes.

Increase in velocity above the optimized value does not produce a corresponding increase in hardness of the coatings. The highest hardness was obtained from the optimized parameters and this is thought to be due to the lower porosity of its coating as explained in Section 3.2, and greater plastic deformation of Al particles at 500 °C. The greater the plastic deformation, the higher is the hardness [46]. Plastic deformation of Al particles will be higher at 500 °C than at 550 °C because thermal softening (which reduces plastic deformation) would have taken place at 550 °C since Al melts at 660 °C.

Ideally, as the gas temperature is increased, the average kinetic energy of the molecules of the gas will increase leading to increase in both particle temperature and impact velocity. This will produce corresponding increase in the momentum of the powder particles and greater particle impact, higher work hardening effect, greater coating densification, and increased coating hardness. However, the highest temperature does not produce the highest coating hardness. This is probably because the heat produced at the jetting regions at the higher deposition temperature would have favored dynamic recrystallization of Al particles which will lead to a reduction in hardness. Moreover, in addition to thermal softening explained above, the heat at the jetting region may also prevent sufficient plastic deformation of the Al, and consequently lessen strain hardening and hardness.

4. Conclusions

The continuity, momentum, and energy equations were solved by transforming the partial differential equations into a single ordinary differential equation. A constitutive equation was introduced to close the system, and also to account for stress and flow viscosity. The solution was used to calculate the optimum velocity for the deposition of SiC on titanium substrate using the LPCGD. CFD analysis of the flow through the nozzle was carried out to determine the input temperature that will yield the calculated velocity at the exit of the nozzle. Although there is no noticeable difference between the phases obtained in the coatings deposited with optimized and non-optimized parameters, coating with optimized parameter yielded less porosity and higher hardness. The higher hardness was traced to greater plastic deformation. Having established the relationship between the velocity, temperature, and pressure of the gas, for SiC, the solution can be used to predict these parameters for other powder materials by substituting the thermo-mechanical properties of the powder into the boundary conditions.

Acknowledgments: This material is based upon work supported financially by the National Research Foundation (NRF). The Cold Spray Laboratory of the University of Witwatersrand, Johannesburg is appreciated for the use of facilities. The authors also acknowledge the support from the African Laser Centre and the Tshwane University of Technology, Pretoria, South Africa which helped to accomplish this work.

Author Contributions: All authors contributed extensively to the work presented in this paper. A.P.P. and I.B. conceived the idea, supervised the work and revised the manuscript. D.I.A. performed the optimization, CFD analysis and experiments, analyzed and interpreted the data, and drafted the manuscript.

Conflicts of Interest: The authors declare no conflict of interest.

Abbreviations

ρ_g	gas (air) density
δ_g	*Kronecker delta*, which is a component of the identity tensor defined such that $\delta_{ij} = 1$ if $i = j$, otherwise $\delta_{ij} = 0$
ν	velocity vector
τ	stress tensor
μ	molecular viscosity

References

1. Adebiyi, D.I.; Popoola, A.P.I.; Pityana, S.L. Phase constituents and microhardness of laser alloyed Ti-6Al-4V alloy. *J. Laser Appl.* **2015**, *27*, S29104. [CrossRef]
2. Adebiyi, D.I.; Popoola, A.P.I. Mitigation of abrasive wear damage of Ti-6Al-4V by laser surface alloying. *Mater. Des.* **2015**, *74*, 67–75. [CrossRef]
3. Sahasrabudhe, H.; Soderlind, J.; Bandyopadhyay, A. Laser processing of *in situ* TiN/Ti composite coating on titanium. *J. Mech. Behav. Biomed. Mater.* **2016**, *53*, 239–249. [CrossRef] [PubMed]
4. Ravindra, D.; Patten, J. Ductile regime single point diamond turning of CVD-SiC resulting in an improved and damage-free surface. In Proceedings of the 4th International Conference on Recent Advances in Materials, Minerals & Environment and 2nd Asian Symposium on Materials & Processing, Penang, Malaysia, 1–3 June 2009.
5. Tului, M.; Giambi, B.; Lionetti, S.; Pulci, G.; Sarasini, F.; Valente, T. Silicon carbide based plasma sprayed coatings. *Surf. Coat. Technol.* **2012**, *207*, 182–189. [CrossRef]
6. Moridi, A.; Hassani-Gangaraj, S.M.; Guagliano, M.; Dao, M. Cold spray coating: Review of material systems and future perspectives. *Surf. Eng.* **2014**, *30*, 369–395. [CrossRef]
7. Villafuerte, J. Recent trends in cold spray technology: Looking at the future. *Surf. Eng.* **2010**, *26*, 393–394. [CrossRef]
8. Goyal, T.; Prince, S.; Walia, R.S.; Sidhu, T.S. Effect of nozzle geometry on exit velocity, temperature and pressure for cold spray process. *Int. J. Mat. Sci. Eng.* **2011**, *2*, 65–72.
9. Tria, S.; Elkedim, O.; Li, W.Y.; Liao, H. Ball milled Ni-Ti powder deposited by cold spraying. *J. Alloy. Compd.* **2009**, *483*, 334–336. [CrossRef]
10. Champagne, V. *The Cold Spray Materials Deposition Process: Fundamentals and Application*; Woodhead: Cambridge, UK, 2007.
11. Ghelichi, R.; Bagherifard, S.; Guagliano, M.; Verani, M. Numerical simulation of cold spray coating. *Surf. Coat. Technol.* **2011**, *205*, 5294–5301. [CrossRef]
12. Gartner, F. Advances in cold spraying. *Surf. Eng.* **2006**, *22*, 161–163. [CrossRef]
13. Schmidt, T.; Gaurtner, F.; Assadi, H.; Kreye, K. Development of a generalized parameter window for cold spray deposition. *Acta Mater.* **2006**, *54*, 729–742. [CrossRef]
14. Assadi, H.; Gaurtner, F.; Stoltenhoff, T.; Kreye, H. Bonding mechanism in cold gas spraying. *Acta Mater.* **2003**, *51*, 4379–4394. [CrossRef]
15. Alkimov, A.P.; Kosarev, V.E.; Papyrin, A.N. A method of cold gas-dynamic deposition. *Dokl. Akad. Nauk. SSSR* **1990**, *318*, 1062–1065.
16. Legoux, J.G.; Irissou, E.; Moreau, C. Effect of substrate temperature on the formation mechanism of cold-sprayed aluminum, zinc and tin coatings. *J. Therm. Spray Technol.* **2007**, *16*, 619–626. [CrossRef]

17. Gao, P.; Li, C.; Yang, C.; Li, Y.; Li, C. Influence of substrate hardness on deposition behavior of single porous WC-12Co particle in cold spraying. *Surf. Coat. Technol.* **2008**, *203*, 384–390. [CrossRef]
18. Li, C.J.; Wang, H.T.; Zhang, Q.; Yang, G.J.; Li, W.Y.; Liao, H.L. Influence of spray materials and their surface oxidation on the critical velocity in cold spraying. *J. Therm. Spray Technol.* **2010**, *19*, 95–101. [CrossRef]
19. Yokoyama, K.; Watanabe, M.; Kuroda, S.; Gotoh, Y.; Schmidt, T.; Gartner, F. Simulation of solid particle impact behaviour for spray processes. *J. Mater. Trans.* **2006**, *47*, 1697–1702. [CrossRef]
20. Henao, J.; Concustell, A.; Cano, I.G.; Cinca, N.; Dosta, S.; Guilemany, J.M. Influence of Cold Gas Spray process conditions on the microstructure of Fe-based amorphous coatings. *J. Alloy. Compd.* **2015**, *622*, 995–999. [CrossRef]
21. Li, C.J.; Li, W.Y.; Liao, H. Examination of the Critical Velocity for Deposition of Particles in Cold Spraying. *J. Therm. Spray Technol.* **2006**, *15*, 213. [CrossRef]
22. Sun, J.G.; Kim, H.D.; Park, J.O.; Jin, Y.Z. A Computational study of the gas-solid suspension flow through a supersonic nozzle. *Open J. Fluid Dyn.* **2012**, *2*, 242–247. [CrossRef]
23. Dmitrienko, G.S.; Uvarova, L.A. Modeling of dendritic growth in two dimensions. *Int. J. Pure Appl. Math.* **2013**, *82*, 663–668. [CrossRef]
24. Vutova, K.; Donchev, V. Electron beam melting and refining of metals: Computational modeling and optimization. *Materials* **2013**, *6*, 4626–4640. [CrossRef]
25. Goyal, T.; Sidhu, T.S.; Walia, R.S. Taguchi Based Optimization for Micro Hardness of Cold Spray Coating Process. *Intl. J. Surf. Eng. Interdiscip. Mat. Sci.* **2014**, *2*, 23–33. [CrossRef]
26. Papyrin, A.N. *Cold Spray: State of the Art and Applications*; Cold spray technology: Albuquerque, NM, USA, 2006; pp. 1–21.
27. Lee, J.C.; Kang, H.G.; Chu, W.S.; Ahn, W.S. Repair of damaged mold surface by cold-spray method. *CIRP Ann. Manuf. Technol.* **2007**, *56*, 577–580. [CrossRef]
28. Lin, J.; Ruan, X.; Chen, B. *Fluid Mechanics*; Tsinghua University Press: Beijing, China, 2005; pp. 72–91.
29. Chuayjan, W.; Pothiphan, S.; Wiwatanapataphee, B.; Wu, Y.H. Numerical simulation of granular flow during filling and discharging of a silo. *Int. J. Pure Appl. Math.* **2010**, *62*, 347–364.
30. Reddy, J.N.; Gartling, D.K. *The Finite Element Method in Heat Transfer and Fluid Dynamics*; CRC Press: London, UK, 1994.
31. Lupoi, R. Current design and performance of cold spray nozzles: Experimental and numerical observations on deposition efficiency and particle velocity. *Surf. Eng.* **2014**, *30*, 316–322. [CrossRef]
32. Lee, M.W.; Park, J.J.; Kim, D.Y.; Yoon, S.S.; Kim, H.Y.; Kim, D.H.; Park, D.S. Optimization of supersonic nozzle flow for titanium dioxide thin-film coating by aerosol deposition. *J. Aerosol. Sci.* **2011**, *42*, 771–780. [CrossRef]
33. Li, C.J.; Li, W.Y.; Liao, H. Examination of the critical velocity for deposition of particles in cold spraying. *J. Therm. Spray. Technol.* **2006**, *15*, 212–222. [CrossRef]
34. Motta, F.V.; Balestra, R.M.; Ribeiro, S.; Taguchi, S.P. Wetting behaviour of SiC ceramics: Part I. E_2O_3/Al_2O_3 additive system. *Mater. Lett.* **2004**, *58*, 2805–2809. [CrossRef]
35. Adebiyi, D.I.; Botef, I.; Popoola, P.A. Computational technique for optimization of the process parameter for cold spray coating of titanium. In Proceedings of the 1st International Conference on Mathematical Methods & Computational Techniques in Science & Engineering, Athens, Greece, 28–30 November 2014; pp. 239–243.
36. Geels, K.; Fowler, D.B.; Kopp, W.U.; Ruckert, M. *Metallographic and Materialographic Specimen Preparation, Light Microscopy, Image Analysis, and Hardness Testing*; ASTM International: West Conshohocken, PA, USA, 2007.
37. ASTM Standard. *E384-11E1, Standard Test Method for Knoop and Vickers Hardness of Materials*; ASTM International: West Conshohocken, PA, USA, 2007. [CrossRef]
38. ASTM Standard. *C1327, Standard Test Method for Vickers Indentation Hardness of Advanced Ceramics*; ASTM International: West Conshohocken, PA, USA, 2008. [CrossRef]
39. Konecna, R.; Nicoletto, G.; Majerova, V. Largest extreme value determination of defect size with application to cast Al-Si alloys porosity. *Metal* **2007**, *16*, 94.
40. Ganvir, A.; Curry, N.; Bjorklund, S.; Markocsan, N.; Nylen, P. Characterization of Microstructure and Thermal Properties of YSZ Coatings Obtained by Axial Suspension Plasma Spraying (ASPS). *J. Therm. Spray Technol.* **2015**, *1*, 1–10. [CrossRef]
41. Moore, E.M.; Papavassiliou, D.V.; Shambaugh, R.L. Air velocity, air temperature, fiber vibration and fiber diameter measurements on a practical melt blowing die. *Int. Nonwovens J.* **2004**, *13*, 43–53.

42. Huang, R.Z.; Fukanuma, H. The Influence of Spray Conditions on Deposition Characteristics of Aluminum Coatings in Cold Spraying. In Proceedings of the International Thermal Spray Conference, 4–7 May 2009; Marple, B.R., Hyland, M.M., Lau, Y.-C., Eds.; Thermal Spray 2009: Las Vegas, Nevada, USA, 2009; pp. 279–284. [CrossRef]

43. Khalid, A.A. The effect of testing temperature and volume fraction on impact energy of composites. *Mater. Des.* **2006**, *27*, 499–506. [CrossRef]

44. Triantou, K.I.; Pantelis, D.I.; Guipont, V.; Jeandin, M. Microstructure and tribological behavior of copper and composite copper+alumina cold sprayed coatings for various alumina contents. *Wear* **2015**, *336*, 96–107. [CrossRef]

45. Medica, L.A. *Guidelines for Safe Handling of Powders and Bulk Solids*; Wiley: Hoboken, NJ, USA, 2004.

46. Smith, G.T. *Industrial Metrology: Surfaces and Roundness*; Springer Science & Business Media: New York, NY, USA, 2013; p. 202.

metals

MDPI

Article

Microstructure Evolution and High-Temperature Compressibility of Modified Two-Step Strain-Induced Melt Activation-Processed Al-Mg-Si Aluminum Alloy

Chia-Wei Lin, Fei-Yi Hung *, Truan-Sheng Lui and Li-Hui Chen

Department of Materials Science and Engineering, National Cheng Kung University, Tainan 701, Taiwan; qqkm0526@gmail.com (C.-W.L.); luits@mail.ncku.edu.tw (T.-S.L.); chenlh@mail.ncku.edu.tw (L.-H.C.)
* Correspondence: fyhung@mail.ncku.edu.tw; Tel.: +886-6-275-7575 (ext. 31395); Fax: +886-6-234-6290

Academic Editor: Soran Birosca
Received: 3 February 2016; Accepted: 10 May 2016; Published: 13 May 2016

Abstract: A two-step strain-induced melt activation (TS-SIMA) process that omits the cold working step of the traditional strain-induced melt activation (SIMA) process is proposed for 6066 Al-Mg-Si alloy to obtain fine, globular, and uniform grains with a short-duration salt bath. The results show that increasing the salt bath temperature and duration leads to a high liquid phase fraction and a high degree of spheroidization. However, an excessive salt bath temperature leads to rapid grain growth and generates melting voids. The initial degree of dynamic recrystallization, which depends on the extrusion ratio, affects the globular grain size. With an increasing extrusion ratio, the dynamic recrystallization becomes more severe and the dynamic recrystallized grain size becomes smaller. It results in the globular grains becomes smaller. The major growth mechanism of globular grains is Ostwald ripening. Furthermore, high-temperature compressibility can be improved by the TS-SIMA process. After a 4 min salt bath at 620 °C, the high-temperature compression ratio become higher than that of a fully annealed alloy. The results show that the proposed TS-SIMA process has great potential.

Keywords: aluminum alloy; strain-induced melting activation (SIMA); semi-solid metal processing; high-temperature compression

1. Introduction

6xxx series Al alloys, a series of precipitation-hardened Al alloys, are widely used. The present study considers 6066 Al alloy, whose strength is higher than that of the great majority of other alloys in this series due to its Cu and Mn addition and excess Si [1,2]. According to previous research, adding Cu can improve strength and hardness by refining the precipitated phases during artificial aging [3]. Adding Mn enhances corrosion resistance, improves mechanical strength, increases the recrystallization temperature, and inhibits grain growth [4]. Even though adding Cu and Mn increases strength, they decrease formability. In order to overcome this problem, the strain-induced melt activation (SIMA) process is used in forming alloys at high temperatures.

The SIMA process is a semi-solid process, in which the materials are manufactured at the temperature of the mushy zone. The finished products have a near-net-shape advantage [5]. Due to its low cost and high stability, the SIMA process is useful. Figure 1a shows the procedure of the traditional SIMA process [6–12]. The steps are: (1) casting, which produces a dendritic structure; (2) hot work, which disintegrates the initial structure; (3) cold work, which introduces strain energy into the alloy; and (4) heat treatment, which makes the material recrystallize and partially melt at the temperature of solid-liquid coexistence. It is defined as a three-step process because casting materials is done via three steps to obtain globular grains. This study proposes an improved SIMA process that has two steps after casting. The two major differences between the traditional SIMA process and the proposed two-step SIMA (TS-SIMA) process are: (1) the proposed TS-SIMA process uses severe hot extrusion instead of cold work to introduce a large amount of strain energy; (2) the proposed SIMA process uses a salt bath instead of an air furnace to improve heating uniformity and reduce heating time. The procedure of the modified TS-SIMA process is shown in Figure 1b.

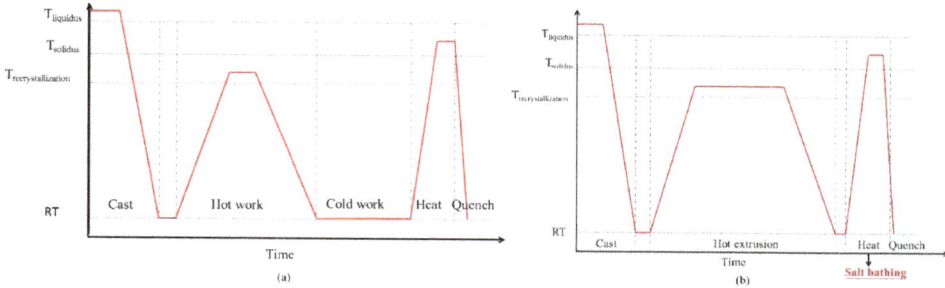

Figure 1. Procedures of (**a**) traditional three-step strain-induced melt activation (SIMA) process and (**b**) modified two-step SIMA process. RT: room temperature.

The aims of this research are to determine the effect of TS-SIMA process conditions on microstructural evolution, discuss the formation mechanism of globular grains in the TS-SIMA process, and confirm that the TS-SIMA process can improve the high-temperature compressibility.

2. Materials and Methods

The material used in this study was extruded 6066 Al alloy supplied from cooperative aluminum cooperation. Its composition, determined using a glow discharge spectrometer, is shown in Table 1. Two thicknesses (3 and 9 mm) were selected to determine the effect of strain energy induced by hot extrusion. The 6 in diameter casting materials were extruded with two dimensions of 52 mm (width) × 3 mm (thickness) and 75 mm (width) × 9 mm (thickness), respectively. The extrusion ratios of two materials with thicknesses of 3 and 9 mm (abbreviated as Ex3 and Ex9) were 117:1 and 27:1 (the true strains were 4.8 and 3.3), respectively. Two materials are designated as Ex3 and Ex9.

Table 1. Composition of 6066 aluminum alloy.

Element	Mg	Si	Cu	Mn	Fe	Cr	Al
Mass%	1.02	1.29	0.98	1.02	0.19	0.18	Balanced

The first phase transformation temperature of 6066 Al alloy was about 565 °C, as shown in Figure 2, which is measured by a differential scanning calorimeter (DSC). In order to determine the effects of salt bath temperature, salt bath duration, and stored strain energy on the microstructure evolution, specimens with two thicknesses (3 and 9 mm) were heated at 550–630 °C within 1–60 min in salt bath. The codes of salt bath specimens with two thicknesses are interpreted by the following example: SB3/620-10 means a material which is 3 mm in thickness was heated by salt bath at 620 °C for 10 min.

Microstructural characteristics and grain size were analyzed using an optical microscope (OM) (BX41M-LED, Olympus, Tokyo, Japan). The identification of phases and distribution of elements were detected using an electron probe micro-analyzer (EPMA) (JEOL, Peabody, MA, USA). The liquid fraction of lower-melting-point second phases was measured using ImageJ software (National Institetes of Health, Java 1.8.0_60, New York, NY, USA). Two shape parameters, x and z, were defined for the degree of spheroidization [7]. In Figure 3, *a*, *b*, *c*, and *A* represent the major axis, minor axis, perimeter, and area of a grain, respectively. According to the definitions $x = (b/a)$ and $z = (4\pi A)/c^2$, x is the ratio of the minor axis to the major axis and z becomes closer to 1 as the shape becomes more circular. When x and z are closer to 1, the grains are more equiaxial and the degree of spheroidization is higher.

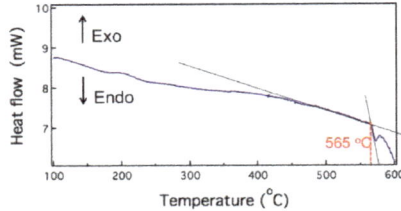

Figure 2. Differential scanning calorimeter (DSC) data of 6066 aluminum alloy. Exo: exothermic; Endo: endothermic.

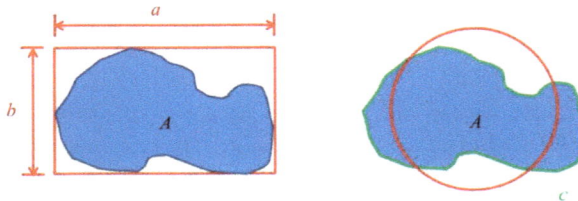

Figure 3. Parameters of spheroidization degree definition. *a*, *b*, *c*, and *A* represent the major axis, minor axis, perimeter, and area of a grain, respectively.

Nine-mm-thick sheets were used in the high-temperature compression tests. As-extruded alloys, fully annealed alloys, and salt bath alloys were tested to compare their high-temperature compressibility. The compression ratio is defined as $R\% = (t_0 - t_f)/t_0$, where t_0 is the thickness of the initial sheet (9 mm) and t_f is the thickness after compression. The compression temperatures were set as 550 and 600 °C (abbreviated as C550 or C600). The compression specimens had dimensions of 20 mm (length) × 20 mm (width) × 9 mm (thickness). The compression rate was 20 mm/min and the compressive loading was 60 kg/cm^2. The compression ratios of the above specimens were measured and compared. A higher compression ratio indicates that the resistance of deformation at high temperature is smaller and thus high-temperature formability is better [13]. The compression ratio was calculated and the microstructure of the cross-section parallel to the compression direction was observed.

3. Results and Discussion

3.1. Microstructural Evolution

Figure 4 shows the microstructure of as-extruded alloys with two thicknesses. A fiber-shape microstructure can be seen in both Figure 6a,b. Additionally, a lot of the second phases are distributed uniformly and a large number of very fine dynamic recrystallized grains appear in the Ex3 alloy as shown in Figure 4a. In contrast, recrystallization only occurred in parts of Ex9, and its recrystallized grain size was larger than that of Ex3, as shown in Figure 4b. The differences between the two materials resulted from the different extrusion ratios. Because the extrusion ratio of Ex3 is higher than that of Ex9, it led the dynamic recrystallization degree of Ex3 to also be higher than that of Ex9.

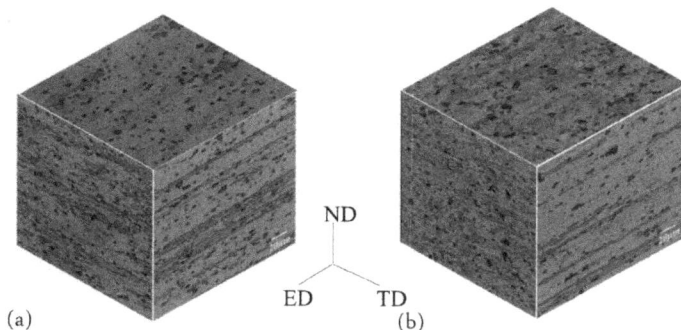

Figure 4. Microstructures of extruded alloys: (**a**) 3-mm-thickness extruded material (Ex3) and (**b**) 9-mm-thickness extruded material (Ex9). ED: extruded direction; ND: normal direction; TD: transverse direction.

The effect of salt bath temperatures on microstructure is shown in Figure 5. After a 30 min salt bath, the grains were not spheroidized uniformly when the temperature was lower than 610 °C, and the material deformed severely or was partially melted severely when the temperature was higher than 630 °C. When the temperature of the salt bath reached 630 °C, melting voids and cracks were probably generated within 7 min of the salt bath, as shown in Figure 5f.

Figure 5. Microstructure of Ex3 after immersion in salt bath at various time periods (represented by signals in Table 2): (**a**) SB3/550-30; (**b**) SB3/570-30; (**c**) SB3/590-30; (**d**) SB3/610-30; (**e**) SB3/620-30; (**f**) SB3/630-7. SB3: salt bath material with 3 mm thickness.

Figure 6 shows the microstructure evolution of 6066 alloys in the salt bath at 610, 620, and 630 °C with various time periods. With the increasing salt bath duration, the grain growth increased. For a given duration, a higher salt bath temperature led to higher grain growth except for SB3/610-1 (Figure 7a) and SB3/620-1 (Figure 7d). The average grain sizes of these two specimens are almost identical due to the duration not being long enough for the grain to grow obviously at 610 and 620 °C. The grain growth rates for various salt bath temperatures were calculated based on the Lifshitz-Slyozov-Wagner (LSW) theory [5,7,14–17]. The formula is $d^n - d_0{}^n = Kt$, where d is the average grain size that depends on the salt bath duration, d_0 is the initial grain size, t is the salt bath duration, K (units: $\mu m^3 \cdot min^{-1}$) is the coarsening rate constant, and exponent n is determined by the diffusive mechanism of grain growth. For instance, $n = 2$ indicates surface diffusion and $n = 3$ indicates volume diffusion. In this study, $n = 3$ can be used. The theoretical formula can be rewritten as $d^3 = Kt + d_0{}^3$, which is plotted in Figure 7a. The results of linear fitting are shown in Table 2. The coefficients of determination (R^2) are above 0.98, indicating that the three results were highly linear. The slopes of the three fitting lines represent the K value (*i.e.*, grain coarsening rate). The K value at 630 °C was much higher than those at the other two temperatures. The results show that grain growth was very rapid at 630 °C.

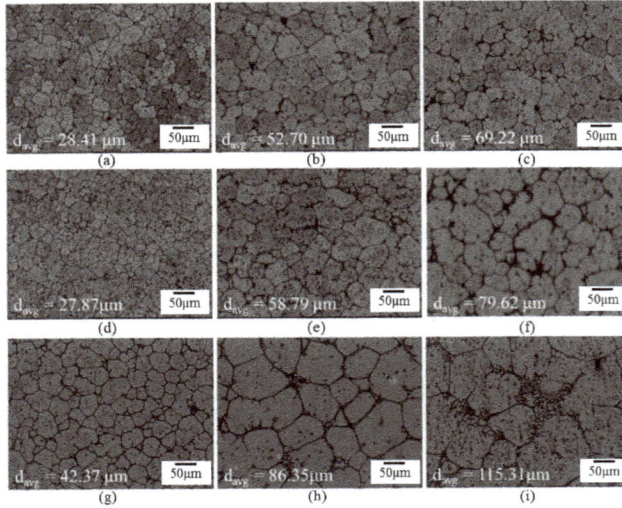

Figure 6. Microstructures of several materials: (**a**) SB3/610-1; (**b**) SB3/610-10; (**c**) SB3/610-30; (**d**) SB3/620-1; (**e**) SB3/620-10; (**f**) SB3/620-30; (**g**) SB3/630-1; (**h**) SB3/630-10; and (**i**) SB3/630-30.

Table 2. K values and coefficients of determination obtained via linear fitting using Lifshitz-Slyozov-Wagner (LSW) theory for SB3 with various salt bath temperatures. SB3: salt bath material with 3 mm thickness.

Specimen	SB3/610	SB3/620	SB3/630
K ($\mu m^3 \cdot min^{-1}$)	10751	16806	50820
R^2	0.9852	0.9926	0.9818

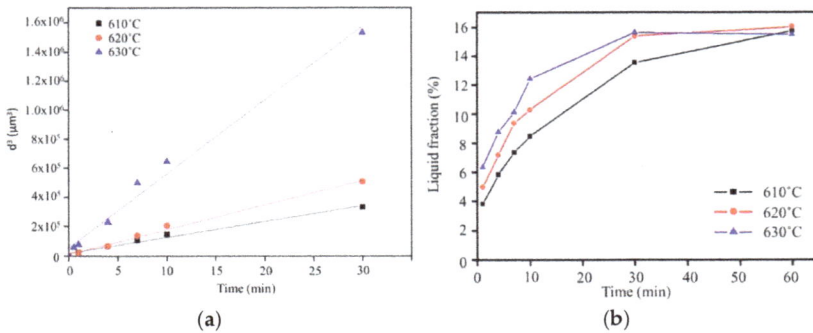

Figure 7. (**a**) Plot of grain size *versus* salt bath duration from Lifshitz-Slyozov-Wagner (LSW) theory (Ex3) and (**b**) liquid fraction for various salt bath temperatures (Ex3).

Grain boundary broadening was found for all salt bath temperatures, as shown in Figure 6. This broadening was due to the lower-melting-point second phases melting [18], and the liquid penetrating into the grain boundaries. After quenching, the low-melting-point second phases solidified. When the heating duration of the salt bath was increased, liquid pools formed because large amounts of liquid phases aggregated. The liquid fraction obtained using ImageJ (National Institetes of Health, Java 1.8.0_60, New York, NY, USA) is shown in Figure 7b. The liquid fraction increased with the increasing salt bath temperature and duration. It saturated within 30 min at the three salt bath temperatures. With the increasing salt bath temperature, the liquid fraction saturation becomes faster. In addition, cracks appeared in the specimen at a salt bath temperature of 630 °C, as shown in Figure 5f, at which the second phases melted and the volume shrank too severely during quenching. The results show that even though 630 °C was suitable for liquid phase formation, it likely led to the generation of defects. Therefore, 620 °C was the most suitable salt bath temperature, due to fast liquid formation and high stability.

The influence of the thickness of the extruded alloy was considerable. It represents the effect of strain energy. Figure 8 shows the microstructures of sheets with two thicknesses (3 and 9 mm). The globular grain size of the SB9 specimen is much larger than that of the SB3 specimen. According to the recrystallization and partial melting mechanism [8,12], the existence of recrystallized grains is necessary for providing high-angle (high-energy) grain boundaries for liquid penetration at high temperatures. In the SB3 specimen, all fine grains were generated all around the material through severe dynamic recrystallization as hot extrusion. In the SB9 specimen, the dynamic recrystallized grains formed only in parts of the material and some static recrystallization occurred within the salt bath period. These dissimilarities resulted in different grain sizes after a given salt bath duration. They also prove that both dynamic recrystallization and static recrystallization provide high-energy grain boundaries for liquid penetration and globular grain formation.

Figure 8. Microstructures of extruded alloys with two thicknesses after salt bath at 620 °C: (**a**) SB3/620-1; (**b**) SB3/620-10; (**c**) SB3/620-30; (**d**) SB9/620-1; (**e**) SB9/620-10; and (**f**) SB9/620-30.

For the degree of spheroidization, three representative conditions are plotted in Figure 9. The figure shows that x and z become closer to 1 as the salt bath duration is increased. This indicates that an increase in salt bath duration makes the grains more globular and equiaxial. Increasing the salt bath temperature also improved the degree of spheroidization determined by comparing z values obtained at 610 and 620 °C. The extrusion ratio did not affect the spheroidization determined by comparing SB3 and SB9 at the same salt bath temperature of 620 °C. The results show that the liquid formation enhanced the degree of spheroidization.

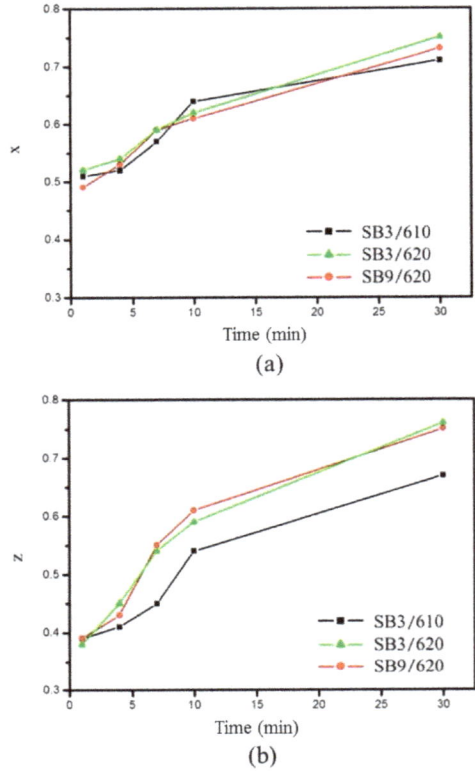

Figure 9. Variation of degree of spheroidization with salt bath conditions: (**a**) *x* and (**b**) *z*.

The evolution of elemental distribution was analyzed using EPMA. The results are shown in Figure 10. Mg, Si, Cu, and Mn, the four major elements added, were distributed uniformly in the as-extruded alloy. After the salt bath, Mg, Si, and Cu were located at the grain boundaries and formed a network structure, but Mn aggregated as compound particles. This was due to the melting points of Mn-rich phases being higher than 620 °C [19]. Figure 11 shows the phase analysis using the wavelength-dispersive X-ray spectrometer (WDS) of EPMA. The brighter plate-like grain boundary particles, those marked by "d", are mainly composed of Si and Cu. It is composed of the eutectic phase of Al and Al_2Cu and the eutectic phase of Al and Si. The darker particles marked by "c" on the grain boundaries and liquid pools are composed of mostly Si and Mg. It is speculated to be the phase aggregation of Al, Mg_2Si, and excess Si. Moreover, Si, Mn, Cr, and Fe are the major elements of the brighter equiaxed particles on the grain boundaries and in the grain interior (see those marked by "a"). These particles should be the $Al_{15}(Fe,Mn,Cr)_3Si_2$ phase with a melting temperature close to 660 °C. Its particle size increases with increasing the salt bath duration. The coarsening mechanism of the particles probably follows Ostwald ripening, a Gibbs-Thompson effect to reduce the total surface energy.

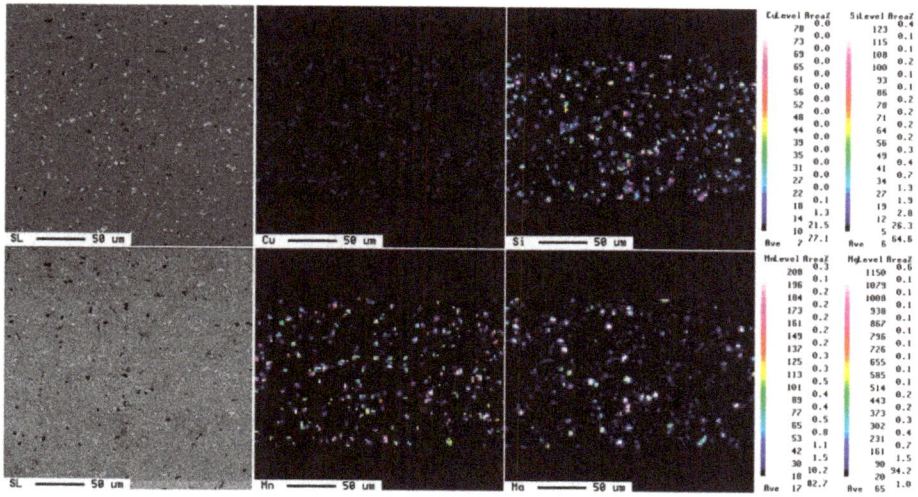

(a)

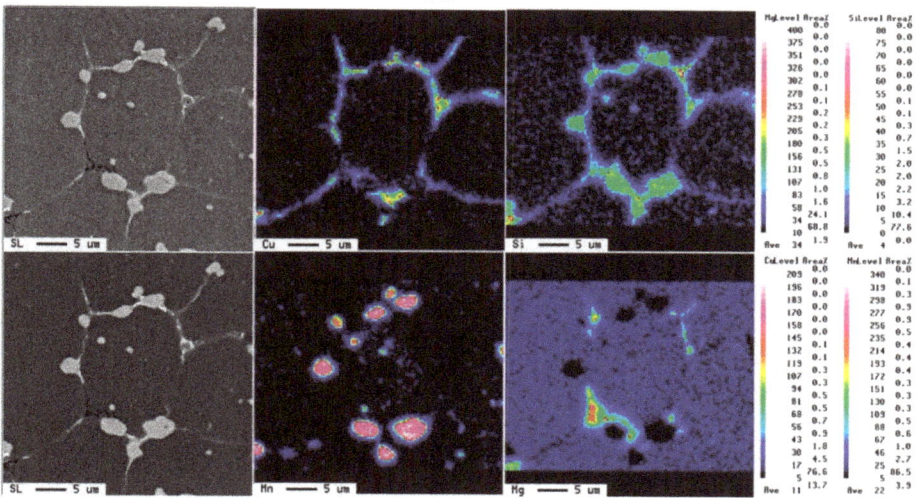

(b)

Figure 10. Elemental distribution obtained using electron probe micro-analyzer (EPMA): (**a**) Ex3 and (**b**) SB3/620-1.

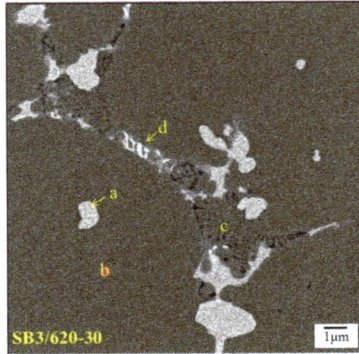

a	Cu	Si	Mn	Mg	Fe	Cr	Al
	0.74	7.50	10.57	0.05	3.85	3.44	Bal.

b	Cu	Si	Mn	Mg	Fe	Cr	Al
	0.18	0.33	0.20	0.77	0.00	0.26	Bal.

c	Cu	Si	Mn	Mg	Fe	Cr	Al
	0.61	5.30	0.07	11.46	0.02	0.00	Bal.

d	Cu	Si	Mn	Mg	Fe	Cr	Al
	10.33	16.14	0.00	1.70	0.00	0.31	Bal.

Figure 11. Phase analysis of SB. a: $Al_{15}(Fe,Mn,Cr)_3Si_2$ particle; b: Al matrix; c: dark grain boundary or liqud pool; d: bright grain boundary; red number: major element.

3.2. Formation Mechanism for the Two-Step Strain-Induced Melt Activation (TS-SIMA) Process

The mechanisms of globular grain formation for the traditional SIMA process and the TS-SIMA process are shown in Figure 12a,b, respectively. The mechanism of TS-SIMA, shown as Figure 12b, works on the assumption that the grains recrystallize over the whole alloy after hot extrusion (as for Ex3). The procedure for the TS-SIMA process is: (1) a suitable composition alloy is cast with a dendritic microstructure and uniformly distributed secondary phases; (2) the initial dendritic microstructure disintegrates, sufficient strain energy is introduced, and grains are dynamically recrystallized over the material through severe hot extrusion; (3) for the materials subjected to a salt bath, the lower-melting-point second phases melt and the liquid penetrates into grain boundaries and then surrounds the grains; (4) the salt bath makes the grains grow and become globular. Compared to the traditional SIMA process, the proposed TS-SIMA process can produce globular grains faster, and the grains spheroidize uniformly and are finer.

After the salt bath, the microstructure of the Ex9 specimen can also be spheroidized. It indicates that even though the lower stored energy from hot extrusion could not make Ex9 recrystallize fully, the static recrystallized grains in the salt bath also provided high-energy grain boundaries for liquid penetration. In other words, both dynamic and static recrystallization generated grain spheroidization.

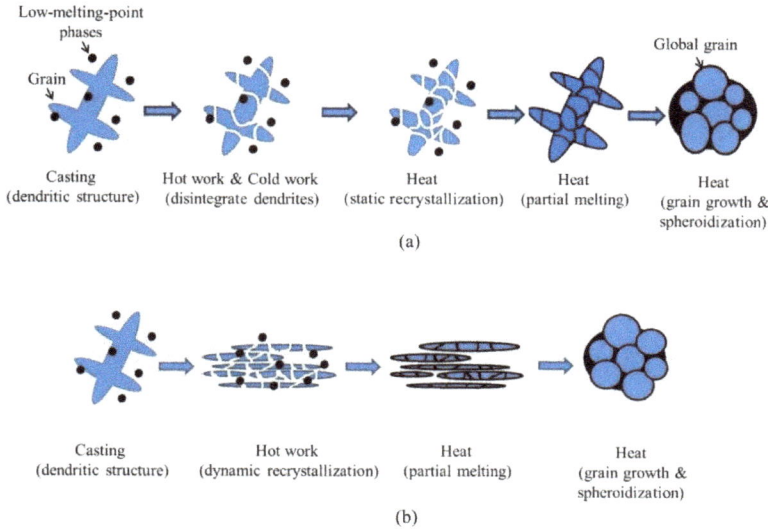

Figure 12. Evolution of globular grain formation: (**a**) traditional three-step SIMA process and (**b**) proposed two-step SIMA process.

For the growth mechanism of globular grains, when alloys are heated to the temperature of solid-liquid coexistence, two coarsening mechanisms are considered [11,20–23], namely grain coalescence and Ostwald ripening. Many studies have indicated that the two mechanisms occur at the same time and do not affect each other. Ostwald ripening dominates for long-duration heating and higher liquid fractions, and grain coalescence dominates for short-duration heating and lower liquid fractions. In this study, the growth mechanism of globular grains followed Ostwald ripening, since the coefficients of determination, shown in Table 2, are very close to 1 for $n = 3$ (volume diffusion) in the LSW theory. This theory indicates that as n becomes closer to 3, Ostwald ripening increasingly dominates. Grain coalescence did not dominate majorly because the method of heating was the salt bath, and thus the sample reached the required temperature quickly and the liquid formed almost immediately.

Ostwald ripening is a dissolution-precipitation diffusion-controlled mechanism [9,14]. Large grains become larger and small grains become smaller (even disappearing). As shown in Figure 6, when the salt bath duration and temperature were increased, liquid pools formed. It resulted from small Al grains dissolving gradually and Al atoms re-precipitating at neighboring larger grains; therefore, larger grains continued to grow but smaller grains lost Al atoms until they disappeared. Finally, grain boundaries filled with liquid phases and created liquid pools. Therefore, the appearance of liquid pools is also evidence of Ostwald ripening.

3.3. High-Temperature Compression of TS-SIMA-Processed Alloy

For the compression test with a given compression loading, Figure 13 shows microstructures of high-temperature compression samples. The level of grain deformation increased with the increasing compressive temperature. The grains after compression at 600 °C were flatter than those after compression at 550 °C. Moreover, liquid pools disappeared and grain boundaries became thinner, as shown in Figure 13e (after compression) and Figure 13f (before compression). This phenomenon became more obvious when the salt bath duration was increased.

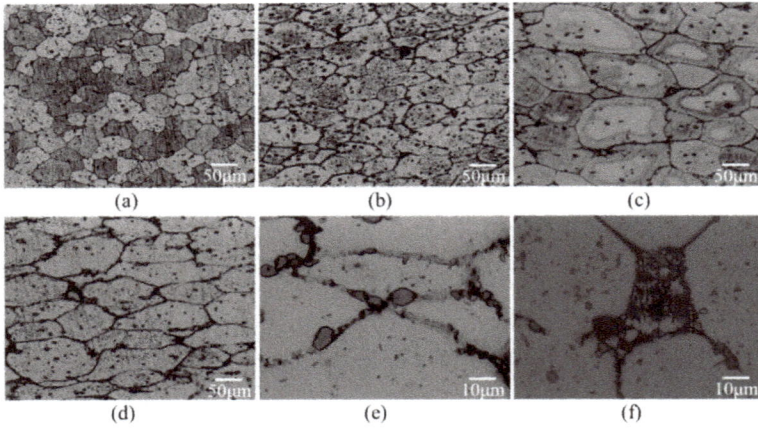

Figure 13. Microstructural characteristics of grain boundaries before and after high-temperature compression (**a**) SB9/620-1 + C550; (**b**) SB9/620-1 + C600; (**c**) SB9/620-30 + C550; (**d**) SB9/620-30 + C600; (**e**) SB9/620-30 + C600 (high magnification); and (**f**) SB9/620-30 (high magnification).

The compression ratios of as-extruded (Ex9), fully annealed (O9), and salt bath samples with various durations (SB9/620-time) are shown in Figure 14. The compression ratio at 600 °C, which is above the solid-liquid coexistence temperature, is much higher than that at 550 °C, which is close to the solid-liquid coexistence temperature. The compression ratio increased with the increasing salt bath duration. The compression ratio of SB9 was higher than that of Ex9 for all salt bath durations and higher than that of O9 when the salt bath duration was over 4 min. This proves that the high-temperature compressibility can be improved by the TS-SIMA process.

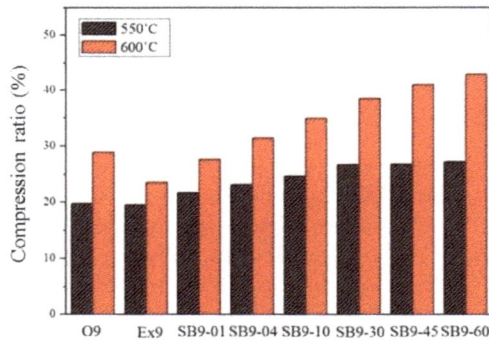

Figure 14. Compression ratios for samples.

The results show that a higher compression temperature (such as 600 °C) leads to a higher compression ratio than a lower compression temperature (such as 550 °C). Theoretically, the compression ratio is affected by the degree of grain deformation and grain boundary slip due to liquid filling at grain boundaries. The curve of the compression ratio *versus* salt bath duration, shown in Figure 15, can be divided into two parts: (1) shorter salt bath duration, where the slope of the curve continued to become smaller (resulting from liquid fraction variation, as shown in Figure 7b); (2) longer salt bath duration, where the slope of the curve was almost fixed. In part II, the liquid fraction reaches saturation. However, even though the liquid fraction does not increase, grains continue to grow. The grain growth increases the compression ratio according to the Hall-Petch theory [22,23].

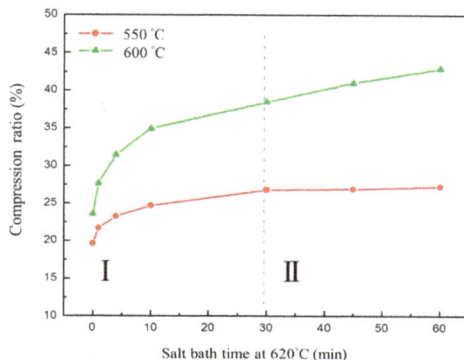

Figure 15. Relationship between compression ratio, salt bath duration, and salt bath temperatures.

4. Conclusions

(1) Globular grains were obtained using the TS-SIMA process. The TS-SIMA process has one fewer step compared to the traditional SIMA process, but the degree of spheroidization and uniformity are good. The most suitable salt bath temperature is 620 °C.

(2) Both dynamic recrystallization and static recrystallization provide high-energy grain boundaries for liquid penetration and spheroidized grain formation. The grain growth mechanism of globular grain was Ostwald ripening.

(3) The formation of globular grains increased high-temperature compressibility. After a 4 min salt bath, the high-temperature compressibility of SIMA alloys is higher than that of fully annealed alloys. High-temperature formability can be improved by the TS-SIMA process.

Acknowledgments: The authors are grateful to the Instrument Center of National Cheng Kung University and the National Science Council of Taiwan (NSC MOST103-2221-E-006-056-MY2) for their financial support.

Author Contributions: Chia-Wei Lin, Fei-Yi Hung and Truan-Sheng Lui conceived and designed the experiments; Chia-Wei Lin performed the experiments, analyzed the data and wrote the paper; Fei-Yi Hung, Truan-Sheng Lui and Li-Hui Chen gave suggestions for improving experiments and analysis.

Conflicts of Interest: The authors declare no conflict of interest.

References

1. Hatch, J.E. *Aluminum: Properties and Physical Metallurgy*; ASM International: Materials Park, OH, USA, 1984; Volume 1, p. 50.
2. Zhen, L.; Fei, W.D.; Kang, S.B.; Kim, H.W. Precipitation behavior of Al-Mg-Si alloys with high silicon content. *J. Mater. Sci.* **1997**, *32*, 1895–1902. [CrossRef]
3. Laughlm, D.E.; Miao, W.F. The effect of Cu and Mn content and processing on precipitation hardening behavior in Al-Mg-Si-Cu alloy 6022. *Miner. Met. Mater. Soc.* **1998**, 63–78.
4. Mondolfo, L.F. *Aluminum Alloys Structure & Properties*; Butterworths: London, UK, 1976; pp. 806–842.
5. Fan, Z. Semisolid metal processing. *Int. Mater. Rev.* **2002**, *47*, 49–85. [CrossRef]
6. Song, Y.B.; Park, K.T.; Hong, C.P. Recrystallization behavior of 7175 Al alloy during modified strain-induced melt-activated (SIMA) process. *Mater. Trans.* **2006**, *47*, 1250–1256. [CrossRef]
7. Tzimas, E.; Zavaliangos, A. A comparative characterization of near-equiaxed microstructures as produced by spray casting, magnetohydrodynamic casting and the stress induced, melt activated process. *Mater. Sci. Eng. A* **2000**, *289*, 217–227. [CrossRef]

8. Paes, M.; Zoqui, E.J. Semi-solid behavior of new Al-Si-Mg alloys for thixoforming. *Mater. Sci. Eng. A* **2005**, *406*, 63–73. [CrossRef]

9. Parshizfard, E.; Shabestari, S.G. An investigation on the microstructural evolution and mechanical properties of A380 aluminum alloy during SIMA process. *J. Alloy. Compd.* **2011**, *509*, 9654–9658. [CrossRef]

10. Akhlaghi1, F.; Farhood, A.H.S. Characterization of globular microstructure in NMS processed aluminum A356 alloy: The role of casting size. *Adv. Mater. Res.* **2011**, *264–265*, 1868–1877. [CrossRef]

11. Tzimas, E.; Zavaliangos, A. Evolution of near-equiaxed microstructure in the semisolid state. *Mater. Sci. Eng. A* **2000**, *289*, 228–240. [CrossRef]

12. Emamy, M.; Razaghian, A.; Karshenas, M. The effect of strain-induced melt activation process on the microstructure and mechanical properties of Ti-refined A6070 Al alloy. *Mater. Des.* **2013**, *46*, 824–836. [CrossRef]

13. Lee, K.S.; Kim, S.; Lim, K.R.; Hong, S.H.; Kim, K.B.; Na, Y.S. Crystallization, high temperature defroemtaion behavior and solid-to-dolid formability of a Ti-based bulk metallic glass within supercooled liquid region. *J. Alloy. Compd.* **2016**, *663*, 270–278. [CrossRef]

14. Tang, H.; Cheng, Z.; Liu, J.; Ma, X. Preparation of a high strength Al-Cu-Mg alloy by mechanical alloying and press-forming. *Mater. Sci. Eng. A* **2012**, *550*, 51–54. [CrossRef]

15. Katayama, T.; Nakamachi, E.; Nakamura, Y.; Ohata, T.; Morishita, Y.; Murase, H. Development of process design system for press forming multi-objective optimization of intermediate die shape in transfer forming. *J. Mater. Process. Technol.* **2004**, *155–156*, 1564–1570. [CrossRef]

16. Wang, Z.; Ji, Z.; Hu, M.; Xu, H. Evolution of the semi-solid microstructure of ADC12 alloy in a modified SIMA process. *Mater. Charact.* **2011**, *62*, 925–930. [CrossRef]

17. Yan, G.; Zhao, S.; Ma, S.; Shou, H. Microstructural evolution of A356.2 alloy prepared by the SIMA process. *Mater. Charact.* **2012**, *69*, 45–51. [CrossRef]

18. Lin, C.W.; Hung, F.Y.; Lui, T.S.; Chen, L.H. High-temperature deformation and forming behavior of two-step SIMA-processed 6066 ally. *Mater. Sci. Eng. A* **2016**, *659*, 143–157. [CrossRef]

19. Bolouri, A.; Shahmiri, M.; Kang, C.G. Coarsening of equiaxed microstructure in the semisolid state of aluminum 7075 alloy through SIMA processing. *J. Mater. Sci.* **2012**, *47*, 3544–3553. [CrossRef]

20. Hardy, S.C.; Voorhees, P.W. Ostwald ripening in a system with a high volume of coarsening phase. *Metall. Trans. A* **1988**, *19*, 2713–2721. [CrossRef]

21. ASM International Alloy Phase Diagram and the Handbook Committees. *Alloy Phase Diagram, ASM Handbook 3*; ASM International: Materials Park, OH, USA, 1992; pp. 307–308.

22. Zhang, L.; Liu, Y.B.; Cao, Z.Y.; Zhang, Y.F.; Zhang, Q.Q. Effects of isothermal process parameters on the microstructure of semisolid AZ91D alloy produced by SIMA. *J. Mater. Process. Technol.* **2009**, *209*, 792–797. [CrossRef]

23. Qin, Q.D.; Zhao, Y.G.; Xiu, K.; Zhou, W.; Liang, Y.H. Microstructure evolution of *in situ* Mg_2Si/Al-Si-Cu composite in semisolid remelting processing. *Mater. Sci. Eng. A* **2005**, *407*, 196–200. [CrossRef]

MDPI AG
St. Alban-Anlage 66
4052 Basel, Switzerland
Tel. +41 61 683 77 34
Fax +41 61 302 89 18
http://www.mdpi.com

Metals Editorial Office
E-mail: metals@mdpi.com
http://www.mdpi.com/journal/metals

www.ingramcontent.com/pod-product-compliance
Lightning Source LLC
Chambersburg PA
CBHW051722210326
41597CB00032B/5568